Rubber-Modified Thermoset Resins

ADVANCES IN CHEMISTRY SERIES **208**

Rubber-Modified Thermoset Resins

C. Keith Riew, EDITOR
The BFGoodrich Company

John K. Gillham, EDITOR
Princeton University

Based on a symposium sponsored by
the Division of Polymeric Materials Science and Engineering
at the 186th Meeting
of the American Chemical Society,
Washington, D.C.,
August 28–September 2, 1983

American Chemical Society, Washington, D.C. 1984

Library of Congress Cataloging in Publication Data

Rubber-modified thermoset resins.
(Advances in chemistry series, ISSN 0065-2393; 208)

Bibliography: p.
Includes indexes.

1. Thermosetting plastics—Additives—Congresses.
2. Rubber—Congresses.

I. Riew, C. Keith. II. Gillham, John K. III. American Chemical Society. Division of Polymeric Materials Science and Engineering. IV. American Chemical Society. Meeting (186th: 1983: Washington, D.C.) V. Series.

QD1.A355 no. 208 540 s [668.4′22] 84-21566
[TP1180.T55]
ISBN 0-8412-0828-X

PRINTED IN THE UNITED STATES OF AMERICA

Advances in Chemistry Series

M. Joan Comstock, *Series Editor*

FOREWORD

ADVANCES IN CHEMISTRY SERIES was founded in 1949 by the American Chemical Society as an outlet for symposia and collections of data in special areas of topical interest that could not be accommodated in the Society's journals. It provides a medium for symposia that would otherwise be fragmented, their papers distributed among several journals or not published at all. Papers are reviewed critically according to ACS editorial standards and receive the careful attention and processing characteristic of ACS publications. Volumes in the ADVANCES IN CHEMISTRY SERIES maintain the integrity of the symposia on which they are based; however, verbatim reproductions of previously published papers are not accepted. Papers may include reports of research as well as reviews since symposia may embrace both types of presentation.

ABOUT THE EDITORS

C. KEITH RIEW is an R&D Fellow at The BFGoodrich Company, Research and Development Center, Brecksville, Ohio. He received his M.S. and Ph.D. in organic chemistry from Wayne State University, Detroit, Michigan, and his B.S. in Chemistry from Seoul National University, Seoul, Korea. He served as research scientist or manager for the New Products' R&D, Chemical Group, The BFGoodrich Company for more than 15 years. He has presented more than 30 technical papers and owns 15 patents on the synthesis and application of telechelic polymers, including rubber-toughened or -modified thermosets, hydrophilic polymers, and latexes. His latest research is in the synthesis, characterization, compounding, and performance evaluation of polymeric additives, especially impact modifiers for engineering thermoplastics. His research interests include correlating polymer chemistry and physics, morphology, processing, and static and dynamic thermomechanical properties to the failure mechanisms of impact-modified thermosets and thermoplastics.

JOHN K. GILLHAM received a B.A. from Cambridge University, England, and a Ph.D. in Chemistry from McGill University, Canada. He is a professor in the Polymer Materials Program of the Department of Chemical Engineering at Princeton University. His research with polymers includes thermosetting systems, transitions, spiral fractures in composite materials, and pyrolysis. He was the recipient of the 1978 Borden Award of the American Chemical Society in the Chemistry of Plastics and Coatings, the 1978 Mettler Award of the North American Thermal Analysis Society, and corecipient of the Doolittle Award for the best paper presented in 1979 to the American Chemical Society Division of Organic Coatings and Plastics Chemistry. He has served on National Materials Advisory Board Committees on "Characterization of Organic Matrix

Composites" and "High-Performance, Low-Energy-Curing Resins". In 1983 he was chairman of the Gordon Research Conference on Thermosets, was corecipient of the First Prize in the Roon Foundation Awards Competition of the Federation of Societies for Coatings Technologies for the best technical paper, and presented a series of lectures in Japan sponsored by the Japan Society for the Promotion of Science. In 1984 he presented a series of lectures in the People's Republic of China at the invitation of the Chinese Academy of Sciences. He is the developer of the torsion braid analysis (TBA) technique and the Time-Temperature-Transformation (TTT) cure diagram for thermosetting systems.

CONTENTS

PREFACE

THE DEVELOPMENT OF RUBBER-TOUCHENED THERMOSET and thermoplastic resins is an important contribution to the commercial polymer industry. A small amount of discrete rubber particles in a glassy plastic can greatly improve the crack and impact resistance of the normally brittle plastic. Generally, this improvement should be accomplished without significant deterioration of the thermomechanical properties inherent to the unmodified original polymer.

Unlike the modification of thermoplastics, which often requires only the simple physical blending of a particular elastomeric modifier, the modification of thermoset resins is achieved almost exclusively through chemistry and is a challenge to the expertise of the polymer chemist. A useful, final product with improved durability and applicability is a major concern of materials scientists and engineers who are responsible for answering this challenge.

Therefore, thermoset resin systems require a team effort among multidisciplinary scientists to tackle the various challenges and attain successful results. Thus, science and technology can be applied to produce the products that will be acceptable to various industrial and consumer groups because of greatly improved performance and simplified design combined with reduced cost.

Although many papers that discuss rubber-modified thermoplastics have been published in journals or books and have been presented in symposia, no symposium on rubber-modified thermoset resins had previously been sponsored by the American Chemical Society. Therefore, the symposium upon which this book is based is the first comprehensive coverage of rubber-modified thermoset resins.

We are indebted to each contributor for the cooperation and enthusiasm we have received, and we thank The BFGoodrich Company for providing administrative support.

C. KEITH RIEW
The BFGoodrich Company

JOHN K. GILLHAM
Princeton University

August 1984

CHEMISTRY AND PHYSICS

1

Cross-linking of Epoxy Resins

KAREL DUŠEK

Institute of Macromolecular Chemistry, Czechoslovak Academy of Sciences, 162 06 Prague 6, Czechoslovakia

The theoretical and experimental investigation of network build-up in the curing of epoxy resins is reviewed. The application of the statistical theory of branching processes based on cascade substitution and probability generating functions is outlined. The network build-up is dependent on reactivities of functional groups, functionality of monomers, composition of the system, and reaction paths as a function of conversion. Results obtained on simple diepoxide–diamine systems agree well with the theory. For acid curing (polycarboxylic acids, cyclic anhydrides), progress has been reached in the elucidation of complicated reaction paths (reaction mechanisms) and their statistical treatment. The effect of chemical and diffusion control of the curing reaction on network structure and the applicability of the theory for predicting variations in the network structure are discussed.

THE PROCESSING AND FINAL PHYSICAL PROPERTIES of epoxy resin–curing agent systems applied in adhesives, sealants, coatings, or laminates depend primarily on their chemical composition and degree of cure (*1*) as well as on the details of network structure. The network structure determines the processing characteristics below the gel point, the position of the gel point, and the mechanical behavior of cured systems in the rubbery state. It also affects—sometimes indirectly and in a less explicit way—the physical properties and durability of cured epoxy resins in the glassy state. This chapter reviews the present theoretical treatment and experimental characterization of network formation and structure in the curing of epoxy resins and the use of the results of these studies in the elucidation of curing processes.

Characterization of Network Formation

Of the main theories for multiplicative growth of network structures (*2*)—statistical, kinetic, and percolation—only the statistical theories that are based on tree-like models, which can absorb the majority of chemical information and are applicable beyond the gel point, have been used to describe the curing of epoxy resins. In the sixties,

0065-2393/84/0208-0003/$06.00/0

authors (*1*) used the Flory–Stockmayer combinatorial approach to describe the pregel state (degree of polymerization) and the gel point. However, the results are somewhat invalidated because the reaction schemes used were too simplified or even incorrect. The importance of deviations from an ideal network structure in cured epoxy resins that are a result of stoichiometric imbalance or incompleteness of reaction was recognized by Bell et al. (*3, 4*) who used semiempirical corrections. The applicability of these corrections was limited, however, to rather small deviations from perfectness.

A complex approach to network formation in curing was made possible by the development of modern statistical theories, in particular the theory of branching processes employing the cascade substitution and the formalism of the probability generating functions as mathematical tools (*2, 5–8*). The underlying model was derived from the Flory–Stockmayer tree-like model with random circuit closing beyond the gel point. This model allows one to consider the different reactivities of reacting groups, the specific rules of bond formation (reaction mechanism), and a perturbation treatment of conformationally determined cyclization. It also offers a great amount of structural information on pregel and postgel states. A fully equivalent but somewhat less developed theory employing the same model and yielding the same results is the recursive theory by Macosko and Miller (*9–11*).

A brief review of the theoretical treatment of curing and of the corresponding experimental studies includes the pregel and postgel studies of diamine–diepoxide and diamine–diepoxide–monoepoxide systems (molecular weights, critical conversion, sol fraction, and the concentration of elastically active network chains and its relationship to the stress–strain behavior in the rubbery state) by the Prague group (*12–18*), which has also studied the problems of acid curing (epoxide–carboxyl and epoxide–anhydride reactions) (*19–22*). Burchard's group has concentrated on the pregel and critical regions of the polymerization (polyetherification) of diepoxides released by diphenols (molecular weight averages, radii of gyration, and diffusion coefficients) (*23–26*). The studies headed by Rozenberg and Irzhak were concerned with aromatic diamine–diepoxide systems (molecular weights, gel points, and sol fractions) (*27–29*). Using the conditional probability approach, Charlesworth (*30, 31*) analyzed the experimental results he obtained for the reaction of diamines and diepoxides (molecular weights, and the composition of the pregel system and the sol fraction). In a somewhat simplified way, Bokare and Ghandi (*32*) extended the conditional probability approach to include epoxide homopolymerization as a consecutive reaction, but they did not compare their results with experiments. The diepoxide used in these studies was almost exclusively diglycidyl ether of bisphenol A (DGEBA). The only exception was the diglycidyl ether of resorcinol used in the Soviet studies. In the previously mentioned studies, the agreement between the theoretical predictions and experiments was relatively good. This observation challenges the often assumed inhomogeneous cross-linking and heterogeneity of all cured epoxy resins. This problem will be briefly discussed in the last section.

Curing and Network Structure

The following factors affect the network formation and structure of cured epoxy resins:

1. Functionality of monomers—usually expressed by the number of functional groups per monomer molecule. However, the functionality expressed as the maximum number of bonds by which the monomer unit can be bound with other units depends also on the reaction mechanism (reaction paths).
2. Relative reactivities—expressed as the ratios of rate constants of functional groups participating in the curing. The reactivities are either intrinsically different or their reactivity changes. The change depends on whether or not the neighboring groups have reacted (substitution effect).
3. Initial composition of the cured system—molar fractions of the epoxide, curing agent, and possibly the initiator.
4. Reaction paths (reaction mechanism)—usually resulting in conversion-dependent fractions of bonds of a given type issuing from a given monomer unit. This information also includes the consumption of existing reactive groups and the formation and consumption of newly formed reactive groups.

Studies of model reactions using low functionality monomers and yielding a small number of products that can be identified and determined are very helpful, particularly with respect to factors 2 and 4. The results cannot be, however, mechanically transposed to systems with polyfunctional components.

Curing with Polyamines

In the addition reaction with amino groups, the epoxy group is monofunctional, but becomes bifunctional if the hydroxyl group formed reacts with another epoxy group. Although the reactivities of epoxy groups in DGEBA are independent (*12–18*, *23–26*, *30*, *31*), this observation need not be so for other polyepoxy compounds; various epoxy groups have intrinsically different reactivity (*1*). If groups that differ in reactivity participate in curing, the network structure may be strongly affected by this difference.

A primary amino group is bifunctional and a secondary amino group is monofunctional. Substitution in the amino group is to be expected; that is, the reactivities of the hydrogens in the primary (rate constant k_1) and secondary (rate constant k_2) amino groups may be different. Steric and substitution effects play a role and the reactivity is thus dependent on the nature of the substituent. In the ideal case, the ratio k_2/k_1 is equal to unity, if the rate constants are related to moles of amine hydrogens, or to 1/2, if moles of the amino groups are used. The absolute values of the rate constants do not matter in the network build-up, only their ratios do. In principle, three methods are available for determining this ratio: (1) reaction kinetics obtained by monitoring the time change in the concentration of epoxy and/or amino groups, (2) chromatographic determination of distribution of free reactants and reaction products, and (3) critical conversion at the gel point (or a modification of the method to determine the limiting excess of one component to reach gelation at all) and possibly other parameters that characterize the development of a network structure (molecular weight, sol fraction, or equilibrium modulus in the rubbery state).

An analysis of kinetic data is not easy because of the complex time dependence of the concentrations of reactive groups—acceleration from the OH groups formed and retardation from hydrogen bonding between the amine and hydroxyl groups (*33*). Therefore, bulk kinetics does not yield unambiguous results (*34*), and a large excess of OH groups is to be added (*35–37*). The results, however, are somewhat affected by the polarity of the medium. Quantitative liquid chromatographic (LC) separation of reactants and reaction products obtained at different molar ratios of amine to epoxy groups (*14*) gives better data. The existing results are somewhat controversial, but an analysis of all data shows that the ratio of the rate constants (k_2/k_1) per mole of amino groups is 0.4–0.6 for aliphatic amines and 0.1–0.3 for aromatic amines (the ideal value is 0.5). Thus, the substitution effect is weak in aliphatic amines and negative in aromatic amines ($k_2/k_1 < 0.5$). This effect is well documented by observation of the shift of the gel point to higher conversions (*13*).

With some curing agents, the substitution effect is highly negative. Two examples can be mentioned: 2,5-dimethylhexane-2,5-diamine (*38*) (a result of steric effect, which can be exploited for the preparation of stable epoxy–amine adducts) and diaminodiphenyl sulfone (DDS). The observed very low reactivity of the secondary amino group formed in DDS in the curing reaction (*39*, *40*) can be ascribed to the induction effect (possible conjugation with the sulfone group). However, such a drastic reduction of reactivity was not observed in model reactions with *p*-tolylglycidyl ether in solution or even in bulk (*36*, *37*). Therefore, the observed low reactivity (*39*, *40*) of the secondary amino group may be tied to the high functionality epoxy resin used—tetraglycidyldiaminodiphenylmethane—and the resulting more densely cross-linked system. Indeed, a study of the stoichiometric amount of DDS and diglycidyl ether of butanediol indicated a complete reaction (*41*).

The negative substitution effect delays gelation on the conversion scale and shifts the network build-up to higher conversions. No positive substitution effect has been reported as yet. A theoretical analysis shows that, in a stoichiometrically equivalent system of diepoxide and diamine, the gel point conversion can vary between 0.5 (very positive substitution effect), 0.577 (equal reactivity of hydrogen atoms), and 0.618 (very negative substitution effect) (*12*). Beyond the gel point, where the increase in the effective cross-linking density as a function of conversion is more curved upward, the substitution effect is more negative, that is, fewer effective cross-links are formed just beyond the gel point and more numerous ones appear at high conversions than with zero or positive substitution effects.

In diamines (i.e. diaminodiphenylmethane, diaminodiphenyl sulfone, or hexamethylenediamine) the reactivity of amino groups seems to be practically independent, but may not be in polyamines in which the distance between the amino groups is short. Diethylenetriamine may serve as an example: the distribution of adducts with phenyl glycidyl ether could not be described by the ratios of three rate constants by assuming only the different reactivities of the primary and secondary amino groups and of the middle secondary amino group. This result is presumably due to interaction between the amino groups (*42*).

The deviations from stoichiometry caused by adding an excess of amino or epoxy groups lower the cross-linking density, and distinct minima or maxima are observed for the sol fraction and equilibrium swelling or equilibrium modulus, respectively (*13, 15–18, 43–45*). In addition, the glass transition temperature (T_g) passes through a maximum at this point, but the glass transition width passes through a minimum (*45*). Curing with polyether triamines and diamines yields somewhat softer products that are useful for elasticity studies (*43, 45*), because measurements can be performed well above the T_g. The assumption of a strictly alternating polyaddition is, however, fully justified only for stoichiometric and amine-rich samples, because polyetherification may interfere if epoxy groups are in excess.

Base-catalyzed polyetherification following or concurrent with (if active hydrogens are present) amine–epoxide addition is much slower and becomes operative only if almost all the amine hydrogen atoms have reacted and epoxy groups are still present. The relative reaction rate is at least by an order of magnitude smaller (*40*) than that of amine–epoxide addition and depends on the basicity of the catalytically active tertiary nitrogen atom present. The amino alcohols formed in the addition phase are relatively weak; formation of some polyether oligomers was observed in systems with aliphatic amines and epoxy groups in excess (*14*), but not in systems with aromatic amines (*46*). Interference of polyetherification has been offered as a reason for the asymmetry of modulus, swelling, and sol fraction curves (*42, 43*) and has been confirmed experimentally (*44*). Polyetherification can be more favored in the presence of a more basic tertiary nitrogen atom, such as *N*,*N*-diethyl-1,3-diaminopropane (*47*) or tetraglycidyldiaminodiphenylmethane (*39*, *40*, *46*), or its bifunctional analogue, diglycidylaniline. For the analogue, in addition to intermolecular polyetherification, an intramolecular etherification may be sterically favored and lead to the formation of a morpholine ring (*45*). (*See* Scheme 1).

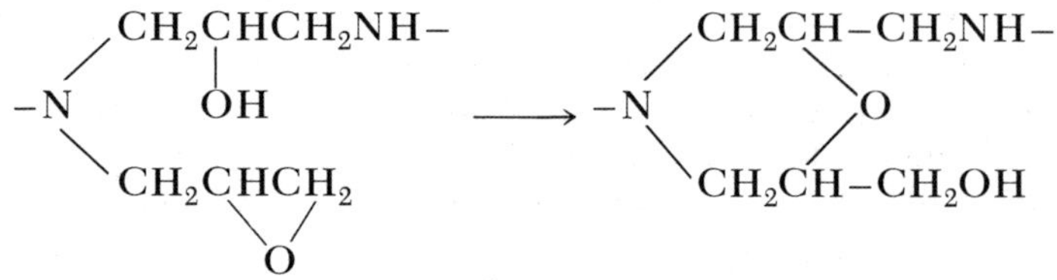

Scheme 1

A statistical treatment of simultaneously occurring amine–epoxide addition and etherification is possible, but conclusive studies of network formation are still missing. In the statistical treatment, one has to take into account the initiating role of amine–epoxide addition in producing a hydroxyl group that is itself an initiation point for addition of epoxy groups. In theory, this feature is taken into account by using oriented graphs such as the polyetherification released by the reaction of carboxyls with epoxides (*19*, *21*, *22*) or diphenols (*23*–*26*).

Impurities in epoxy resins and/or curing agents may considerably affect network formation and processing characteristics. Commercial grade tetraglycidyldiaminodiphenylmethane (TGDDM) can be used as an example. The admixtures in TGDDM have a significant effect on the stability and reactivity of the resin and on the reliability and performance of fiber-reinforced composites as well (*48*, *49*). In Araldite MY-720 (Ciba-Geigy Corporation); the TGDDM oligomers are the main by-product in addition to a small amount of reactive chlorohydrin groups in some lower functionality compounds. Although the content of epoxy groups decreases by oligomerization, the average functionality of the monomer increases. This effect is especially reflected in the weight average derived from the second moment of the functionality distribution (this average may increase, even if the number average decreases because of the presence of some nonfunctional or low functional impurities). This weight average is also decisive in gelation and in the increase in viscosity during curing. A functionality distribution has also been found in the polyoxypropylenetriamine, Jeffamine T-403, and has to be taken into account in calculating the network parameters (*45*).

All these previously discussed parameters affect the development of the network structure, and quantitative information about the magnitude of these effects can, in principle, be employed in the statistical treatment. More well-designed experimental work is needed, however.

The network characteristics determine the processing and physical properties of systems subjected to cure, but these relationships are not always clear. The main emphasis of the previous work was the cross-linking of pure components to find out whether network formation theories can predict the real behavior. For the DGEBA (possibly mixed with PGE)–diamine systems, the critical conversion and the change in molecular weights in the pregel regions were in good agreement with the predicted values. The main purpose of the postgel studies (sol fraction, w_s, and the concentration of elastically active network chains, ν_e, and its relationship to rubber elasticity and swelling) has been to find out whether the theory is reliable in predicting the network structure beyond the gel point. Detailed studies of the dependence of the sol fraction on conversion as a function of the previously mentioned parameters revealed very good agreement between the theory and experiment. In addition, the determination of the sol fraction may be a more reliable measure of the degree of cure than a direct determination of conversion. The dependence of ν_e calculated from the equilibrium modulus of elasticity in the rubbery state (high temperature of measurement or in swollen systems) on conversion or w_s over several decades of ν_e has proved to be a good application of the theory to these epoxy–amine systems. The strictly alternating character of the epoxy–amine addition and its exclusive occurrence, at least in the stoichiometric and amine-rich systems, makes the epoxy–amine networks suitable as a model for studies of the validity of the rubber elasticity theories (*15–18*).

Acid Curing

Network formation in the curing of polyepoxides with polycarboxylic acids and cyclic anhydrides is much less understood, mainly because of the much more complex reaction mechanisms.

In the carboxyl–epoxide reaction used mainly for cross-linking of carboxyl-terminated reactive polymers or preparation of unsaturated polyesters, the addition esterification is followed by a transesterification reaction in which the hydroxy ester unit is transformed into a diester and a diol unit (*50*) (*see* Scheme 2).

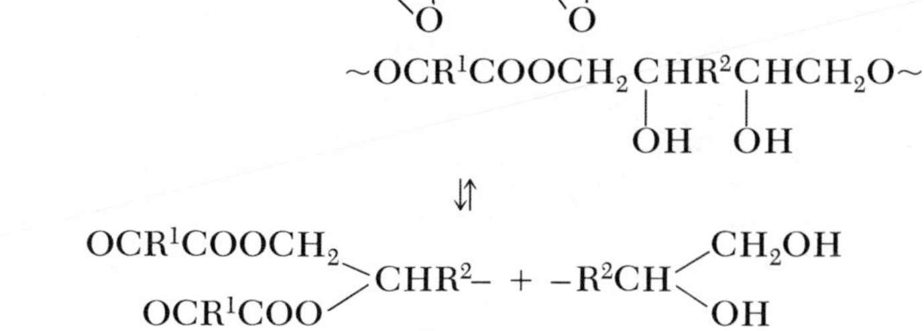

Scheme 2

Although the total number of ester bonds does not change, their redistribution (randomization of distribution) causes gelation and network formation of linear poly(hydroxyesters). The theory is in satisfactory agreement with experiments on DGEBA–azelaic acid and carboxyl-terminated polybutadiene–DGEBA systems (*21, 22*). A special feature of this type of cross-linking is an increase in the cross-linking density while the sol fraction remains high and changes little with the extent of transesterification. With an excess of epoxide, additional esterification is followed by polyetherification, which occurs simultaneously with transesterification. Quantitative studies are not yet available.

The mechanism of curing with cyclic anhydrides is less understood than that with polyamines. Recently, the reaction studied originally by Tanaka and Kakiuchi (*51–53*) was reanalyzed (*54–57*), and, in the presence of tertiary amines, the reaction was found to proceed in the absence of proton donors, most probably via a zwitterion (*58*). The simplified mechanism is shown in Scheme 3.

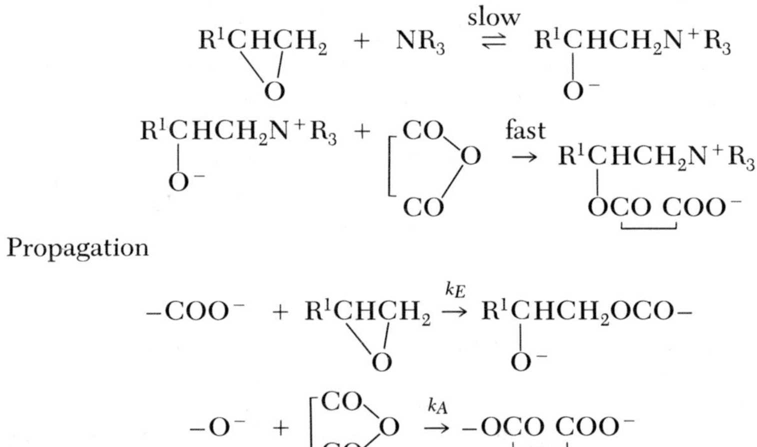

Scheme 3

The tertiary amine is thus irreversibily bound in the structure, and further chain growth occurs by an anionic mechanism (*58*). Consequently, the number of growing chains is equal to the number of reacted amine molecules, which is an example of an initiated cross-linking reaction. The network formation greatly differs from the random stepwise alternating polyaddition that is typical of curing with polyamines, and an analogous statistical treatment (*58*) is incorrect. By using an oriented graph model, the gel point and network structure have been shown to depend on the relative concentration of tertiary amine, an observation that has been confirmed experimentally (*20*).

Other Types of Curing

Of the other types of curing, only the polyetherification of diepoxides released by diphenols has been treated statistically and studied experimentally in the region up to the gel point (*23–26*). The situation is similar to other types of initiated polyetherifications such as the polyetherification initiated by the addition of amine. The observed change in molecular weights and conformational averages derived from light-scattering experiments agreed well with the theory and brought interesting information on the excluded volume and chain stiffness of polyepoxides.

The well-established ionic mechanism of curing with imidazoles and BF_3 complexes (*59–61*) seems to make possible a theoretical treatment of network formation in these systems. The extent to which the initiator may be split off from the chain ends and regenerated so that it can initiate the chain growth several times is a necessary and important piece of information.

Recently, the mechanism of the curing with dicyanodiamide has been elucidated (*62*). The main product of the model reaction of dicyanodiamide with phenylglycidyl ether (PGE) is 2-aminooxazolidine, which is obtained by the reaction of dicyanodiamide with three PGE molecules (*see* **I**).

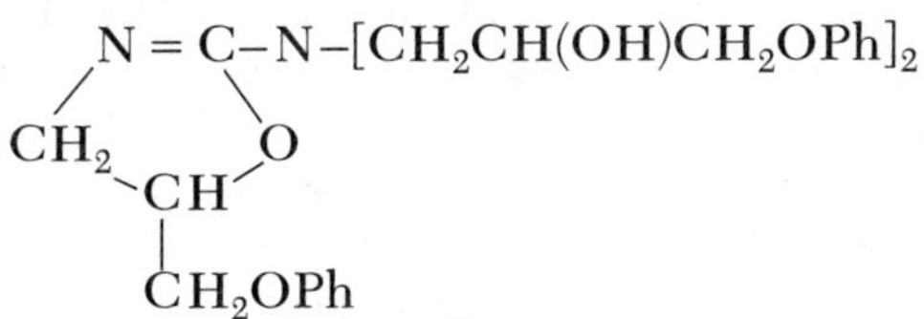

I

Dicyanodiamide is thus trifunctional with respect to the formation of bonds to epoxy units. At the same time, the product is basic and catalyzes the homopolymerization of the epoxide released by the hydroxyls formed; thus, the product is bifunctional as an initiator. In this respect, structure **I** is bifunctional. The relative rates of these reactions and the concentration of some by-products are still to be determined as a prerequisite for the statistical treatment.

Applicability of Statistical Methods to Curing

The treatment is based on an assumption of the chemical control of curing (i.e. on the independence of the apparent reactivity of a group of the size of the molecule to which it is attached) as well as on the assumption of a uniform distribution of reactive groups in the volume to make the mass action law applicable. Both assumptions have been questioned, but they seem to be valid in the majority of simple epoxy systems composed of compatible units where alternating stepwise polyaddition controls curing.

The first assumption is concerned with the rejection of a specific diffusion control; this control makes the reactivity of a group dependent on the size of the macromolecule to which the group is attached. Chemical reactions proceed via formation of an activated complex. Specific diffusion control is operative only in the case of rapid reactions (free radical recombination, fluorescence quenching) for which the rate of formation of the activated complex is the rate-determining step and is affected by the diffusion of the reacted species. This diffusion is dependent on the size of the reacted species. For slow reactions, such as stepwise polyaddition, polycondensation, and the majority of organic reactions, the transition of the activated complex to the product is rate determining, and the formation of the activated complex is controlled by equilibrium and is not size dependent (*62*). The curing reactions for epoxies belong to this category, and the independence of the reactivity from size is commonly accepted.

The overall diffusion control that sets in when the reacting system enters the glass transition region because of an increase in T_g with increasing conversion is very important for the curing of the epoxy group and strongly affects the curing rate (*63*). The reaction essentially stops when the T_g of the system (given by the maximum of mechanical losses) reaches a value 25–30 K above the reaction temperature. Although the overall diffusion control has a pronounced effect on the curing rate, it does not seem to affect the network formation on the conversion scale, for example, the molecular weight distribution and the gel point conversion (*64*). Further work is needed, however.

A widely accepted view is that cross-linked epoxy resins are inhomogeneous and inhomogeneously cross-linked in particular. This assumption is based mainly on the examination of electron micrographs that show nodular structures. However, similar nodular structures can also be seen in other polymers—cross-linked and uncross-linked—in the bulk. Leaving the problem of the structure of the glassy state open, we have no convincing evidence of inhomogeneous cross-linking from the analysis of reaction rates, network formation, small-angle X-ray scattering, or electron microscopy (*65*, *66*). In all papers on curing kinetics, the validity of the mass action law is assumed. The conclusion about the homogeneous cross-linking within the limits of statistical fluctuation of cross-link densities does not apply, however, to all cured epoxy resins. Systems with less compatible components and formed by parallel reactions producing chain sequences composed of different kind of units (e.g. epoxy–amine addition and subsequent polyetherification) may exhibit more pronounced fluctuations in composition and cross-linking density. Each case should be carefully analyzed with direct methods for determination of the spatial distribution of components and bonds.

Literature Cited

1. May, C. A.; Tanaka Y. "Epoxy Resins Chemistry and Technology"; Dekker: New York, 1973.
2. Dušek, K. *Makromol. Chem. Suppl.* **1979,** *2,* 35.
3. Bell, J. P. *J. Polym. Sci.* **1970,** *A-2,* 417.
4. Lin, C. J.; Bell, J. P. *J. Appl. Polym. Sci.* **1972,** *16,* 1721.
5. Gordon, M. *Proc. R. Soc. London,* **1962,** *A 268,* 240.
6. Gordon, M.; Malcolm, G. N. *Proc. R. Soc. London* **1966,** *A295,* 29.
7. Dušek, K. *Rubber Chem. Technol.* **1982,** *55,* 1.
8. Burchard, W. *Adv. Polym. Sci.* **1982,** *48,* 1.
9. Macosko, C. W.; Miller, D. R. *Macromolecules* **1976,** *9,* 199.
10. Miller, D. R.; Macosko, C. W. *Macromolecules* **1976,** *9,* 206.
11. Miller, D. R.; Macosko, C. W. *Macromolecules* **1978,** *11,* 656.
12. Dušek, K.; Ilavský, M.; Luňák, S. *J. Polym. Sci. Polym. Symp.* **1975,** *No. 53,* 29.
13. Luňák, S.; Dušek, K. *J. Polym. Sci. Polym. Symp.* **1975,** *No. 53,* 45.
14. Dušek, K.; Bleha, M.; Luňák, S. *J. Polym. Sci. Polym. Chem. Ed.* **1977,** *15,* 2393.
15. Dušek, K.; Ilavský, M. *Colloid Polym Sci.* **1980,** *258,* 605.
16. Dušek, K.; Ilavský, M. In "Elastomers and Rubber Elasticity"; Mark, J. E.; Lal, J., Eds.; ACS SYMPOSIUM SERIES No. 193, American Chemical Society: Washington, D.C., 1982.
17. Dušek, K.; Ilavský, M. *J. Polym. Sci. Polym. Phys. Ed.* **1983,** *21,* 1323.
18. Ilavský, M.; Bogdanova, L. M.; Dušek, K. *J. Polym. Sci. Polym. Phys. Ed.* **1984,** *22,* 265.
19. Dušek, K.; Matějka, L.; Bouchal, K.; Pokorný, S.; Ryska, M.; Vojta, V. *Proc. Int. Rubber Conf.* **1978,** *A1,* 18.
20. Dušek, K., Luňák, S.; Matějka, L. *Polym. Bull.* **1982,** *7,* 145.
21. Matějka, L.; Dušek, K. *Prepr. Div. Org. Coat. Plast. Chem.* Am. Chem. Soc. Meet., Washington, D.C. 1983.
22. Dušek, K.; Matějka, L. Chap. 2 in this book.
23. Burchard, W.; Bantle, S.; Müller, M.; Reiner, A. *Pure Appl. Chem.* **1981,** *53,* 1519.
24. Burchard, W.; Bantle, S.; Zahir, S. A. *Makromol. Chem.* **1981,** *182,* 143.
25. Bantle, S.; Burchard, W. *Prepr. IUPAC Macro 82,* Amherst **1982,** 531.
26. Zahir, S. A.; Bantle, S. *Prepr. Div. Org. Coat. Plast. Chem.* **1982,** *46,* 651.
27. Topolkaraev, V. A.; Zhorina, L. A.; Vladimirov, L. V.; Berlin, Al.Al.; Zelenetskii, A. N.; Prut, E. V.; Enikolopyan, N. S. *Vysokomol. Soedin.* **1979,** *A21,* 1655.
28. Bogdanova, L. M.; Domashneva, G. S.; Irzhak, V. I.; Sapozhnikova, E. M.; Rozenberg, B. A. *Vysokomol. Soedin.* **1976,** *A-18,* 1100.
29. Irzhak, V. I.; Rozenberg, B. A.; Enikolopyan, N. S. "Setchatye Polimery" (Network Polymers); Nauka: Moscow, 1979.
30. Charlesworth, J. M. *J. Polym. Sci., Polym. Phys. Ed.* **1979,** *17,* 1557.
31. Charlesworth, J. M. *J. Polym. Sci., Polym. Phys. Ed.* **1979,** *17,* 1571.
32. Bokare, V. M.; Ghandi, K. S. *J. Polym. Sci. Polym. Chem. Ed.* **1980,** *18,* 857.
33. Arutunyan, Kh. A.; Tonoyan, A. O.; Davtyan, S. P.; Rozenberg, B. A.; Enikolopyan, N. S. *Dokl. Akad. Nauk SSSR* **1973,** *212,* 1128.
34. Charlesworth, J. M. *J. Polym. Sci. Polym. Chem. Ed.* **1980,** *18,* 621.
35. Dobáš, I.; Eichler, J. *Collect. Czech. Chem. Commun.* **1973,** *38,* 2602.
36. Eichler, J.; Dobáš, I. *Collect. Czech. Chem. Commun.* **1973,** *38,* 3461.

37. Dobáš, I.; Eichler, J.; Klaban, J. *Collect. Czech. Chem. Commun.* **1975,** *40,* 2989.
38. Buckley, L.; Roylance, D. *Polym. Eng. Sci.* **1982,** *22,* 166.
39. Mones, E. T.; Morgan, R. J.; *Polym. Prepr. Am. Chem. Soc. Div. Polym. Chem.* **1971,** *22(2),* 249.
40. Gupta, A.; Cizmecioglu, M.; Coulter, D.; Liang, R. H.; Yavronian, A.; Tsay, F. D.; Moacanin, J. *J. Appl. Polym. Sci.* **1983,** *28,* 1011.
41. Chang, T. D.; Carr, S. H.; Brittain, J. O. *Polym. Eng. Sci.* **1982,** *22,* 1213.
42. Dušek, K.; Bleha, M.; Luňák, S. unpublished results.
43. Morgan, R. J.; Kong. F. -M.; Walkup, C. M. *Polymer* **1984,** *25,* 375.
44. LeMay, J. D.; Swetlin, B. J.; Kelley, F. N. *Adv. Chem. Series* in press.
45. Luňák, Jr., S. Thesis, Charles University, Prague 1983.
46. Byrne, C. A.; Schneider, N. S.; Hagnauer, G. L. *Proc. IUPAC Macro 82,* Amherst, 686.
47. Whitting, D. A.; Kline, D. E. *J. Appl. Polym. Sci.* **1974,** *18,* 1043.
48. Hagnauer, G. L.; Pearce, P. J. *Prepr. Div. Org. Coat. Appl. Polym. Sci.* **1982,** *46,* 580.
49. Hagnauer, G. L.; Pearce, P. J.; LaLiberte, B. R.; Roylance, M. E. Manuscript in press.
50. Matějka, L.; Pokorný, S.; Dušek, K. *Polym. Bull.* **1982,** *7,* 123.
51. Tanaka, Y.; Kakiuchi, H. *J. Appl. Polym. Sci.* **1963,** *7,* 1063.
52. Tanaka, Y.; Kakiuchi, H. *J. Polym. Sci.* **1964,** *A2,* 3405.
53. Tanaka, Y.; Kakiuchi, H. *J. Macromol. Sci. Chem.* **1966,** *1,* 307.
54. Antoon, M. K.; Koenig, J. L. *J. Polym. Sci. Polym. Chem. Ed.* **1981,** *19,* 549.
55. Lustoň, J.; Maňásek, Z. *J. Macromol. Sci. Chem.* **1978,** *12,* 983.
56. Lustoň, J.; Maňásek, Z.; Kuličková, M. *J. Macromol. Sci. Chem.* **1978,** *12,* 995.
57. Matějka, L.; Lövy, J.; Pokorný, S.; Dušek, K. *J. Polym. Sci., Polym. Chem. Ed.* **1983,** *21,* 2873.
58. Tanaka, Y.; Kakiuchi, H. *J. Polym. Sci. Part A* **1965,** *3,* 3279.
59. Fischer, M.; Lohse, F.; Schmid, R. *Makromol. Chem.* **1980,** *181,* 1251.
60. Lohse, F.; Schmid, R. *Proc. 5th Int. Conf. Org. Coat. Sci. Technol., Athens* **1979,** 31.
61. Ricciardi, F.; Romanchick, W. A.; Joullié, M. M. *J. Polym. Sci., Polym. Chem. Ed.* **1983,** 21, 1475.
62. Zahir, S. A. *Proc. 6th Int. Conf. Org. Coat. Sci. Technol., Athens* **1981,** 83.
63. Enns, J. B.; Gillham, J. K.; Small, R. *Polym. Prepr. Am. Chem. Soc. Div. Polym. Chem.* **1981,** *22,* 123.
64. Luňák, S.; Vladyka, J.; Dušek, K. *Polymer* **1978,** *19,* 931.
65. Dušek, K.; Lednický, F.; Pleštil, J.; Luňák, S. *Polymer* **1978,** *19,* 393.
66. Oberlin, A.; Ayache, J.; Oberlin, M.; Guigon, M. *J. Polym. Sci. Polym. Phys. Ed.* **1982,** *20,* 579.

RECEIVED for review November 18, 1983. ACCEPTED April 25, 1984.

2

Transesterification and Gelation of Polyhydroxy Esters Formed from Diepoxides and Dicarboxylic Acids

KAREL DUŠEK and LIBOR MATĚJKA

Institute of Macromolecular Chemistry, Czechoslovak Academy of Sciences, 162 06 Prague 6, Czechoslovakia

The base-catalyzed addition esterification reaction between an epoxy and a carboxyl group is followed by a transesterification reaction, in which the hydroxy ester is transformed into a diester and 1,2-diol. This reaction is reversible. In the reaction of a diepoxide (diglycidyl ether of bisphenol A) and a dicarboxylic acid (azelaic acid or a carboxyl-terminated polybutadiene), gelation occurs by transesterification; the cross-linking density of the gel increases and reaches a steady value corresponding to the transesterification equilibrium. A statistical theory based on the theory of branching processes has been developed for gelation and network formation by addition esterification and transesterification. A relatively high sol fraction is predicted as a result of simultaneously occurring bond formation and scission. Both the experimental sol fractions and degrees of cross-linking calculated from the elasticity modulus are in good agreement with the theory.

ADDITION ESTERIFICATION RESULTING IN THE FORMATION of a hydroxy ester is the primary reaction in the base-catalyzed reaction between carboxyl and epoxy groups. In the excess of epoxy or carboxyl groups, addition esterification can be followed, respectively, by polyetherification involving the hydroxyl group formed and the epoxy group, or condensation polyesterification involving the hydroxyl group formed and the carboxyl group. Recently (*1*), we found in the model system phenylglycidyl ether–hexanoic acid that the resulting hydroxy ester (2-hydroxy-3-phenyloxypropyl hexanoate) was transformed by a transesterification reaction (called disproportionation) into a diester and a diol (*see* Scheme 1).

0065-2393/84/0208-0015/$06.00/0

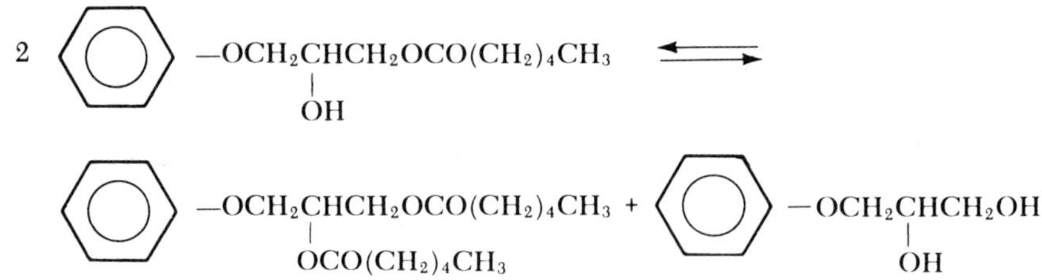

Scheme 1

Notably, this transesterification reaction is the only possible reaction to follow polyaddition esterification in stoichiometric systems. Under conditions of catalysis by bases, etherification or condensation esterification is much slower than addition esterification and, when transesterification sets in, practically no unreacted epoxy and carboxyl groups are available. The transesterification reaction occurs also in nonstoichiometric systems, but then etherification or condensation esterification interferes.

The transesterification reaction is expected to have a great effect in polyfunctional systems, because the epoxide unit in the poly(hydroxy ester) is monofunctional. However, in the diester the epoxide unit is bifunctional and the diol (glycol) unit has a functionality of zero (no bond issues from this unit). Therefore, if a diepoxide and a diacid are involved in the reaction, the reacted diepoxide units participate in the number of bonds (*i*) ranging from zero to four and gelation is possible. For example, for 100% conversion of epoxy groups, the following diepoxide units are present in the system:

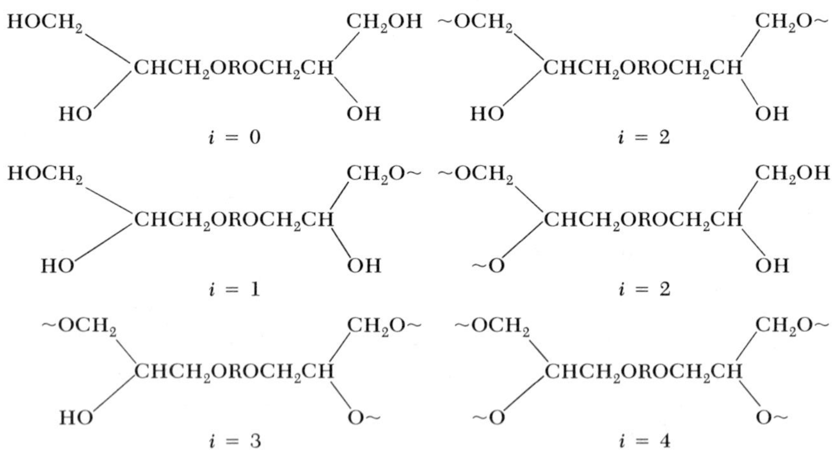

The unit with $i = 0$ (a tetrol) is contained fully in the sol, but the sol also contains structures obtained by combination of two units with $i = 1$ and other low molecular weight oligomers.

The occurrence of the transesterification reaction is important in many systems of practical interest, such as the curing of epoxy resins with polyacids, the preparation of (functional) polyesters, and the cross-linking of carboxyl-terminated liquid rubbers.

In this chapter, we give experimental evidence of gelation caused by transesterification in systems composed of a diepoxide and a dicarboxylic acid. Also, the addition esterification and transesterification are treated theoretically with the theory of branching processes. In addition, the calculated and experimentally determined gel fractions and concentrations of elastically active network chains are compared.

Experimental

Chemicals. Azelaic acid (AA, nonanedioic acid) (99.5%) was obtained from Fluka, Switzerland; redistilled phenylglycidyl ether (PGE) was of 99.7% purity (gas chromatography); diglycidyl ether of bisphenol A (DGEBA) contained 5.70 mmol epoxy groups per gram and its molecular weight (VPO) was 350, which corresponded to the number average functionality (f) 1.99. The tertiary amines (tributylamine, triethylamine) were boiled with acetic anhydride and redistilled. The sample of carboxyl-terminated polybutadiene (CTPB) was prepared by anionic polymerization and its composition was 17% 1,4-*trans*, 7% 1,4-*cis*, and 76% vinyls; $M_n = 3500$, $M_w/M_n = 1.26$, and 0.48 mmol COOH/g; and the functionality distribution was 1% $f = 0$, 18% $f = 1$, and 81% $f = 2$ (liquid chromatography) (*2*).

Reactions. All reactions were carried out in sealed glass ampules in nitrogen atmosphere. For the preparation of cylindrical samples used for determining the sol fraction and modulus of elasticity, the inner surfaces of the ampules were hydrophobized with trimethylchlorosilane to suppress adhesion of the polymer to the glass surface. DGEBA and azelaic acid were mixed in stoichiometric proportion in the absence of solvent at elevated temperatures above the melting point of the acid. To facilitate good mixing of the CTPB and DGEBA, DGEBA was first dissolved in toluene, and mixing proceeded at room temperature; the amount of toluene was 15 wt% based on the total weight of the mixture.

Analytical Methods. The concentration of carboxyl groups was determined by potentiometric titration with methanolic KOH. The samples were extracted with chloroform at room temperature until a constant weight was reached. The stress–strain dependence was determined in the compression régime with an inductive force transducer (Hottinger–Baldwin Messtechnik, Stuttgart, FRG). The compression measurements were carried out at 120 °C—about 50 K above the glass transition temperature (T_g) of the sample. The equilibrium value was obtained by monitoring the time dependence of stress.

Statistical Treatment of Network Formation

The theory of branching processes based on the tree-like model and employing the cascade substitution (*3, 4*) has been used to describe the network formation from diepoxides and diacids by addition esterification and transesterification. In this approach to network formation, the covalently bound structures are constructed (generated) from monomer units distributed according to the number of issuing bonds (reaction states) by observing the reaction paths given by the chemical mechanism. We used the same theory for the treatment of curing of diepoxides with polyamines (5–7) or gelation in diepoxide–cyclic anhydride systems (8).

We will consider a diacid and a diepoxide in which the reactivities of the carboxyl and epoxy groups are independent. Such a situation is true in a majority of cases where the reactive groups are sufficiently well separated, such as DGEBA (9). Therefore, we will consider the reaction states of the carboxyl and epoxy groups and the distribution of reaction states of the diepoxide and diacid units obtained by squaring the set of probabilities for the reaction states of the respective unit. Cyclization is not considered in the treatment, because it is believed to be weak. The smallest cycle expected to be formed with the highest probability is the 13-membered ring formed by esterification of two vicinal OH groups with one molecule of azelaic acid

The carboxyl group can occur in two reaction states—reacted and unreacted—with probabilities given by the molar conversion of the carboxyl group, ξ_C. Thus the probability of finding an unreacted carboxyl is $1 - \xi_C$ and the probability of finding the reacted carboxyl is ξ_C. The epoxy groups also occur in the reacted or unreacted states with probabilities ξ_E and $1 - \xi_E$, respectively; ξ_E is the molar conversion of the epoxy group. The reacted epoxy group can extend one chain if it occurs in the hydroxy ester unit (conditional probability $1 - \xi_T$), or it can issue two bonds if it occurs in the diester unit (conditional probability, ξ_T). The conditional probability ξ_T is defined as the molar ratio of epoxy groups in diester over all reacted epoxy groups and is, therefore, equal to the extent of transesterification reaction; ξ_T is zero if the polymer is composed exclusively of hydroxy ester units.

The set of probabilities for the diacid and diepoxide units in different reaction states can be described by the probability generating functions (PGF) F_{OC} and F_{OE}, respectively,

$$F_{OC}(\theta) = (1 - \xi_C + \xi_C\theta_E)^2 \tag{1}$$

$$F_{OE}(\theta) = \{1 - \xi_E + \xi_E[(1 - \xi_T)\theta_C + \xi_T(1/2 + \theta_C^2/2)]\}^2 \tag{2}$$

where θ_C and θ_E are variables of PGFs. The coefficients at θ_E^i or θ_C^i in the power expansions of F_{OC} and F_{OE} are the probabilities of finding a diacid or a diepoxide unit issuing bonds to, respectively, i diepoxide or diacid units. The construction of F_{OE} is in accordance with the fact that a hydroxy ester unit is found in the system with a probability $\xi_E(1 - \xi_T)$, whereas for a diester or glycol unit this probability is $\xi_E\xi_T/2$.

The PGFs F_{OC} and F_{OE} describe the distribution of probabilities of finding a unit in a given reaction state located in the root of the probability trees (i.e., a unit in the whole system including the unreacted monomer). The PGFs F_{1C} and F_{1E} for the number of bonds issuing from a unit in the first or higher generation are obtained by differentiating F_{OC} and F_{OE} with respect to the variables θ followed by renormalization.

$$F_{1C}(\theta) = 1 - \xi_C + \xi_C\theta_E \tag{3}$$

$$F_{1E}(\theta) = \{1 - \xi_E + \xi_E[(1 - \xi_T)0_C + \xi_T(1/2 + \theta_C^2/2)]\} \times (1 - \xi_T + \xi_T\theta_C) \tag{4}$$

Because of the alternating character of the reaction, the gel point condition is given by

$$F'_{1C}F'_{1E} = 1 \tag{5}$$

where F'_{1C} and F'_{1E} are values of the first derivatives of F_{1C} and F_{1E} with respect to θ_C^3 and θ_E^2 for $\theta_C = \theta_E = 1$.

From Equations 3, 4, and 5, the critical values of conversions are bound by

$$\xi_C(\xi_E + \xi_T) = 1 \tag{6}$$

and for $\xi_C = \xi_E$ (stoichiometric equivalence)

$$\xi_T = (1 - \xi_C^2)/\xi_C \tag{7}$$

The postgel parameters are determined by the extinction probabilities v_C and v_E given by

$$v_C = F_{1C}(v_E) \text{ and } v_E = F_{1E}(v_C) \tag{8}$$

After substituting v_E and v_C for θ_E and θ_C in Equations 3 and 4 and eliminating the trivial root $v_E = v_C = 1$, one obtains

$$v_C = \frac{1/\xi_C\xi_E - (1 - \xi_T/2 + \xi_T^2/2)}{\xi_T(1 - \xi_T/2)} \tag{9}$$

$$v_E = \frac{v_C - (1 - \xi_C)}{\xi_C} \tag{10}$$

The sol fraction w_s is given by

$$w_s = m_CF_{OC}(v_E) + m_EF_{OE}(v_C) = m_E\{1 - \xi_E + \xi_E[(1 - \xi_T)v_C + (\xi_T/2)(1 + v_C^2)]\}^2 + m_C(1 - \xi_C + \xi_Cv_E)^2 \tag{11}$$

where m_C and m_E, respectively, are the weight fractions of diacid and diepoxide units in the system, $m_C = n_CM_C/(n_CM_C + n_EM_E)$ and $m_E = 1 - m_C$.

The number of elastically active network chains (EANC) per monomer unit, N_e, is derived from the number of active cross-links. An active cross-link is a monomer unit issuing at least three paths to infinity. A bond issuing from a monomer unit has a finite continuation with probability v and an infinite continuation with probability $1 - v$. In this case, only the epoxide units can contribute to N_e; N_e can be obtained from the PGF $T_E(\theta)$ defined as

$$T_E(\theta) = F_{OE}[v_C + (1 - v_C)\theta] \equiv \sum_i t_i\theta^i \tag{12}$$

In this equation, the probability that a bond exists is weighted by the probability that it has an infinite continuation. Of the distribution of units given by t_i, only units with three and four infinite paths, whose fractions are given by t_3 and t_4, can contribute to N_e. The contribution is equal to half the number of issuing infinite paths. Therefore,

$$N_e = (1/2)n_E(3t_3 + 4t_4) \tag{13}$$

By combining Equations 2, 9, and 12 and selecting coefficients at θ^3 and θ^4 of Equation 12, one obtains

$$N_e = (1/2)n_E[3\xi_E^2\xi_T^2(1 - \xi_T)(1 - v_C)^3 + 3\xi_E^2\xi_T^2 v_C(1 - v_C)^3 + \xi_E^2\xi_T^2(1 - v_C)^4] \tag{14}$$

which yields

$$N_e = 1/2 n_E\xi_E^2\xi_T^2(1 - v_C)^3[3 - 2\xi_T(1 - v_C)] \tag{15}$$

The concentration of EANC per unit volume of nonextracted sample is

$$\nu_e = N_e d/\overline{M} \tag{16}$$

where $\overline{M} = n_C M_C + n_E M_E$, $w_g = 1 - w_s$, and d is the density of the sample.

The special effect of the transesterification reaction on network formation can be demonstrated by the change in the gel fraction, w_g, and the number of EANCs, N_e (Figure 1). Because the transesterification reaction causes both chain cross-linking and scission, the sol fraction remains always finite. In Figure 1, the sol fraction $w_s = 1 - w_g$ does not change much in a broad range of ξ_T, if the conversion of carboxyl and epoxy groups is high. It may even pass through a minimum, while N_e monotonically increases. This special feature is

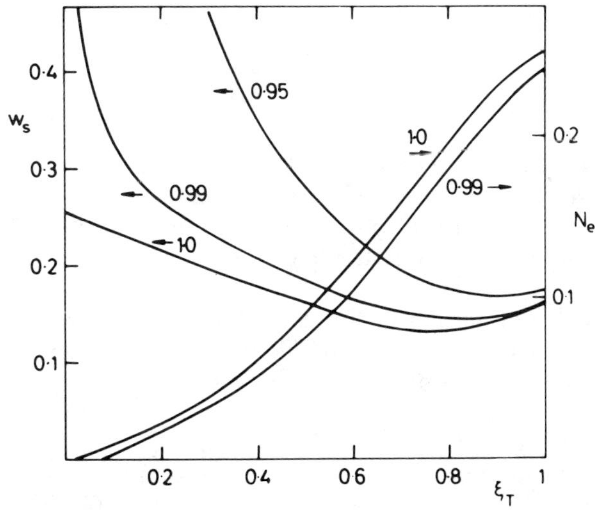

Figure 1. The calculated dependence of the sol fraction, w_s*, and the number of EANCs per monomer unit,* N_e*, on the extent of transesterification reaction,* ξ_T *in a stoichiometric mixture of diepoxide (M = 340) and dicarboxylic acid (M = 188). The extent of addition esterification* $\xi_C = \xi_E$ *(1.0, 0.99, 0.95) is indicated.*

in contrast to the situation in the usual cross-linking processes (e.g., vulcanization or polyfunctional condensation) in which w_g and N_e always increase simultaneously and w_s may be brought down to very small values.

Results and Discussion

Reversibility of the Transesterification (Disproportionation) Reaction. Transesterification of poly(hydroxy esters) is expected to be a reversible reaction. Indeed, the model reaction of PGE with hexanoic acid in the presence of tertiary amines yielding the hydroxy ester, 2-hydroxy-3-phenyloxypropyl hexanoate (*M*), in the first step and a diester, 3-phenyloxy-1,2-propanediyl dihexanoate (*D*), and a glycol, 3-phenyloxypropanediol (G), in the next step was found to be reversible. The equilibrium constant, *K*, is related to the extent of the transesterification reaction, ξ_T.

$$K = \frac{[G]\,[D]}{[M]^2} \tag{17}$$

For $[G] = [D]$ (stoichiometric system), $[M]/[M]_0 = 1 - \xi_T$ and $[G]/[M]_0 = \xi_T/2$, and Equation 17 becomes

$$K = \frac{\xi_T^2}{4(1 - \xi_T)^2} \tag{18}$$

or

$$\xi_T = \frac{2\sqrt{K}}{1 + 2\sqrt{K}} \tag{19}$$

The value of K is of the order of 10^{-1}–10^0, because only steric effects and some differences in hydrogen bonding may play a role ($\xi_T = 1/2$ would yield $K = 1/4$). From our measurement, a value of $K = 0.15 \pm 0.05$ was obtained for the reaction catalyzed with tributylamine (TBA) at 110–120 °C.

Network Formation in the AA–DGEBA System. When equimolar mixtures of AA and DGEBA are heated under catalysis with TBA, the concentrations of epoxy and carboxyl groups decrease at the same rate; and, when these concentrations are almost zero, gelation occurs. A typical example is shown in Figure 2. Continuous heating makes the network denser and the sol fraction smaller (Figure 3); a steady state is reached after 60–100 h (110 °C). The concentration of EANC, ν_e, has been calculated from equilibrium compression measurements, $\sigma = ART\nu_e(\Lambda - \Lambda^{-2})$, where σ is the equilibrium stress (negative in compression), R is the gas constant, T is the temperature in K, Λ is the deformation ratio, and A is the front factor. The concentration of EANCs, ν_e, is related to nonextracted samples; from Equation 16, $N_e = \nu_e \overline{M}/d$. Generally, A should vary between $(f_e - 2)/f_e$ and 1 (f_e is the effective functionality of an active cross-link), and for compression measurements A should be close to unity (*10*). An additional contribution to ν_e may come from trapped entanglements (*11*). In our calculation of ν_e, $A = 1$ has been used.

As shown in Figures 3 and 4, the predicted trends in w_g and N_e

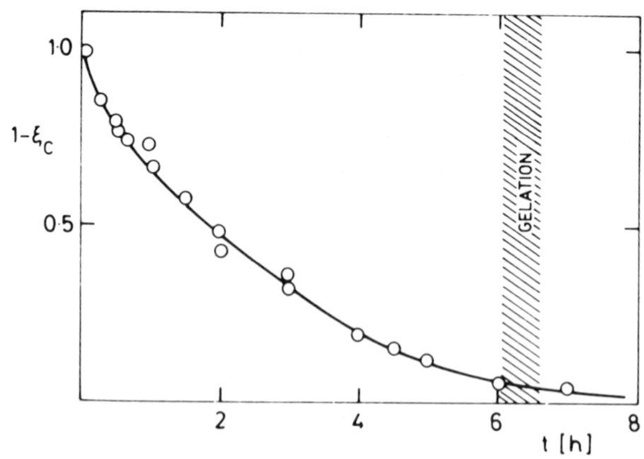

Figure 2. Time dependence of the fraction of unreacted carboxyl groups in the stoichiometric mixture of AA and DGEBA catalyzed with 1 wt% tributylamine at 110 °C.

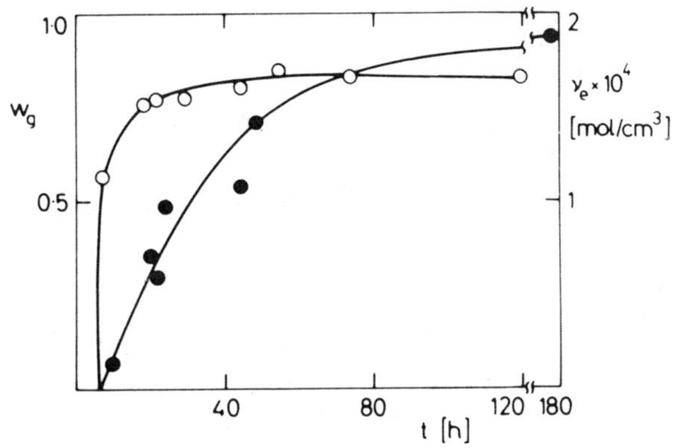

Figure 3. Time dependence of the gel fraction, w_g (○), and concentration of EANC, ν_e (●), in the stoichiometric mixture of AA and DGEBA catalyzed with 1% tributylamine at 110 °C.

are obeyed: w_g increases steeply just after gelation sets in and later changes only little while the concentration of EANCs continues to increase. A quantitative comparison with the theory is, however, somewhat difficult because of the great sensitivity of w_g and N_e to the extent of the addition esterification, $\xi_C = \xi_E$. Although ξ_C is close to unity just at the gelation threshold, 1–2% unreacted groups may remain beyond the gel point. Figure 4 shows the predicted dependence of N_e on w_g for $\xi_E = \xi_C = 1$ and 0.99, as well as the experimental data. The predicted trend is well fulfilled. The steady (equilibrium) value of $w_g = 0.83$ corresponds to $\xi_T = 0.46$ ($\xi_C = 1$) or 0.57

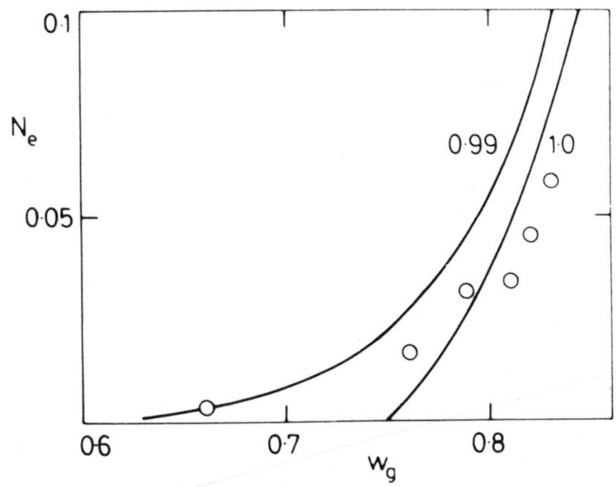

Figure 4. The number of EANCs per monomer unit, N_e, as a function of the gel fraction, w_g, in the stoichiometric mixture of AA and DGEBA catalyzed with 1% tributylamine at 110 °C. The curves are calculated for the value of the extent of the addition esterification $\xi_C = \xi_E$ indicated; experimental points were calculated from the compression modulus.

($\xi_C = 0.99$), and these values give a K of 0.18 and 0.45, respectively. The higher value seems more probable, because a small deviation (of the order of 1%) from completeness and stoichiometry is likely.

Network Formation in the CTB–DGEBA System. The average functionality of anionically polymerized CTB is less than two and, therefore, a higher sol fraction is expected in comparison with the strictly bifunctional AA. Indeed, this expectation is confirmed by the data in Figure 5.

To compare the results with theory, a slight modification of the PGFs to take into account the functionality distribution of CTB has been necessary. The C (carboxylic) component is composed of molecules of functionalities 0, 1, and 2, so that the PGFs F_{OC} and F_{1C} become

$$F_{OC}(\theta) = n_0 + n_1(1 - \xi_C + \xi_C\theta_E) + n_2(1 - \xi_C + \xi_C\theta_E)^2 \quad (20)$$

$$F_{1C}(\theta) = \phi_1 + \phi_2(1 - \xi_C + \xi_C\theta_E) \quad (21)$$

where n_i is the mole fraction of molecules with i carboxyl groups and $\phi_i = in_i/\Sigma_i in_i$. Equations 9 and 10 now become

$$v_E = \frac{v_C - (1 - \phi_2\xi_C)}{\phi_2\xi_C} \quad (22)$$

$$v_C = \frac{1/\xi_C\xi_E\phi_2 - 1 + \xi_T - \xi_T^2/2}{\xi_T(1 - \xi_T/2)} \quad (23)$$

and the sol fraction is given by

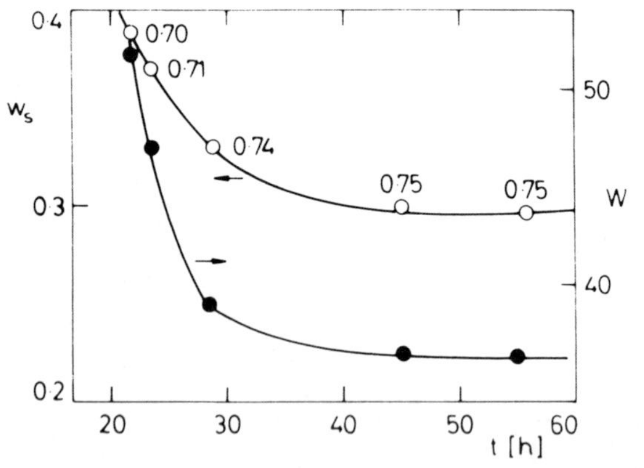

Figure 5. The sol fraction, w_s, and the weight degree of swelling, W, in chloroform as a function of time for the stoichiometric system CTB–DGEBA.

The mixture was diluted with 16% toluene to decrease its viscosity; 1% tributylamine, 110 °C; W = weight of sample swollen divided by weight of dry extracted sample; the extent of transesterification (ξ_T) calculated from w_s (0.70, 0.71, 0.74, 0.75, 0.75) is indicated.

$$w_s = m_E\{1 - \xi_E + \xi_E[(1 - \xi_T)v_C + (\xi_T/2)(1 + v_C^2)]\}^2 + m_C\{n_0 + n_1(1 - \xi_C + \xi_C v_E) + n_2(1 - \xi_C + \xi_C v_E)^2\} \tag{24}$$

In these equations the molecular weight distribution is assumed to be independent of functionality distribution. If it were not so, the number fractions (n_i) would have to be replaced by weight fractions.

The time dependence of the sol fraction and weight degree of swelling in chloroform are given in Figure 5. Equation 24 has been used for calculation of ξ_T from w_s by using the functionality distribution data given in the experimental part. The equilibrium value of ξ_T is now ~0.75, which corresponds to $K = 2.2$. This value is higher by a factor of five than that for the system DGEBA–AA and by an order of magnitude than that for the model system hexanoic acid–PGE. The increase in K may be due to the decreasing polarity of the system, which may enhance self-association by hydrogen bonding and stabilization of the vicinal OH groups in the glycol unit (formation of a five-membered ring).

Summary

Transesterification of poly(hydroxy esters) formed by addition esterification of diepoxides with dicarboxylic acids leads to gelation and formation of cross-linked structures. In stoichiometric systems catalyzed by tertiary amines, transesterification is the only reaction succeeding addition esterification. However, transesterification is also operative in nonstoichiometric systems where it competes with polyetherification or condensation esterification. Further work is needed in this direction, because the use of an excess of the epoxy component is the routine procedure for preparation of in situ reinforced rubbers.

Although transesterification is considerably slower than addition esterification, it interferes whenever a high conversion of epoxy and carboxyl groups is required. Such a situation arises, for instance, in the synthesis of unsaturated polyesters. The relative extent of transesterification seems to depend largely on the catalyst. Chromium complexes catalyze more specifically the addition esterification (*1*), so that the tendency to gelation can be suppressed (*12*).

List of Symbols

A	Front factor
AA	Azelaic acid (nonanedioic acid)
CTPB	Carboxyl-terminated polybutadiene
[D],[G],[M]	Molar concentrations of diester (3-phenyloxy-1,2-propanediyl hexanoate), glycol (3-phenyloxypropanediol), and monoester (3-phenyloxy-2-hydroxypropyl hexanoate), respectively
DGEBA	Diglycidyl ether of bisphenol A
EANC	Elastically active network chain

$F_{OC}, F_{OE}, F_{1C}, F_{1E}$	Probability generating functions for diacid and diepoxy units
K	Equilibrium constant
M_C, M_E	Molar weights of the acid and epoxy monomer, respectively ($\overline{M} = n_C M_C + n_E M_E$)
N_e	Number of elastically active chains per monomer unit
T_E	Probability generating function for active crosslinks
TBA	Tributylamine
W	Weight degree of swelling
d	Density
m_E, m_C	Weight fractions of diepoxy and diacid units, respectively
n_i	Mole fractions of molecules with i carboxyl groups
t_i	Fraction of units with i bonds having infinite continuation
v_C, v_E	Extinction probabilities
w_s, w_g	Sol and gel fractions, respectively
ϕ_i	Fraction of groups belonging to molecules with i carboxyl groups
Λ	Deformation ratio
ν_e	Concentration of elastically active chains
σ	Equilibrium stress
θ_C, θ_E	Dummy variables of probability generating functions
ξ_c, ξ_E	Conversion of the carboxyl and epoxy groups, respectively
ξ_T	Extent of the transesterification

Literature Cited

1. Matějka, L.; Pokorný, S.; Dušek, K. *Polym. Bull.* **1982,** *7*, 123.
2. Pokorný, S.; Janča, J.; Mrkvičková, L.; Turečková O.; Trekoval, J. *J. Liq. Chromatogr.* **1981,** *4*, 1.
3. Dušek, K. *Makromol. Chem., Suppl.* **1979,** *2*, 35.
4. Dušek, K. *Rubber Chem. Technol.* **1982,** *55*, 1.
5. Dušek, K.; Ilavský, M.; Luňák, S. *J. Polym. Sci. Polym. Symp.* **1975,** *53*, 29.
6. Dušek, K.; Ilavský, M. *Colloid. Polym. Sci.* **1980,** *258*, 605.
7. Dušek, K.; Ilavský, M. *J. Polym. Sci. Polym. Phys.* in press.
8. Dušek, K.; Luňák, S.; Matějka, L. *Polym. Bull.* **1982,** *7*, 145.
9. Luňák, S.; Dušek, K. *J. Polym. Sci. Polym. Symp.* **1975,** *53*, 45.
10. Erman, B.; Flory, P. J. *J. Chem. Phys.* **1978,** *68*, 5363.
11. Gottlieb, M.; Macosko, C. W.; Benjamin, G. S., Meyers, K. O.; Merrill, E. W. *Macromolecules* **1981,** *14*, 1039.
12. *Czech. Patent* 193354, 1982.

RECEIVED for review November 18, 1983. ACCEPTED April 25, 1984.

3

The Toughening of Epoxy Resins with Reactive Polybutadienes

KAREL DUŠEK[1], FRANTIŠEK LEDNICKÝ[1], STANISLAV LUŇÁK[2], MILOSLAV MACH[2], and DAGMAR DUŠKOVÁ[3]

[1]Institute of Macromolecular Chemistry, Czechoslovak Academy of Sciences, 162 06 Prague 6, Czechoslovakia
[2]Research Institute for Synthetic Resins and Lacquers, Pardubice, Czechoslovakia
[3]Research Institute of Production Cooperatives, Prague 5, Czechoslovakia

The modification of epoxy resins with carboxyl-terminated and hydroxyl-terminated liquid polybutadienes is described. The effect of this modification on the toughening of cured resins is compared with the effect of carboxyl-terminated butadiene–acrylonitrile copolymers (CTBN). Carboxyl-terminated polybutadienes (CTPB) are less compatible with the resin than CTBNs; resins modified with a prereacted CTPB of molecular weight ~2000 separate into two phases at room temperature, but may form a homogeneous solution at the elevated temperatures used in curing with anhydrides. The miscibility of the polybutadienes (PBD) was increased by attaching to the hydroxyl-terminated polybutadiene (HTPB) or CTPB polyester end-blocks formed by an in situ reaction of tetrahydrophthalic anhydride and phenyl glycidyl ether. The toughness of cured resins modified with prereacted CTPBs or PBDs with polyester end-blocks was comparable to those modified with CTBNs, although their morphology was usually much finer and had diffuse phase-separated regions.

Carboxyl-terminated butadiene–acrylonitrile (CTBN) copolymers have been used most often for improving fracture properties of epoxy resins (*1–4*). The existence of phase-separated rubber particles 10^2–10^3 nm in diameter is believed to be the necessary condition for a substantial increase in fracture energy. The modified resins should be stable prior to curing over an extended period of time. The two-phase structure should develop during polymerization at a time when the system is viscous enough to prevent macroscopic phase separation, but still below the gel point. If these conditions are not met, the chemical network would prevent the rubber phase from forming particles of a desirable size (*5*).

0065-2393/84/0208-0027/$06.00/0

For some purposes, however, the acrylonitrile (AN) copolymers are not suitable. Monomeric AN is carcinogenic and butadiene–acrylonitrile (B–AN) copolymers always contain traces of the AN monomer. Existing regulations do not permit the use of this material in products coming into direct or indirect contact with the human body. Also, when making liquid rubbers by anionic polymerization, AN cannot be used as a comonomer. For these reasons, an investigation of a modification using telechelic PBDs is of interest. No systematic study of this modification has been carried out as yet, obviously because of the general lack of compatibility of PBD and epoxy resins.

In this chapter, we report our results on the modification of epoxy resins of diglycidyl ether of bisphenol A (DGEBA) with liquid PBDs. Their miscibility with the resins was increased (a) by prereaction of the carboxyl groups with epoxy groups of the resins and (b) by attaching to the OH or COOH groups of the PBDs a polyester block formed by an in situ reaction of monoepoxide and cyclic anhydride.

Experimental

Resins and Rubbers. Several commerical low molecular weight epoxy resins of the DGEBA type containing 5.07–5.28 mmol/g epoxy groups and 0.15–0.40% Cl were used. The HYCARs (PBD) are commercial products of The BFGoodrich Company. The development samples of HTPB (BH) were prepared by free radical polymerization using functional initiators. The development samples of CTPB (LBH) and HTPB (LBC), research products of the Research Institute for Synthetic Rubber (Kralupy, Czechoslovakia), were prepared by anionic polymerization. Their characteristics are given in Table I.

Polybutadienes with Polyester End-blocks. HTPB or CTPB was reacted with hexahydrophthalic anhydride (HHPA) and phenyl glycidyl ether (PGE) under specific catalysis with Cr(III) complexes (*see* Reactions 1 and 2).

Table I. Characteristics of Liquid Rubbers

Rubber	*Reactive Groups*	*AN (%)*	c *(mmol/g)*	M	f	*Microstructure (%)* 1.4	*Microstructure (%)* 1.2
HYCAR CTBN							
1300X8	COOH	19	0.53	3200	1.7	88.0	12.0
LBC 264	COOH	0	1.17	1550	1.9	31.2	68.8
LBC 266	COOH	0	0.79	2200	1.7	33.9	66.1
LBC 272E	COOH	0	0.91	1980	1.8	38.6	61.4
HYCAR HTB							
2000X166	OH	0	0.41	3800	1.6	—	—
BH	OH	0	1.03	2200	2.3	79.0	21.0
LBH 283E	OH	0	0.77	2150	1.7	44.0	56.0

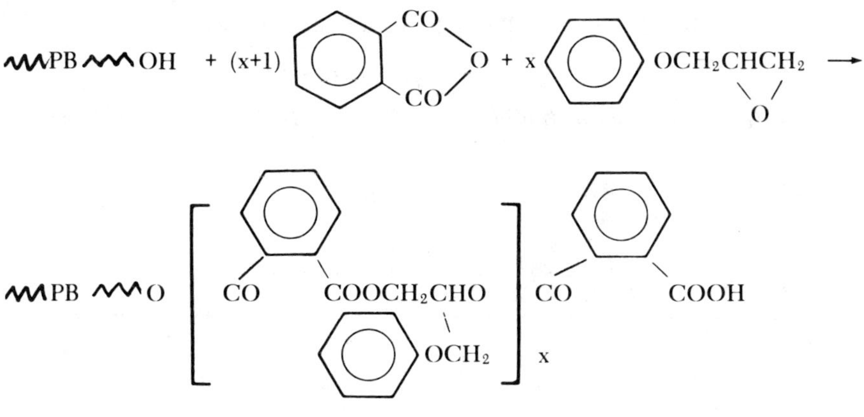

Reaction 1

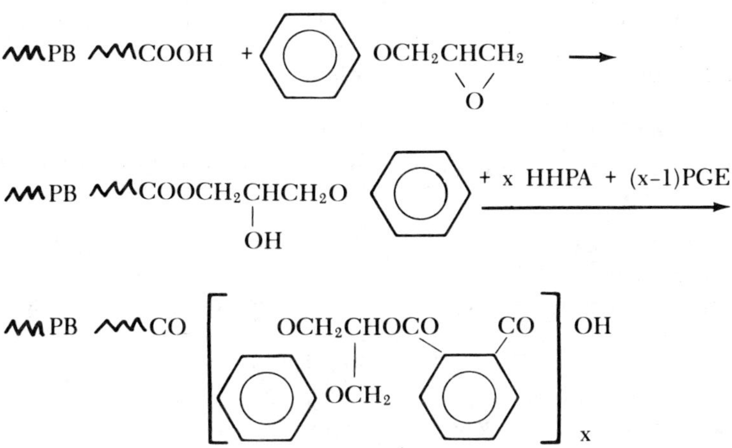

Reaction 2

Typically, the HTPBs were stirred with HHPA and the Cr catalyst ATC-3 (Cordova Chemical, Sacramento, CA), and the temperature was raised to 60–80 °C. After 20 min, PGE was added, and the mixture was kept at 80 °C for 4 h. The CTPBs were first reacted with an equimolar quantity of PGE (catalyst AMC-2) at 80 °C for 2 h. HHPA and possibly PGE were then added, and the mixture was stirred at 80 °C for another 2 h. The block copolymers are listed in Table II. The copolymers were diluted with the epoxy resin, and the mixture was stirred at 80 °C for 2 h; the HYCAR HTB 2000X166 was stirred at 100 °C.

Prereaction. All carboxyl-terminated rubbers were prereacted with the epoxy resins either in the absence of a catalyst or in the presence of Cr(III) catalyst AMC-2. In the absence of the catalyst, the components were reacted for 2 h at 150 °C. (*6*). In the presence of the Cr(III) catalyst, the reaction time was 2 h at 80 °C. The rubbers with polyester end-blocks terminated with a carboxyl group were also prereacted with the resin. The prereaction was catalyzed with the Cr complexes already present in the system to catalyze the formation of polyester as described in the previous section.

Curing and Testing of Cured Resins. Bulk samples were prepared from resins cured with HHPA in presence of 0.5% benzyldimethylamine at 100 °C

Table II. Polybutadiene–Polyester Block Copolymers

Sample Designation	Parent Rubber	Moles per Group		wt% PBD in Copolymer
		PGE	HHPA	
BH/P–1	BH	0	1	88
BH/P–2	BH	1	2	68
BH/P–3	BH	2	3	57
HY/P	HYCAR HTB	2	3	77
LH/P	LBH 283E	1	2	74
LC/P	LBC 272E	1	1	78

for 2 h and at 130 °C for 16 h in Teflon-plated molds. Standard methods were used for determination of thermomechanical properties. Fracture toughness was measured in three-point bend tests on single-notched specimens (*7*, *8*).

Curing with diethylenetriamine at room temperature was used for determination of lap shear strength. The surfaces of an aluminium alloy (Dural) were first degreased with acetone and then pre-etched with acid chromate solution. The curing took place at room temperature for 5 days.

Results and Discussion

In this preliminary study, we have attempted to answer the following questions:

1. Can prereacted CTPB form stable solutions in the resin at room temperature?
2. If these solutions phase separate at room temperature, can they be used in high-temperature curing?
3. How long should the polyester end-block be to secure the solubility of PBD in the resin?
4. How much does the toughening efficiency differ from that produced with CTBN?
5. Is the morphology the same as for CTBN-modified and cured resins, and is the presence of segregated spherical rubber particles a necessary condition for toughening?

The choice of functional rubbers as far as the variations in composition are concerned was rather limited. Samples with a sufficient range of molecular weights were represented by anionically polymerized PBD with a rather high vinyl content and a narrow molecular weight distribution. The samples prepared by free radical polymerization were of different origin and/or the chain terminal units to which the OH or COOH groups were attached were of different composition. In addition to the rubbers listed in Table I, HYCAR 2000X162 (higher molecular weight and low vinyl content) and Nisso G–3000 (very high vinyl content) were modified, and their miscibility with the epoxv resin was tested.

Stability of the Modified Resins. Similarly to B–AN copolymers (*9*), the PBDs also exhibit an upper critical solution temperature, but the immiscibility gap is shifted to considerably higher temperatures. All resins containing 5–20% rubber not prereacted with the resin formed two-phase systems; the mixtures separated more or less rapidly into two bulk layers. The prereaction helps, but only the prereacted PBD with the lowest molecular weight (1550) was compatible with the resin at room temperature (Table III). For these anionically polymerized rubbers, the threshold molecular weight for the appearance of phase separation is estimated to be between 1600 and 1900.

Attachment of the polyester block helps to increase the compatibility as is demonstrated by the results of the modification of the BH polybutadiene (Table IV). The minimum length of the end-block required to produce a stable system is equal to two anhydride and one epoxide unit. The anionically polymerized LBH and LBC rubbers modified by the same procedure behave somewhat untypically in that they are slightly turbid after modification. The turbidity is stable, and no phase separation is observed. Also, the viscosity increases much more than it does as a result of modification of the BH rubber. Originally, the turbidity was presumed a result of a high vinyl content. Experiments with the Nisso rubber (much higher molecular weight) disproved this assumption, however, because the modified rubber was clear. The possibility of impurities such as residual lithium that could catalyze the formation of unattached polyester could be ruled out because of the very low concentration of Li (several parts per million). A reason still to be checked is the relatively narrow distribution of molecular weights.

Although the long-term stability of the modified resins is an important factor, its absence does not prevent the rubber from being used in the modification at higher curing temperatures at which the system is homogeneous again. The prereacted LBC rubbers of mo-

Table III. Appearance and Morphology of Cured Modified Resins

	10% Rubber			*20% Rubber*		
Rubber	*mr*	*anc*	*amc*	*mr*	*anc*	*amc*
HYCAR CTBN 1300X8	c	t,P	—	c	t,P	—
LBC 264	c	c,S	c,S	c	c,S–R	h
LBC 266	t,s	c,h,S–R	—	t,s	h,F	—
BH/P–2	c	c,S–R	h,S–R	c	c,S–R	c,h,F
BH/P–3	c,h		c,h,R	c	—	c,h,F
HY/P	c	c,S–R	c,R	c	c,R	c
LC/P	t,s	c,S–R	h	h,t	c,S–R	h,R
LH/P	t,s	c,S–R	—	h,t	c,S–R	—

Key to abbreviations: mr, modified resin; anc, anhydride cured; amc, amine cured; c, clear; t, turbid; h, hazy; s, separates upon storage into two layers, room temperature. Morphology—P, separated particles; S, smooth surface; F, pronounced fluctuations without a sharp boundary; R, rippled surface.

Table IV. Effect of Length of the Polyester End-block on the Stability of Modified Resins and Lap Shear Strength (σ_s) of Resins Cured with DETA at 25 °C

Rubber Type	*Modified Resin*			*Epoxy (mmol/g)*	*Cured Resin σ_s (MPa)*
	Appearance	*Stability*	*Viscosity*[a]		
Unmodified	clear	stable	13.5	5.3	10.5
Prereacted	turbid	unstable	—	—	—
BH/P–1	turbid	slow separation	27	—	—
BH/P–2	hazy	stable	34	4.2	23
BH/P–3	clear	stable	48	4.1	22

[a] Pascal seconds.

lecular weight up to 2200–2300 form a clear solution in the resin at 100 °C curing temperature for the HHPA–DGEBA systems, but a prereacted LBC rubber of molecular weight 2500 was not compatible even at this temperature.

Appearance and Morphology of Cured Resins. Unlike the situation with typical AN copolymers—transition of clear and homogeneous systems into turbid and phase-separated systems during curing—the clear PB-modified resins usually yield cured systems without the formation of any distinctly phase-separated rubber particles. This observation points to the special role of AN units in their interaction with the partially cured resins. Some cured resins have a morphology not much different from the unmodified resins. They show a more or less smooth fracture surface—type S in Table III—that means the rubber is dissolved in the resin.

Distinctly separated particles, such as in CTBN (type P in Table III), are not usually observed in PBD-modified systems, although the mechanical properties are favorably affected. A characteristic morphology in this case has no sharp phase boundary but has pronounced fluctuations (type F in Table III) in composition (Figure 1) or a rippled surface (R fracture surface in Table III) with striations that are finer than the F-type surface and point to partial demixing. The F- or R-type morphology is sometimes observed for systems modified with rubber without a polyester block, but it is more typical of rubber with a polyester block.

Thermomechanical and Fracture Properties. Comparison of the properties of modified and unmodified resins (Table V) shows that the highest toughening effect is indeed obtained with the classical HYCAR CTBN rubber. However, with PBD rubbers a comparable toughening can be obtained as well without a loss in the heat distortion temperature in spite of the different morphology corresponding to a much less distinct phase separation. A similar observation has been made with modification of the resins with amine-terminated rubbers (3). As expected, the fracture toughness is the quantity most

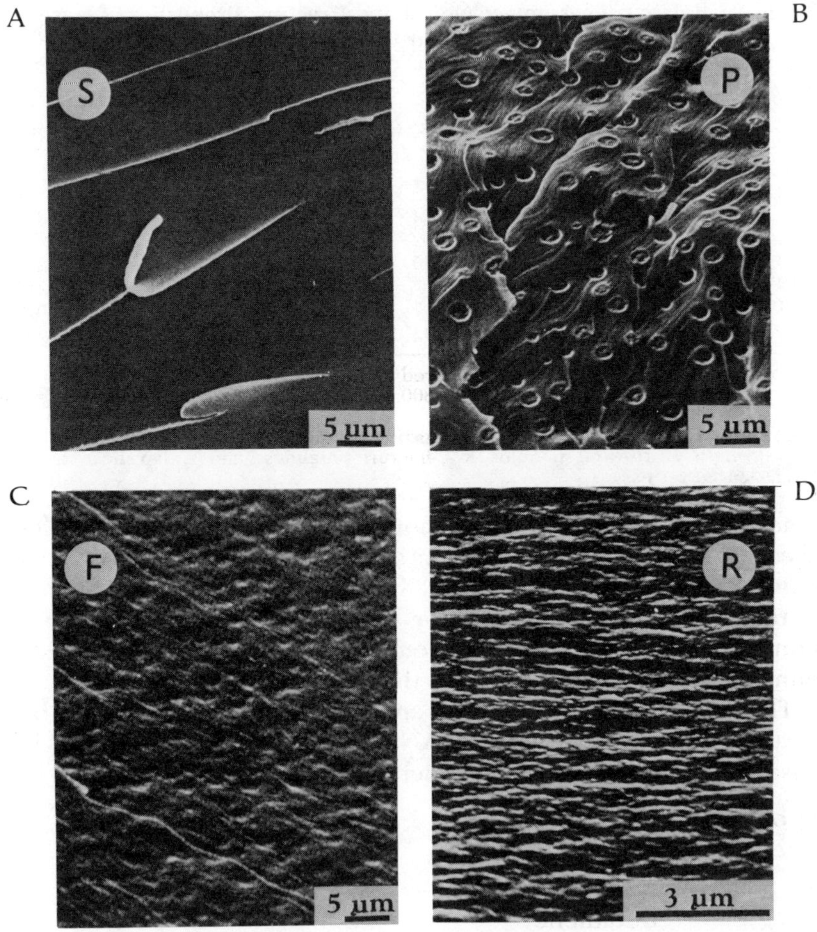

Figure 1. Typical fracture surfaces of cured, modified resins. (A) Smooth fracture surface of unmodified, anhydride-cured resin. (B) Separated particles in the fracture surface of a modified (10% HYCAR CTBN), anhydride-cured resin. (C) Pronounced fluctuations in the fracture surface of a modified (20% LBC 266), anhydride-cured resin. (D) Rippled fracture surface in a modified (10 BH/P–1), amine-cured resin.

sensitive to modification. The highest values correspond to the R-type morphology.

Conclusions

Although the systems and conditions for modification of epoxy resins with functional PBDs have not yet been definitely optimized, clearly the PBDs can replace the B–AN copolymers whenever necessary. The main drawback of PBD is its poor compatibility with the resins. This drawback can be overcome by an increase in temperature and optimization of the molecular weight in high-temperature curing or

Table V. Relative Thermomechanical and Fracture Properties of Cured Modified Resins

Rubber in Resin	*10% Rubber*						*20% Rubber*					
	HDT	τ_b	*A*	*E*	K_{IC}	σ_s	*HDT*	τ_b	*A*	*E*	K_{IC}	σ_s
HYCAR												
CTBN	92	90	130	90	215	—	86	75	170	75	289	—
LBC 264	92	95	115	95	174	195	76	80	150	85	184	160
LBC 266	102	90	170	85	162	—	87	70	200	70	205	—
BH/P–2	104	95	100	70	220	205	94	85	190	60	202	185
BH/P–3	—	—	—	—	—	200	—	—	—	—	—	185
HY/P	105	60	140	50	207	160	85	40	145	30	181	125
LC/P	101	65	150	60	192	—	90	—	—	—	120	195

NOTE: All values are for anhydride-cured systems. Unmodified resin: HDT = 131 °C; τ_b = 130 MPa; *A* = 1.8 J/cm^2; *E* = 2800 MPa; K_{IC} = 0.53 MPa m$^{1/2}$; and σ_s = 10 MPa.

Key to abbreviations: HDT, heat distortion temperature; τ_b, flexural strength; *A*, impact strength; *E*, flexural modulus; K_{IC}, fracture toughness; and σ_s, lap shear strength of DETA-cured resins.

by the chemical attachment of a more polar block. A polyester block can easily be attached by reaction of a cyclic anhydride and monoepoxide to both HTPB and CTPB. For functional PBDs obtained by free radical polymerization and for low-temperature curing, the optimum molecular weight range seems to be 2000–2500, and the optimum molecular weight range of the block is 450–600.

The morphology of modified and cured resins differs from that typical for modification with B–AN copolymers. It is finer and has diffuse boundaries with less distinct segregation of the rubber.

List of Symbols

A	Impact strength
AN	Acrylonitrile
B	Butadiene
BH	Development sample of HTPB prepared by free radical polymerization using functional initiators
c	Concentration, mmol/g
CTBN	Carboxyl-terminated butadiene–acrylonitrile copolymer
CTPB	Carboxyl-terminated polybutadiene
DETA	Diethylenetriamine
DGEBA	Diglycidyl ether of Bisphenol A
E	Flexural modulus, MPa
F	Fluctuation-type morphology of fracture surfaces
f	Number-average functionality of the liquid rubber
HDT	Heat distortion temperature, °C
HHPA	Hexahydrophthalic anhydride
HTPB	Hydroxyl-terminated polybutadiene
K_{IC}	Fracture toughness, MPa m$^{1/2}$
LBC,LBH	Development samples of CTPB and HTPB, respectively

M	Molecular weight
P	Particle-type morphology of fracture surfaces
PBD	Polybutadiene
PGE	Phenyl glycidyl ether
R	Rippled-type morphology of fracture surfaces
S	Type of the fracture surface morphology with smooth surfaces
τ_b	Flexural strength, MPa
σ_s	Lap shear strength, MPa

Literature Cited

1. Sultan, J. N.; McGarry, F. J. *Polym. Eng. Sci.* **1979,** *13*, 29.
2. Riew, C. K.; Rowe, E. H.; Siebert, A. R. In "Toughness and Brittleness of Plastics"; Deanin, R. D.; Crugnola, A. M., Eds.; ACS ADVANCES IN CHEMISTRY SERIES No. 154, American Chemical Society: Washington, D.C., 1976; p. 326–43.
3. Kunz, S. C.; Sayre, J. A.; Assink, R. A. *Polymer* **1982,** *23*, 1897.
4. Drake, R. S.; Siebert, A. R. Chap. 13 in this book.
5. Manzione, L. T.; Gillham, J. K.; McPhearson, C. A. *J. Appl. Polym. Sci.* **1981,** *26*, 889, 907.
6. Paul, N. C.; Richards, D. H.; Thompson, D. *Polymer* **1977,** *18*, 945.
7. Pavlíková, A.; Luňák, S.; Lednický, F.; Dušek, K. *Chem. Prum.* **1980,** *30*, 591.
8. Pavlíková, A.; Luňák, S.; Mach, M.; Dušek, K. *Chem. Prum.* **1981,** *31*, 28.
9. Wang, T. T.; Zupko, H. M. *J. Appl. Polym. Sci.* **1981,** *26*, 2391.

RECEIVED for review November 18, 1983. ACCEPTED April 25, 1984.

4

Simultaneous Interpenetrating Networks Prepared from Special Functional Group Triglyceride Oils

Lesquerella palmeri and Other Wild Plant Oils

M. A. LINNÉ, L. H. SPERLING, A. M. FERNANDEZ, SHAHID QURESHI, and J. A. MANSON

Polymer Science and Engineering Program, Materials Research Center, Lehigh University, Bethlehem, PA 18015

Lesquerella palmeri *is an oil-bearing wildflower that grows in the deserts of the western United States. The structure of the oil from this plant is similar to that of castor oil and contains hydroxyl groups. These hydroxyl groups can be reacted with sebacic acid (decanedioic acid) or toluene 2,4-diisocyanate to form polyesters or polyurethanes. If both cross-linkers are used, a mixed polyester–polyurethane is formed. Several interpenetrating polymer networks (IPNs) and simultaneous interpenetrating networks (SINs) with divinylbenzene (DVB) were made. The morphology of these samples depends on the method of synthesis; the sequential IPNs have smaller domains than the SINs. Modulus–temperature studies on a number of compositions show that the SINs and IPNs are tough, leathery materials in the midrange compositions.*

PETROLEUM PRICES HAVE DECLINED DURING the 1982–83 period, but present oil supplies are inadequate to meet long-term demands. One way to prevent an economic and standard of living crisis resulting from an oil shortage is to have new raw materials waiting "in the wings." When the economic picture changes sufficiently (a condition that may not be very far off), these new materials can be smoothly put into service as needed.

Most of these "new materials" are not new at all; they are natural products, sometimes called "renewable resources." As agricultural products, various plants can be grown over and over again. Many

0065-2393/84/0208-0037/$06.00/0

renewable resources are materials familiar to most people: cellulose, natural rubber, and wool (*1, 2*). Other renewable resources are virtually unknown, however, including the special functional group triglyceride oils (*1, 3*). Although most triglyceride oils, such as linseed and tung, have only double bond functionality, a few others naturally develop hydroxyl and epoxy functional groups (*3, 4*).

Of the special functional group oils, only castor oil [1,2,3-propanetriyl tris(12-hydroxyoctadec-9-enoate), **I**] has become commercially important. Surprisingly, some 99% of this oil goes into items such as urethane rubber and paints, and only about 1% goes into the medicinal market, although this 1% has become the most well-known.

Although castor oil has the advantage of relatively high purity, the castor plant is grown commercially only in a tropical climate. Therefore, we explored the chemistry of OH-bearing oils from plants that grow within the United States, preferably in climates unsuitable for most other agricultural purposes. High on this list is *Lesquerella palmeri*, a desert flower that grows mainly in Arizona. *L. palmeri*, known locally as Bladder Pods or Pop Weed (because of the brittle seed pod), produces an oil [1,2,3-propanetriyl tris(14-hydroxyeicos-11-enoate), **II**] that bears hydroxyl groups.

The number of methylene groups on the acid residue of *L. palmeri* oil is two more than with castor oil; hence the glass transition temperature (T_g) of *L. palmeri* based polymers should be lower. Of course, the method of polymerization must also be taken into account; for example, degree of hydroxy functionality and the curing agent may affect the T_g.

In addition to the natural occurrence of special functional groups in the renewable resource, they can be added through epoxidation or other reactions on the double bond. For example, oils from plants such as linseed, *Lunaria*, *Lesquerella gracilis*, and *Crambe* have been epoxidized to increase reactivity (*5*). This possibility was also confirmed for *L. palmeri* (and used in this study).

These epoxidized oils, as well as the naturally occurring, epoxy-bearing oils (*Vernonia*) can be polymerized through a variety of reactions. One such reaction involves sebacic acid (decanedioic acid), a derivative of castor oil (*see* Scheme 1). Through this type of reaction, an elastomer can be synthesized from all natural products.

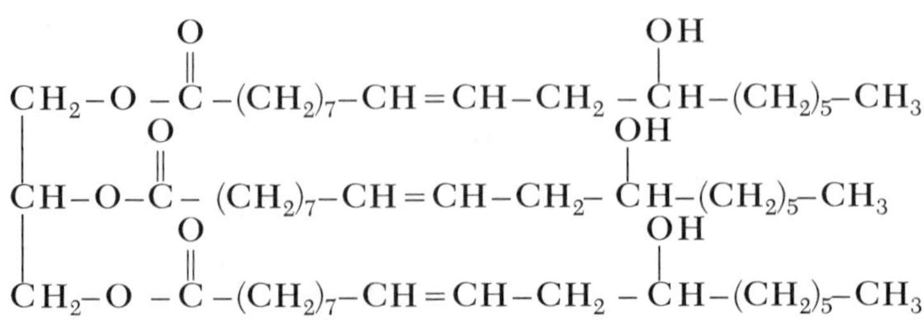

I

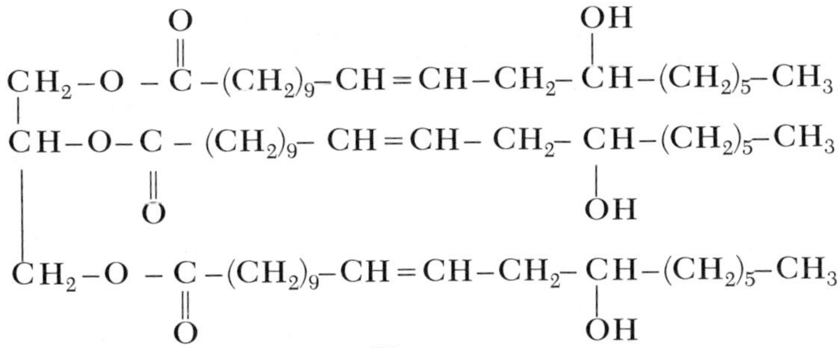

II

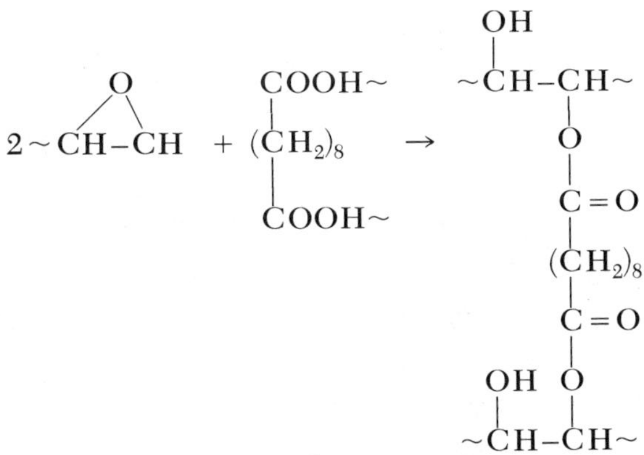

Scheme 1

During polymerization, these oils can be formed into sequential interpenetrating polymer networks (IPNs) or, more commonly, into simultaneous interpenetrating networks (SINs) (*6*). Briefly, an IPN is defined as a combination of two polymers in network form, at least one of which is synthesized and/or cross-linked in the immediate presence of the other. In an SIN, both networks are polymerized simultaneously but by independent reactions.

Both the IPNs and SINs produce toughened elastomers and reinforced plastics in a manner somewhat parallel to block and graft copolymers. However, the IPNs frequently have small domain sizes controlled by the cross-link density (*7*).

This chapter presents new information on the polymerization and SIN formation of novel, special functional group triglyceride oils and describes the morphology and mechanical behavior of these materials.

Experimental

Synthesis. The SIN synthesis from a special functional group oil is illustrated in Scheme 2. The botanical oil plus cross-linker is reacted to form an elastomeric prepolymer. The prepolymer is mixed with the vinyl monomer plus cross-linker. The vinyl monomer is polymerized with stirring until near the gel point. The product is then poured into a cell, heated to network formation, and eventually postcured at 160 °C for 2 h. The oil forms polymer network 1, and the vinyl monomer forms polymer network 2 (*see* Scheme 2).

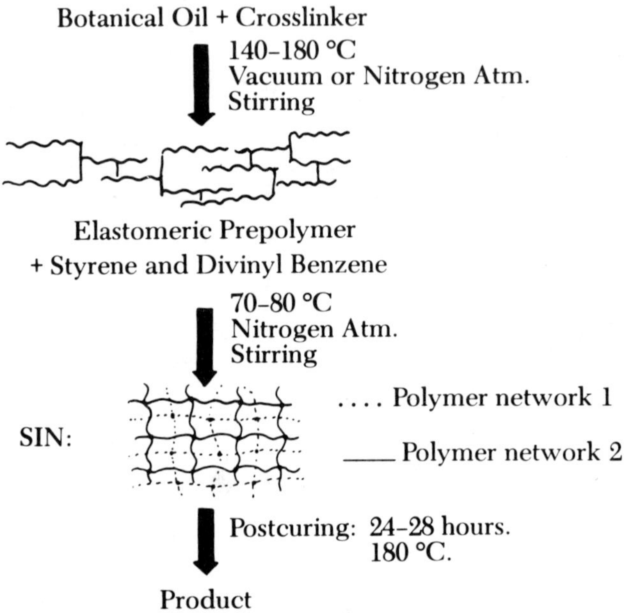

Scheme 2

L. palmeri oil seeds were obtained from R. Kleiman (Northern Regional Laboratory, USDA, Peoria, IL), and the oil was solvent extracted to yield 30% by weight. The oil contains about 51% by weight lesquerolic acid (14-hydroxyeicos-11-enoic acid), which bears hydroxyl groups (*8–10*). This functionality corresponds to 1.73 hydroxyl groups per oil molecule. The remaining triglycerides include both saturated and unsaturated species. IR analysis on our samples, however, yielded a functionality corresponding to 2.25 mol of hydroxyl groups on the basis of $M = 1009.5$ g/mol. This result is somewhat higher than the previous value. Cross-linking the oil by polycondensation with sebacic acid alone at 180 °C under vacuum results in tacky materials. However, suitable polyester or polyester–polyurethane homopolymers were obtained either by increasing the hydroxyl functionality by adding polyols (e.g., glycerol) or, taking advantage of the unsaturation by sulfur vulcanization, by radical-grafting with *tert*-butyl hydroperoxide ($t_{1/2} \sim 5$ h at 180 °C). The glycerol content was kept at 5% on the basis of oil weight, and both sulfur and *tert*-butyl hydroperoxide were 1%.

In order to increase the oil reactivity, the oil was epoxidized to attain a double bond functionality of 3.5/mol of oil. Typical epoxidation procedures involve the use of glacial acetic acid, an ion-exchange cationic catalyst (Dowex 50W.X-8), toluene, oil, and the required amount of hydrogen peroxide. After extraction and cleaning, an 80% yield of epoxidized oil was obtained. The homopolymer was synthesized readily under mild conditions (140 °C); gelation occurred at 70% conversion. The T_g of the homopolymer increased from −65 °C, for the crude oil based elastomer, to −28 °C, for the epoxidized oil based elastomer, because of the higher cross-link density in the epoxidized oil.

Prepolymers were obtained by stopping the homopolymerization well before gelation, and typical SINs were produced from mixtures of the desired amount of prepolymer and styrene with divinylbenzene (DVB) as the polystyrene (PS) cross-linker. Whenever toluene 1,4-diisocyanate (TDI) was used to cure the polyester—polyurethane network, the time of addition to the reaction vessel was critical and was just before the PS gelation.

The term SIN implies that both the networks (one from the polycondensation of the oil and the other from radical polymerization of the styrene mixture) ought to be distinct and interlocked. However, experimental evidence suggests grafting between the two networks. Thus, some grafting is probable because of double bonds in the oil.

Multiple SIN compositions were synthesized by using prepolymers from either crude oil or epoxidized oil. The prepolymers were cured to yield polyester or polyester–polyurethane networks. The SINs included oil/PS ratios of 15/85 and 50/50. The DVB level was 5% molar on the basis of styrene concentration.

In addition to the SIN synthesis, a few sequential IPNs were made. An epoxidized oil was reacted with sebacic acid until the network was complete. Styrene and DVB were then swollen in and polymerized.

The network compositions used are summarized in Table I. Crude *L. palmeri* oil means "as extracted." The epoxidized products are indicated.

Instrumental. Stress–strain curves were obtained on a TTDL model Universal Testing Instrument with a 200-lb tensile load cell (Instron Engineering Corp.). The shear modulus at 10 s, G_{10}, was determined with a Gehman torsion stiffness tester by raising the temperature 1 °C/min.

Transmission electron microscopy (TEM) studies were made on osmium tetroxide stained samples that were cut into thin sections with a Porter-Blum MT-2 ultramicrotome equipped with a diamond knife (Du Pont). A Philips 300 transmission electron microscope was employed.

Results

Because of phase separation, the *L. palmeri* SINs were all opaque; most were light yellow because of the original color of the oil. The

Table I. Network Compositions and Glass Transition Temperatures (T_g)

Composition	T_{g1}[a]	T_{g2}[b]
Elastomer		
Poly[LP(SULF(1%)),SA(1.0)]	−65	—
Poly[LP,SA(0.75),TDI(0.25)]	−55	—
Poly[ELP,SA(0.70)]	−28	—
SIN 50/50		
Poly[(LP(TBHP(1%)),SA(1.0))-SIN-(S,DVB(5%))]	−52	114
Poly[(LP,SA(0.40),TDI(0.60))-SIN-(S,DVB(5%))]	−55	113
Poly[(LP(SULF(1%)),SA(1.0),TDI(0.40))-SIN-(S,DVB(5%))]	−55	120
Poly[(ELP,SA(0.40),TDI(0.60))-SIN-(S,DVB(5%))]	−13	110
Poly[(ELP,SA(0.70))-SIN-(S,DVB(5%))]	−8	112
IPN 50/50		
Poly[(ELP,SA(0.70))-IPN-(S,DVB(5%))]	44	—

NOTE: Key to abbreviations: LP, crude *Lesquerella palmeri* oil; ELP, epoxidized *L. palmeri* oil; SA, sebacic acid; TBHP, *tert*-butyl hydroperoxide; S, styrene; and SULF, sulfur. Polystyrene T_{g2} = 92 °C.

[a] Temperature (°C), plastic phase.

[b] Temperature (°C), elastomer phase.

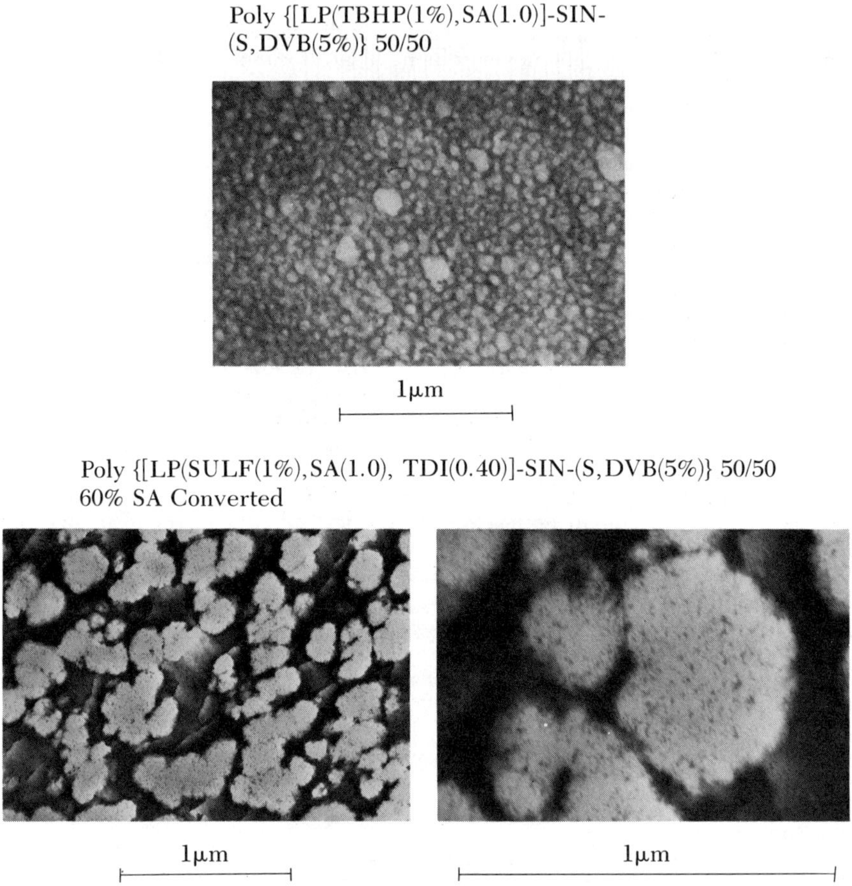

Figure 1. Transmission electron microscopy of OsO_4-stained lesquerella oil based SINs. TDI sharpens the phase boundaries over that produced only by the sebacic acid.

50/50 SINs were rather flexible and had low moduli; therefore, the oil phase may be continuous in these materials. This observation was confirmed by TEM. In contrast, 15/85 SINs were more rigid, although their oil phases also proved to be continuous.

Because of the remaining double bonds, the samples could be stained with osmium tetroxide, which makes the rubber phase appear dark; typical micrographs are shown in Figure 1. When the sample is cured with both sebacic acid and TDI, the phase domain interfaces become more distinct, and the phases become larger. The materials also may have a higher cross-link density because finishing the reaction with TDI is more efficient.

The phase domains of the sequential IPNs were smaller than

Figure 2. The phase domains of an IPN are smaller than those of an SIN; thus, the IPN is more transparent.

those of the corresponding SINs (*see* Figure 2). As a consequence of the smaller domains, the samples were optically clear. This effect is also illustrated in Figure 2.

In the sequential IPNs, the phase domain size of polymer network 2 is controlled by the cross-link density of polymer network 1 (7). Domains on the order of 1000 Å are common. For the SINs, the domain size is controlled primarily by the shear rate, and domains of 1 μm are common.

The modulus–temperature behavior of typical compositions is shown in Figure 3. The crude oil based network forms an elastomer with a T_g of approximately −65 °C. The polystyrene network forms a plastic with a T_g near 92 °C. The SIN shows both T_g values and an intermediate plateau. At room temperature, this composition is in the leathery range.

The glass transitions of several compositions are shown in Table I. The T_g values of the epoxidized oil compositions are higher than

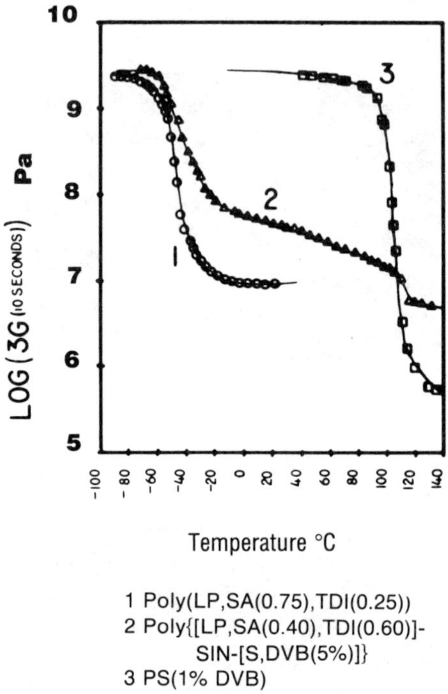

Figure 3. The shear modulus–temperature behavior of the crude oil based elastomer (○), the PS network (□), and an SIN (△).

those of the crude materials, because of the greater level of cross-linking.

The tensile behavior of several SIN and IPN compositions is shown in Table II. The mechanical properties of the crude oil products are better than those of the epoxidized oil and exhibit higher stress at break.

The properties of the crude oil SINs are further illustrated in Figure 4, which shows the tensile energy at rupture for several materials computed from the areas under the stress–strain curves. The crude oil SINs show much higher energy to rupture, in the range of 5 to 10 MJ/m^3 for the better samples.

Discussion

Effect of Grafting and Cross-linking. Although the chemistry of SIN production can be idealized, some of the complexities may be seen in Table I. The T_g values of the epoxidized networks were higher than those of the crude oil networks. Perhaps the present samples had too high a degree of epoxidation. A more moderate level of epoxidation might still allow for better network formation without raising the T_g excessively.

Table II. Tensile Properties of SINs Based on *Lesquerella palmeri* Oil

Composition	*Elastic Modulus (MPa)*	*% Strain at Break*	*Tensile Stress at Break (MPa)*
Homopolymers			
Poly[ELP,SA(0.70)]	33.2	22.1	6.1
PS	2360	2	46.1
Crude Oil			
Poly[(LP(TBHP(2%),SA(0.70),TDI(0.30))–SIN–(S,DVB(5%))]50/50	150.2	59.2	29.2
Poly[(LP,SA(0.40),TDI(0.60))–SIN–(S,DVB(5%)]15/85	161.8	22.3	35.8
Poly[(LP(TBHP)(1%)),SA(1.0))–SIN–(S,DVB(5%))]50/50	19.1	31.5	4.3
Poly[(LP,SA(0.40),TDI(0.60))–SIN–(S,DVB(5%))]50/50	35.9	45.1	14.2
Poly[(LP(SULF(1%)),SA(1.0),TDI(0.40))–SIN–(S,DVB(5%))]50/50	1.2	94.8	1.1
Epoxidized Oil			
Poly[(ELP,SA(0.40),TDI(0.60))–SIN–(S,DVB(5%))]50/50	24.8	40.4	10
Poly[(ELP,SA(0.70))–SIN–(S,DVB(5%))]50/50	68.3	8.9	6.1
Poly[(ELP,SA(0.70))–IPN–(S,DVB(5%))]50/50	33.2	70.2	9.3

NOTE: *See* Table I for key to abbreviations.

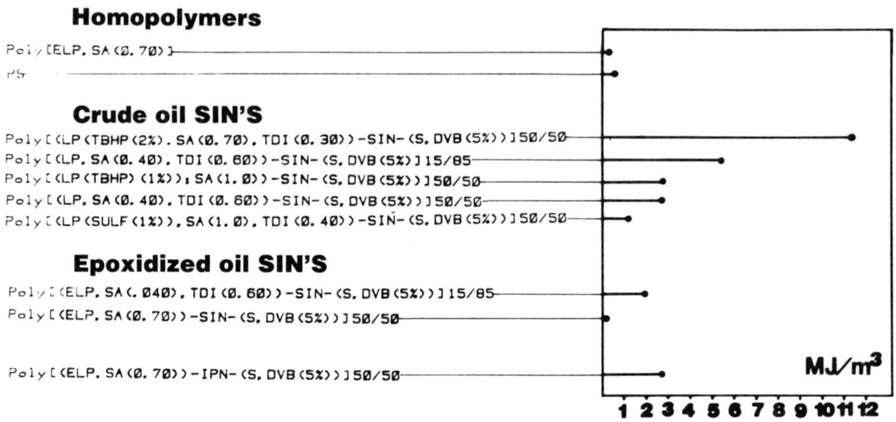

Figure 4. Tensile energy at rupture for various SINs and IPNs.

As can be seen in Table I, the glass transition of the PS is raised in the SINs. If molecular mixing were the chief effect, the T_g values should go down. Two possible mechanisms may cause the increase. First, some sort of antiplasticization may be visualized, where the oil fits into the free volume of the PS. However, this mechanism seems difficult to visualize. Second, one may imagine that the double bonds in the oil are causing extra cross-linking through grafting reactions with the PS; increased cross-linking tends to raise the glass transition of the network.

The lower level of cross-linking in the crude oil products definitely contributes to better mechanical properties. The SINs made from the crude oil 50/50 compositions especially were very tough, leathery materials.

Behavior of Related Materials. As previously mentioned, the naturally occuring triglyceride oils can be divided into two subcategories. Most of the oils contain only a double bond functionality, but, a few, such as vernonia and castor (*11*) as well as lesquerella, contain additional groups. Whereas castor and lesquerella contain hydroxyl groups, vernonia contains epoxy groups (*12*).

However, the functionality of oils with only double bonds can be increased through epoxidation. In the previous sections, we described the epoxidation of lesquerella oil; linseed, crambe, and lunaria oils have also been epoxidized and then polymerized through a variety of reactions. The use of sebacic acid and TDI has already been described but dimer acids (aliphatic, monocyclic, and bicyclic acids with two carboxyl groups) have also been used.

The glass transitions of the polymerized oils are compared in Table III. In general, the oils that were synthetically epoxidized have higher T_g values than those that were not, primarily because of the higher cross-link density. Milder reaction conditions may have per-

Table III. Glass Transition Temperatures of Networks Made from Special Functional Group Oils

Oil	*Cross-linker*	T_g *(°C)*
Not Synthetically Epoxidized		
Castor	Sebacic acid	−66[a]
Castor	Sebacic acid–TDI	−50[a]
Castor	TDI	−4[a]
Lesquerella palmeri	Sebacic acid–TDI	−55
Naturally Epoxidized		
Vernonia anthelmintica	Sebacic acid	−50[b]
Synthetically Epoxidized		
Linseed	Dimer acid	−18
Lunaria annua	Dimer acid	−34
Crambe abyssinica	Dimer acid	−35
Lesquerella gracilis	Dimer acid	−27
Lesquerella palmeri	Sebacic acid	−28

[a] Reference 11.
[b] Reference 12.

mitted lower glass transition temperatures. The aromatic compound TDI also raises T_g above that of the aliphatic sebacic acid.

The T_g values of these elastomers are important because tougher, more impact-resistant plastics can be obtained through the incorporation of lower T_g elastomers. With optimized chemistry, all of these oils could be prepared with T_g values in the range of −40 to −60 °C.

In materials studied so far, the different oils yield SINs with similar morphological features. Below about 15–20% oil, with proper stirring, the system phase inverts to the plastic (PS) phase continuity. The resulting material is a rubber-toughened plastic. Above about 20% oil, phase inversion fails to occur, and the rubber (oil) phase remains continuous. These materials behave as reinforced elastomers. Midrange compositions with the continuous rubber phase have moduli in the leathery range.

Structure of IPNs and SINs. Although the "ideal" structure of an IPN or SIN may be two completely independent networks without any grafting, real IPN and SIN structures have variable but usually low extents of grafting. For example, Widmaier and Sperling (*13*) showed by extraction experiments that less than 5% of the chains were grafted together in a poly(*n*-butyl acrylate)–PS sequential IPN.

The synthetic method clearly places the IPNs as a subclass of the graft copolymers. However, the cross-linking of the two polymers introduces a new structure-influencing element that permits a new dimension in tailor-made morphologies. The important point is that the cross-links outnumber the graft sites and are more important in the final structure, if the material is to be properly termed an IPN (*14*).

Of course, in many cases, a small amount of grafting improves mechanical behavior by increasing the extent of interfacial bonding. The work upon which this chapter is based is part of a continuing study of the interrelationships among synthesis, morphology, and mechanical behavior in IPNs. We have also investigated phase continuity and inversion in castor oil based IPNs and SINs (*15*). In brief, we show that castor oil–sebacic acid–TDI/PS–DVB compositions gel during or immediately after phase inversion, depending on composition. In sequential IPNs, increasing the cross-link density decreases the domain size (*7*). Because the domains in IPNs are probably smaller than optimum for stopping cracks, increasing cross-linking (especially in polymer network 1) beyond a certain point decreases impact resistance (*16*).

Greater cross-linking does increase molecular interpenetration at the phase boundaries, however. Recent experiments combining TEM and small-angle neutron scattering show that a "full" IPN has smaller domains, more irregular interfacial features, and smaller correlation distances (*17*) than the semi-IPNs (polymer network 1, cross-linked; polymer network 2, linear) (*18*). The semi-IPNs are similar to the SINs. Reference 18 is a review of recent oil work.

Summary

The special functional group triglyceride oils research program at Lehigh University started in 1974 as an international program with Colombia, South America, to study and develop tough IPN plastics based on castor oil. The program was then broadened with the study of other oils, such as vernonia and crambe.

This program has now been broadened further to include the study of *Lesquerella palmeri*, a plant native to Arizona. Lesquerella oils can be fashioned into prototype reinforced elastomers and tough plastics, depending on the composition and on which phase is continuous.

Acknowledgment

The authors wish to thank the National Science Foundation for support under Grant No. CPE-8109099.

Literature Cited

1. Carraher, C. E., Jr.; Sperling, L. H., Eds. "Polymer Applications of Renewable-Resource Materials"; Plenum: New York, 1983.
2. MacGregor, E. A.; Greenwood, C. T. "Polymers in Nature"; John Wiley: New York, 1980.
3. Pryde, E. H.; Princen, L. H.; Mukherjee, K. D., Eds. "New Sources of Fats and Oils"; Am. Oil Chem. Soc.: Champaign, Ill., 1981.
4. Sperling, L. H.; Manson, J. A.; Qureshi, S.; Fernandez, A. M. *Ind. Eng. Chem. Prod. Res. Dev.* **1981**, *20*, 163.
5. Qureshi, S.; Manson, J. A.; Sperling, L. H.; Murphy, C. J. In "Polymer

Applications of Renewable Resource Materials"; Carraher, C. E., Jr.; Sperling, L. H., Eds.; Plenum: New York, 1983.
6. Sperling, L. H. "Interpenetrating Polymer Networks and Related Materials"; Plenum: New York, 1981.
7. Yeo, J. K.; Sperling, L. H.; Thomas, D. A. *Polymer,* **1983,** *24,* 307.
8. Smith, C. R., Jr. In "Fatty Acids"; Pryde, E. H., Ed.; *Am. Oil Chem. Soc.:* Champaign, Ill., 1979; Ch. 2.
9. Kleiman, R.; Spencer, G. F.; Earle, F. R.; Nieschlag, H. J.; Barclay, A. S. *Lipids* **1972,** 7, 660.
10. Mikolajczak, K. L.; Earle, F. R.; Wolff, I. A. *J. Am. Oil Chem. Soc.,* **1962,** *39,* 78.
11. Devia, N.; Manson, J. A.; Sperling, L. H.; Conde, A. *Polym. Eng. Sci.* **1979,** 869.
12. Fernandez, A. M.; Murphy, C. J.; DeCrosta, M. T.; Manson, J. A.; Sperling, L. H. In "Polymer Applications of Renewable Resource Materials"; Carraher, C. E., Jr.; Sperling, L. H., Eds.; Plenum: New York, 1983.
13. Widmaier, J. M.; Sperling, L. H. *Macromolecules* **1982,** *15,* 625.
14. Sperling, L. H. *J. Polym. Sci. Macromol. Rev.* **1977,** *12,* 141.
15. Jordhamo, G. M.; Manson, J. M.; Sperling, L. H. submitted, AIChE J.
16. Donatelli, A. A.; Sperling, L. H.; Thomas, D. A. *Macromolecules* **1976,** 9, 671, 676.
17. Fernandez, A. M.; Wignall, G. D.; Sperling, L. H., presented at the 187th Natl. Meet. Am. Chem. Soc., Philadelphia, Pa., August 1984.
18. Sperling, L. H.; Manson, J. A. *J. Am. Oil Chem. Soc.* **1983,** *60,* 1888.

RECEIVED for review November 18, 1983. ACCEPTED April 20, 1984.

5

Rubber Toughening of Oxazolidinone-Modified Epoxy Novolacs

J. A. CLARKE
The Dow Chemical Company, Freeport, TX 77541

Multifunctional resins do not respond as dramatically to elastomer toughening as do their difunctional counterparts. One route to improvement is by partial "advancement" of the resin itself using reactants that provide oxazolidinone linkages. This approach minimizes the downgrading of the thermal properties in cured systems that would occur with bisphenol A advancement. These advanced resins will provide a measurable but not strongly significant improvement in fracture toughness when cured with methylene dianiline. Their main advantage lies in the response to incorporation of carboxyl-terminated nitrile rubber into multifunctional resins. This incorporation brings the measured toughness (via the fracture energy) into the same performance range as typical commercial difunctional epoxy resins.

EPOXY RESINS FORM AN IMPORTANT POLYMER CLASS within the family of thermoset materials. Most types of epoxies exhibit an inherent brittleness that is due to the typical glassy cross-linked nature of thermosets. In many applications, this brittleness is of no importance; in another major application, that of coatings for metals, brittleness is overcome by an advancement reaction of the liquid epoxy with bisphenol A. This advancement produces a solid resin that, upon curing, is substantially tougher than its liquid counterpart. Here, toughness is gained at the expense of high-temperature performance. In other areas, such as structural adhesives, neither brittleness nor loss in glass transition temperature (T_g) can be tolerated. A means of modifying the resin to alleviate brittleness must be found without degrading the important thermal and environmental qualities of the cured epoxy, reasons for selection of this particular polymer type in the first place. Rubber toughening techniques have reached this goal with commercially proven success (*1*). Increases in fracture toughness

0065-2393/84/0208-0051/$06.00/0

on an order of magnitude have become commonplace for the typical, difunctional, bisphenol A based epoxies such as D.E.R.[1] 331.

The solution to the brittleness problem becomes more difficult for multifunctional resins such as D.E.N.[1] 438 epoxy novolac. These high-temperature performance resins ($T_g > 200$ °C) usually combine multifunctionality with partial elimination of flexible linkages to attain their desired properties. The hardener must also be selected to follow these guides in order to develop a highly cross-linked rigid cured network. As a result, the viscoelastic response of the network that is required for energy dissipation (*2*), whatever specific mechanism is applied, is greatly restricted. Although toughening of this class of epoxies is an important need, the ability to respond to the presence of rubber inclusions is much more limited (*3*). In this case, rubber alloying in itself is becoming generally recognized as an inadequate toughening technique. A practical answer might be to combine this alloying with certain principles used in high-temperature thermoplastics. A way to loosen the tight cross-linking to simulate thermoplastic behavior without a simultaneous reduction in the T_g should improve the ability of the network to react in a viscoelastic rather than brittle fashion in the presence of the rubber inclusions. An application of such a principle to a well-established high-temperature epoxy, D.E.N. 438, is the subject of this chapter.

Description of Resin

Two features distinguish D.E.N. 438 epoxy novolac from the bisphenol A based epoxies. First, D.E.N. 438's broad molecular weight distribution is accompanied by an equally broad distribution in epoxide functionality of the individual molecules. Functionality varies from 2 to over 20, as indicated in Figure 1. This variation is expected from the condensation reaction of phenol with formaldehyde to produce an average 3.5-functional novolac after removal of unreacted phenol. Coupling of the phenolic hydroxyls with epichlorohydrin produces the epoxy novolac. The resin makeup includes about 25% diepoxide (diglycidyl ether of bisphenol F), which does not contribute to high-temperature performance, and another 25 wt% with functionality greater than 6. This observation suggests that gelation will occur early in the curing reaction or in any other reaction involving epoxide that we may use to modify the resin.

Second, the solubility parameter (δ) of D.E.N. 438 is higher than that for liquid diglycidyl ether of bisphenol A (DGEBA) epoxies. Small's method of calculation (*4*) estimates δ to be about 1.0 unit higher (10.2 vs. 9.15 if the calculation is based on the pure structural formula). This result suggests a different interaction with solvents and with the nitrile rubber modifier, an important consideration.

[1] Trademark of The Dow Chemical Company.

Molecular Weight Distribution

N-MER	WEIGHT FRACTION	
1	(stripped out)	
2	.25	
3	.175	← wt. avg. = 3.5f
4	.137	
5	.105	
6	.080	
7+	.25	

THEORETICAL STRUCTURE

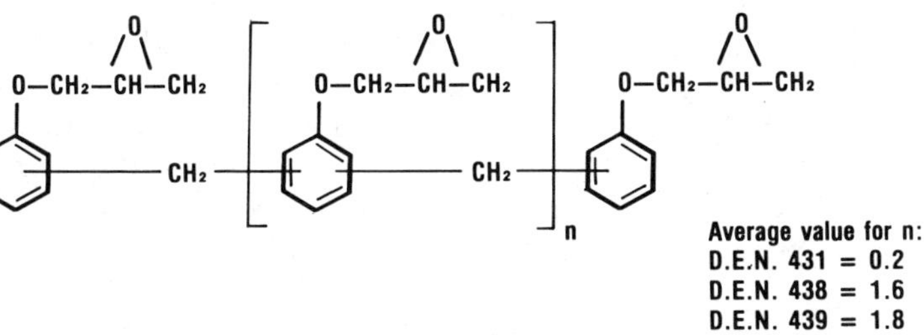

Figure 1. D.E.N. 438 epoxy novolac molecular weight distribution and theoretical structure.

Experimental Approach

The objective of this study is to compare the performance of cured D.E.N. 438 epoxy novolac to a chemically modified D.E.N. 438 epoxy novolac of lower cured cross-link density. This comparison is to be made both with and without rubber toughening to determine the effectiveness of the rubber in each case. The principal criterion is fracture energy, $\mathcal{G}_{Ic}$, which indicates relative resistance to crack propagation.

The obvious way to reduce cross-link density in an epoxy is through "advancement"—reaction of a portion of the epoxide content with a diphenolic such as bisphenol A. This advancement converts a liquid resin into an inherently tougher solid product of reduced epoxide content. D.E.R. 661 solid epoxy resin, epoxide equivalent weight (E.E.W.) = 475–575, is an example. When cured with 4,4′-methylenedianiline (MDA), it has over twice the fracture toughness of its liquid counterpart, D.E.R. 331 epoxy resin, ($\mathcal{G}_{Ic}$ = 0.58 vs. 0.24 kJ/m^2). But at the same time the cured T_g drops from 165 to 115 °C. The result is similar with D.E.N. 438 epoxy novolac—improved toughness (not as great because the amount of advancement allowed is limited by the approach of the gelation point) along with an attendant sharp loss in T_g.

An alternate method to advance an epoxy monomer without a major sacrifice in T_g uses a diisocyanate reactant such as toluene 2,4-diisocyanate (TDI) to direct the reaction so that oxazolidone[2] rings are formed as part of the bridge coupling two epoxy molecules (Figure 2A, where the two original epoxy molecules are depicted as circles). The TDI residue is centered in the bridge—each NCO group has condensed with an opened epoxide (oxirane) to form a five-membered oxazolidone ring. This ring structure has excellent thermal stability in an internal position (Figure 2). It substitutes for the 2-hydroxypropyl diether link, $-O-CH_2-CH(OH)-CH_2-O-$, formed if advancing with bisphenol A. In this reaction, we have avoided the inclusion of this highly flexible aliphatic group that leads to the large drop in T_g.

The chemistry and procedure for this specific, advanced, epoxy novolac is given in the patent literature (5). About 20% of the epoxide content is reacted with the dimethyl carbamate of TDI (formed in situ by reaction of NCO with methanol), to give the structure in Figure 2A. Because the resultant epoxide content is still 17.2%, the product is classed as a modified epoxy resin. Its IR structure is shown in Figure 3.

A second way to include an oxazolidone ring in an epoxide advancement reaction uses isocyanuric acid (ICA) rather than TDI as the coupling reactant. The linkage differs sufficiently to include this product in this discussion. Its structure is shown in Figure 2B. At first glance, this structure has a more flexible appearance than Figure 2A because of the aliphatic chain—one 2-hydroxypropyl ether group is now present. Yet, when the ability to form a hydrogen bond is recognized (confirmed by broadening of the oxazolidone absorption band in the IR scan, Figure 3), a second ring structure develops in this linkage. This development brings the two overall structures in Figures 2A and 2B much closer in appearance. Note that these linkages are identical outside the area enclosed by the wavy lines. The procedure for preparation of the cyanuric acid modified resin is also described in the patent literature (6).

Three resins now have been described: unmodified D.E.N. 438 and D.E.N. 438 modified by reaction of 20% of the epoxide with either TDI or ICA. Properties of these resins are given in Table I. These resins are to be compared in this chapter, before and after incorporation of 10 phr (parts per hundred parts resin) of nitrile rubber.

Experimental

Rubber Incorporation. A carboxyl-terminated butadiene–acrylonitrile (CTBN) rubber containing 26% acrylonitrile (AN) was chosen for optimum

[2] Indexed as oxazolidinone in Chemical Abstracts.

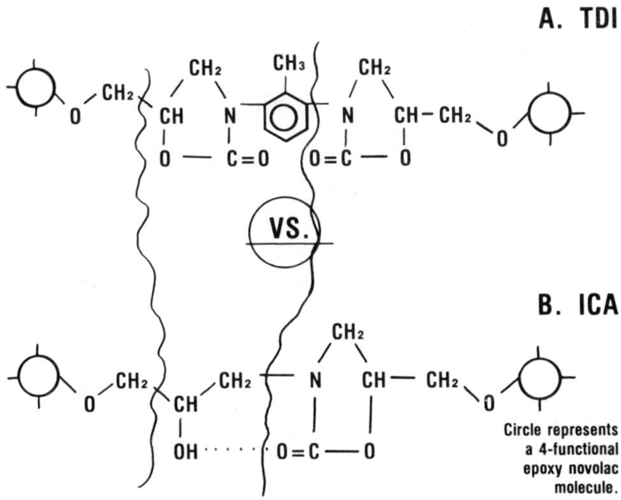

Figure 2. Comparison of bridges joining coupled molecules in modified epoxy novolac: (A) D.E.N. 438–TDI epoxy novolac, (B) D.E.N. 438–ICA epoxy novolac.

compatibility with the epoxy novolac. This liquid, low molecular weight rubber is a product of The BFGoodrich Company and is designated as HYCAR CTBN 1300X13 (7). It is represented structurally as HOOC-$[(CH_2CHCHCH_2)_x(CH_2CHCN)_y]_m$COOH, where the average $x = 3$, $y = 1$, and $m = 15$.

This rubber is soluble in the resins but performance is enhanced by prereaction of the carboxyl endgroups with epoxide. This prereaction was done by mixing for 1 h at 120 °C in the presence of 0.1% DMP-30 catalyst. [DMP-30 is tris(*N*,*N*-dimethylaminomethyl) phenol (Rohm & Haas).]

Preparation of Castings. One hundred grams of the epoxy novolac or the modified epoxy novolac was weighed into a 250-mL beaker and melted and devolatilized at 140 °C in a vacuum oven (160 °C was required for the more viscous CTBN-modified resins).

A stoichiometric amount of molten MDA at 125 °C was added and mixed in for 1 min. The mixture was immediately poured into a heated (125 °C) sheet mold consisting of two 10 × 10 in. aluminum plates coated with a fluorinated release agent. A 1/8 in. silicone gasket was used as a spacer. The cure cycle was 16 h at 125 °C, then 2 h at 175 °C plus 2 h at 225 °C. The mold was cooled slowly in the oven before dissembling.

Physical Testing. Dynamic Mechanical Spectroscopy. A 2 1/2 × 1/2 × 1/8 in. strip was cut from the cured casting to obtain a dynamic mechanical spectroscopy (DMS) curve, Figure 4. These curves were run for the non-CTBN modified resins on a Rheometrics DMS Model 1770 at a frequency of 1 Hz. Values for T_g were taken from the peak of the tan *S* curve for these three specimens and are reported in Table I. T_g values for the three rubber-modified resins were measured by differential scanning calorimetry (DSC).

Differential Scanning Calorimetry. A DuPont 912 thermal analyzer was used with a heating rate of 10 °C/min. The T_g value was calculated automatically through the interactive DSC program by using the inflection point for the break in the heat flow curve.

Durran Softening Point. *See* Reference 8, pp. 4–23.

Fracture Toughness. The compact tension test derived from ASTM E-

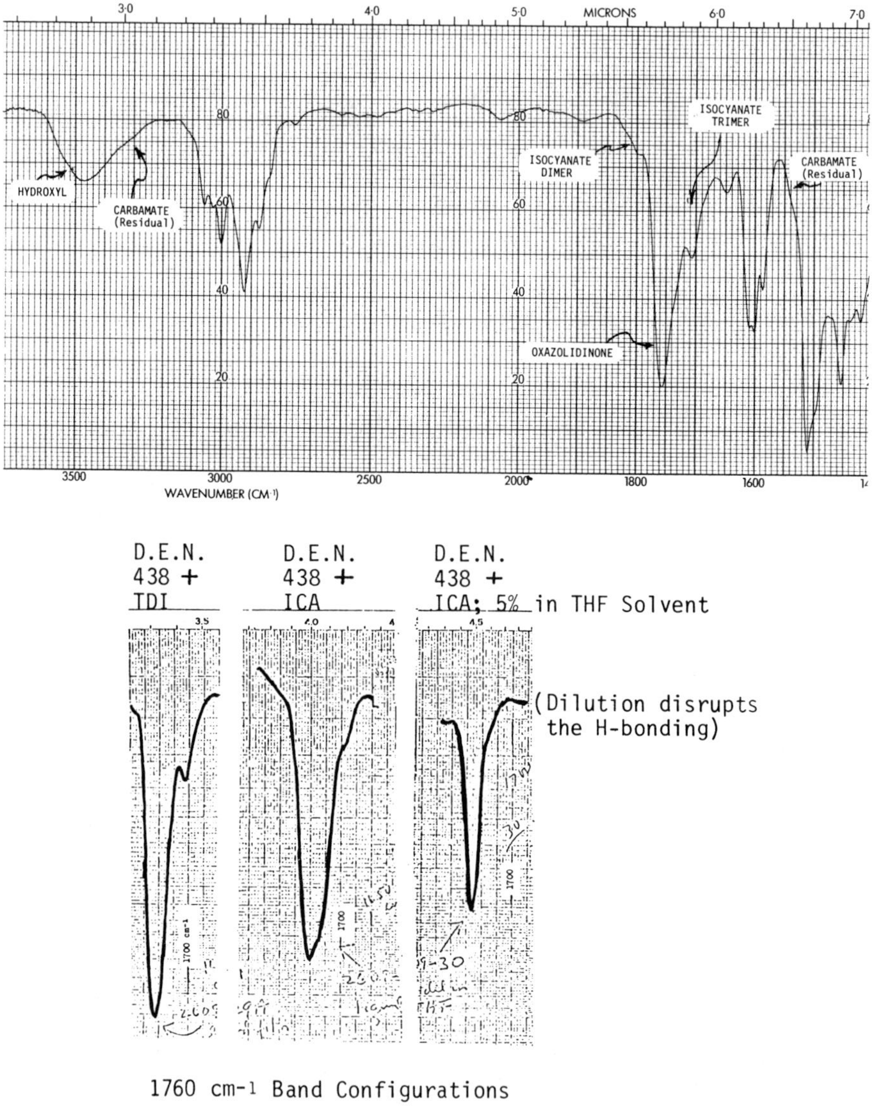

Figure 3. IR scans of modified epoxy novolac.

399 for metals was used (9). This test is one of a variety of crack-propagation tests that are becoming popular for evaluating variations in toughness in polymer systems (*10*).

Test pieces, 1 × 1.04 × 1/8 in., were cut from the castings, notched (9), and then precracked by tapping with a heavy-duty razor blade. The crack length (*a*) varied from piece to piece over a range of 0.06 to 0.25 in. The preliminary test result, K_q, was calculated by combining the force required to start propagation of the precrack with specimen dimensions in the equation given in ASTM

Table I. D.E.N. 438: Comparisons Before and After Cure

Parameter	*Unmodified*	*Modified with TDI*	*Modified with ICA*
Epoxide, %	24.0	17.2	18.4
Softening point, °C (Durran)	40	85	80
DMS Analysis (MDA cure)			
T_g, °C	243	222	231
$\overline{M}_c$	80+	170	145

NOTE: $\overline{M}_c$ is average molecular weight between cross-links

E-399-82. This result was converted to $\mathcal{G}_{Ic}$ by using the measured tensile modulus (E) and an assumed Poisson's ratio (μ) of 0.34

$$\mathcal{G}_{Ic} = \frac{(K_q)^2 (1 - \mu^2)}{E}$$

where $\mathcal{G}_{Tc}$ is defined as the critical strain energy release rate (*11*). A graphical scale relating $\mathcal{G}_{Ic}$ values for various materials is included in Reference 11. Unmodified D.E.N. 438 epoxy novolac is found near the glassy end of this toughness scale.

STRUCTURAL ADHESIVE TESTING. Formulations were prepared as follows: 100 parts resin excluding CTBN rubber; 10 parts dicyandiamide (pulverized DICY); 80 parts aluminum powder (Type Alcoa M-101, 3 mil maximum particle diameter); and 5 phr fumed silica (Cab-o-Sil M-5). The components were hand mixed at 120 °C and devolatilized.

The change in curing agent from the MDA used in preparation of the castings to DICY for the adhesives formulation follows a practice used in this laboratory. MDA is an excellent choice for comparing epoxies of widely differing structure because of its versatility of cure schedule and its good mechanical properties in the cured casting. On the other hand, DICY is the accepted hardener with structural epoxy adhesives and is better adapted to the need for a latent preformulated paste adhesive.

Lab shear samples (ASTM-D-1002) were prepared by using 1 × 4.5 × .062 in. Type 2024-T3 aluminum strips. The strips were treated with hot $Na_2Cr_2O_7$/H_2SO_4/H_2O etchant, 10/2/30 weight ratio. The 1/2 in. overlapped glue lines were cured 1 h at 180 °C. Glue-line thickness was controlled at 0.003 in. by an aluminum filler.

The climbing drum peel test (ASTM D-1781) used flexible type 302 stainless steel strips, 1 × 12 × 0.020 in. The assembly consisted of two adhered strips plus a 1/4 in. thick mild steel backing plate. Surface preparation was simply wiping with an acetone-soaked tissue.

Data and Discussion

Table I gives T_g values for the three resins before rubber inclusion. Note that, although a drop in T_g of 10–20 °C is indicated for the two advanced resins, the T_g values are still substantially above 200 °C.

One advantage to the DMS analysis is that it produces a value for shear modulus in the rubbery plateau region above T_g. This result is clearly evident for the $\mathcal{G}'$ curves in Figure 4 above 250 °C. This

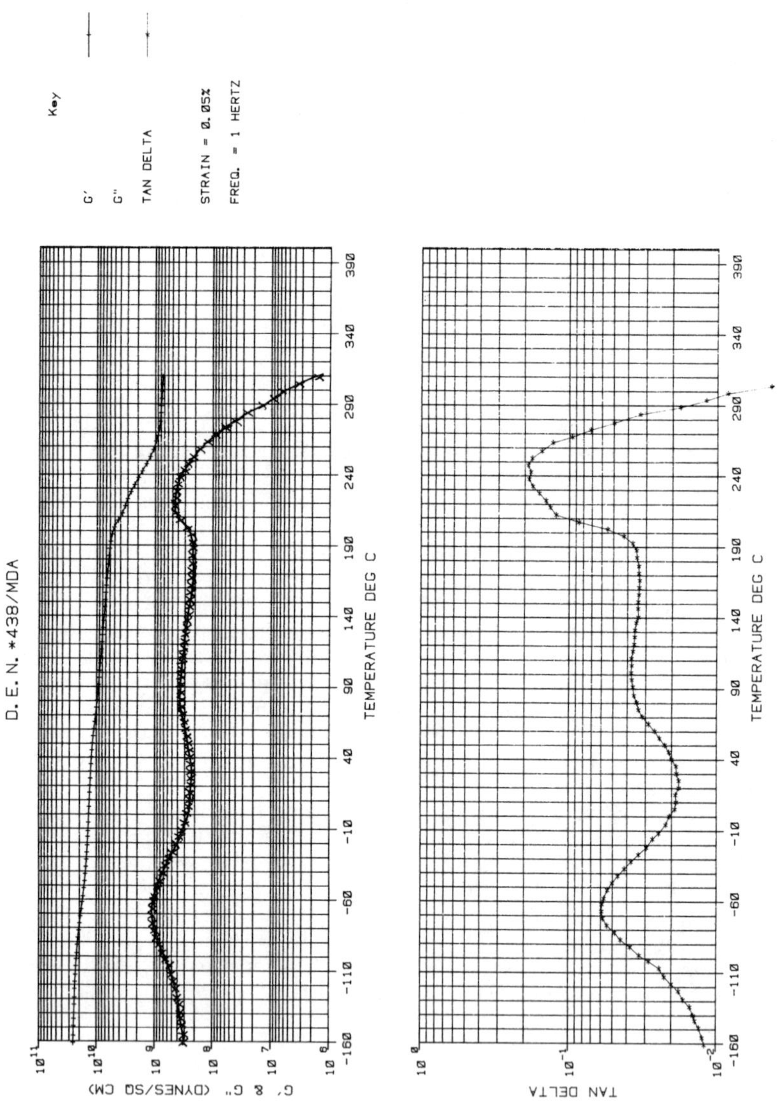

Figure 4. DMA comparison of the three epoxy novolacs.

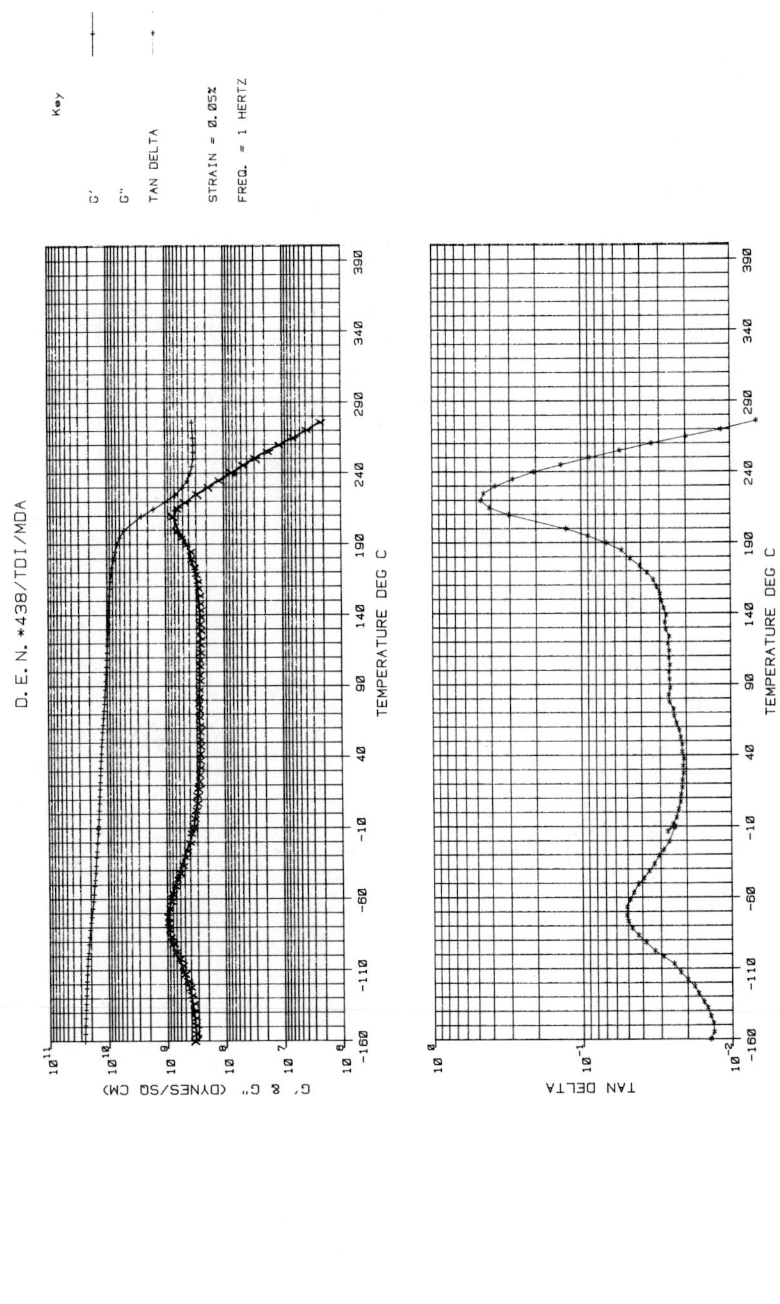

Figure 4. Continued

Continued on next page

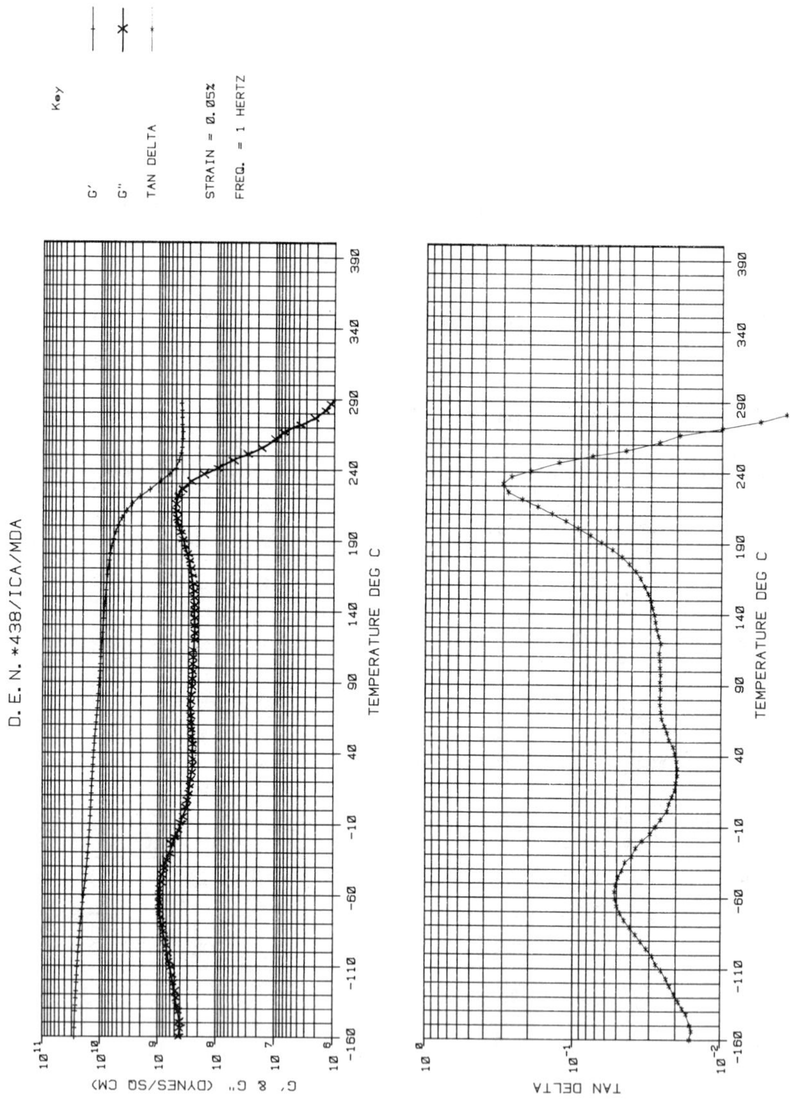

Figure 4. Continued

value may be used to estimate network cross-link density (*12*). The M_c (molecular weight between cross-links) values reported in Table I use this method and indicate a strong reduction in cross-link density for the two modified resins; the reduction is greater for the bulkier TDI-derived linkages. This evaluation of M_c is not absolute because the $\mathcal{G}'$ measured is dependent on test frequency; however, it allows a useful comparison within a single set of data. Also of interest are the low numerical values that were found, especially in comparison to the M_c = 320 value for a D.E.R. 331 epoxy resin–MDA system. These values indicate a tightly cross-linked system that does not undergo a strong drop in rigidity above T_g. This observation is still puzzling because not all the structure between the cross-links is accounted for. However, a satisfactory way was found to match these DMS-derived data to values calculated from the structure of the cross-linked epoxy–MDA network itself; a cross-link site and its adjacent rigid structure were considered as a "point" (*13*), and this weight was excluded from the calculation. The flexible portions of the chains between the cross-links were then used to calculate M_c; the calculations included the total weight of the difunctional epoxide present in the original D.E.N. 438 epoxy novolac (25%). This method gives M_c values quite close (10–20 units higher) to the values reported in Table I and also emphasizes the sharp drop in cross-link density seen in the two modified resins.

Toughness. When defined as resistance to propagation of an already existing crack, toughness can be evaluated by adhesive peel measurements as well as by the $\mathcal{G}_{Ic}$ test. Some correlation should exist between the two tests. The peel values reported in Table II are all very low but indicate improvement in the two modified systems. Lap shear strengths also are higher. The adhesive lap shear test (ASTM D-1002) allows peel forces to be introduced as stress is applied to the sample. The two modified resins resist this test more efficiently, with an advantage to the (more flexible ?) cyanuric acid modifier.

The comparison between peel strength and $\mathcal{G}_{Ic}$ calculated from the compact tension test in Table II is good. The two modified D.E.N. 438 epoxy novolacs are improved in $\mathcal{G}_{Ic}$ but still are more brittle than desired. The cyanuric acid route again showed more improvement than the TDI advancement route. Next, we compared the efficiency of each resin in developing viscoelastic response mechanisms (e.g., microyielding, shear banding) to the presence of rubber inclusions that could result in actual toughening of the matrix. Two points of interest were (1) are the $\mathcal{G}_{Ic}$ values now in an acceptable range and (2) does the reduced cross-link density in the modified systems allow a proportionally greater response.

Table III gives fracture toughness comparisons ($\mathcal{G}_{Ic}$) between the

Table II. D.E.N. 438: Comparisons of Cured Product

Parameter	Unmodified	Modified with TDI	Modified with ICA
Adhesive properties (DICY cure)			
Lap shear, psi	2000	2800	3100
Climbing drum peel, lb/in.	3	7	7.5
$\mathcal{G}_{Ic}$, kJ/m² (MDA cure)	0.10	0.16	0.20

three resins with and without inclusion of 10 phr of CTBN nitrile rubber containing 26% AN.

Although all $\mathcal{G}_{Ic}$ values are increased substantially with the presence of the rubber, the D.E.N. 438 epoxy novolac responds equally as well as the two advanced resins. The latter two were expected to give a higher percentage increase in $\mathcal{G}_{Ic}$ because of their relaxed cross-link density. The data thus do not support one of our original premises. A possible mismatch in solubility parameter between resin and rubber may contribute to this finding. The modified resins display values for δ that are 0.5 to 1.0 unit higher than the value for the unmodified D.E.N. 438 epoxy novolac. The proper relationship of δ is highly important to rubber performance (*14*), and a CTBN with higher AN content might be more useful.

On the positive side, the $\mathcal{G}_{Ic}$ values are now in line with values reported for typical nonmodified liquid epoxies. Table III includes data for D.E.R. 331 epoxy resins cured with dicyandiamide or with aminoethylpiperazine. These systems have had many years of commercial usage. Therefore, the modified rubber-toughened epoxy novolacs should be expected to perform well in established areas for epoxy resins but with a T_g value about 100 °C higher than the 115–120 °C level for the customary liquid systems.

In conclusion, we have shown a way to adapt epoxy novolac resins to rubber toughening in a way that does not harm their high-temperature performance.

Table III. D.E.N. 438: Changes in $\mathcal{G}_{Ic}$ Caused by Incorporation of HYCAR CTBN 1300X13

Parameter	Unmodified	Modified with TDI	Modified with ICA
$\mathcal{G}_{Ic}$, kJ/m²			
Without CTBN	0.10	0.16	0.20
With 10 phr CTBN	0.25	0.34	0.34
T_g (DSC value), °C	216	220	210

NOTE: Comparative values for unmodified liquid epoxy are D.E.R. 331/aminoethylpiperazine—$\mathcal{G}_{Ic}$, 0.41 and T_g, 112 °C and D.E.R. 331/dicyandiamide—$\mathcal{G}_{Ic}$, 0.30 and T_g, 115 °C.

List of Symbols

K_q Stress intensification factor (qualitative), lb/in.^{2} × (in.)$^{1/2}$
$\mathcal{G}_{Ic}$ Critical strain energy release rate, mode I (cleavage), kJ/m^2
M_c Calculated molecular weight between cross-links
T_g Glass transition temperature, °C
E Tensile modulus, psi

Literature Cited

1. Riew, C. K.; Rowe, E. H.; Siebert, A. R. In "Toughness and Brittleness of Plastics"; Deanin, R. D.; Crugnola, A. M., Eds.; ADVANCES IN CHEMISTRY SERIES No. 154, American Chemical Society: Washington, D.C., 1976; p. 326.
2. Bitner, J. L.; Rushford, J. L.; Rose, W. S.; Hunston, D. L.; Riew, C. K. *J. Adhes.* **1981,** *13,* 3.
3. Pearson, R. A.; Yee, A. F. *Prepr. Am. Chem. Soc. Div. Polym. Mater. Sci. and Eng.* **1983,** *49,* 316–20.
4. "Encyclopedia of Polymer Science and Technology"; Interscience: New York; 1965, *3,* 833.
5. Clarke, J. A. U.S. Patent 3 687 897, 1972; Example 1.
6. Clarke, J. A. U.S. Patent 3 676 397, 1972; Example 6.
7. BF Goodrich Product Data Bulletin RLP-1: HYCAR Reactive Liquid Polymers.
8. Lee, H; Neville, K. "Handbook of Epoxy Resins" McGraw-Hill Book Company: New York 1967; Reissue 1982.
9. Jones, W. B.; Lee, C. Y-C *Polym. Eng. Sci.* **1982,** *22,* 1190.
10. Bascom, W. D.; Ting, R. Y.; Moulton, R. J.; Riew, C. K.; Siebert, A. R. *J. Mater. Sci.* **1981,** *16,* 2657.
11. Lee, L-H. In "Physicochemical Aspects of Polymer Surfaces"; Mittal, K. L., Ed.; Plenum Press: New York, 1983; Vol. I.
12. Geil, P. H.; Takahama, T. *J. Polym. Sci. Polym. Lett. Ed.* **1982,** *20,* 453.
13. Katz, D.; Tobolsky, A. V. *J. Polym. Sci. Part A* **1964,** *2,* 1587.
14. Romanchick, W. A.; Sohn, J. E.; Geibel, J. F. In "Epoxy Resin Chemistry II"; Bauer, R. S., Ed.; ACS SYMPOSIUM SERIES No. 221, American Chemical Society: Washington, D.C., 1981; p. 91.

RECEIVED for publication November 18, 1983. ACCEPTED April 9, 1984.

6

Characterization of a Carboxyl-Terminated Polybutadiene

Molecular Weight Distribution, Functionality Distribution, and Microstructure

H. E. DIEM, D. J. HARMON, R. A. KOMOROSKI, J. B. PAUSCH, and R. J. BERTSCH

The BFGoodrich Research and Development Center, Brecksville, OH 44141

A commercial carboxyl-terminated polybutadiene was separated by preparative gel permeation chromatography (GPC) into 12 fractions of relatively narrow molecular weight. Information on the molecular weight, molecular weight distribution, functionality, and microstructure of each fraction was determined. Acid-equivalent weights were obtained from three end-group counting (number of carboxyls) measurements, proton NMR and IR spectroscopies, and titration. $\overline{M}_n$ values were obtained from vapor pressure osmometry and GPC. Results from the end-group counting methods are statistically indistinguishable. By comparing the pooled data from the end-group counting methods with the osmometry data, we calculate the functionality of the polymer to be 1.9 carboxyl groups per molecule. $\overline{M}_n$ values from GPC (polystyrene standards) are about 70% high. The highest molecular weight fraction (3.5% of the polymer) is unique. It consists of branched molecules with an average of about one branch per molecule. We hypothesize that the branching resulted from the reaction of a terminal nitrile on one chain end with a terminal carboxyl on another chain end. The microstructure is uniform across the molecular weight distribution. No "cage product" from coupling of the azo catalyst fragments is found, although a small loss in carboxyl is indicated by the end-group accounting. This discrepancy is assigned to the necessity to estimate carboxyl equivalents for one fraction, plus accumulated errors.

REACTIVE LIQUID POLYMERS (RLP), such as HYCAR CTB, are used in applications where they are actually monomers for condensation polymerizations. Monofunctional or polyfunctional constituents in di-

0065-2393/84/0208-0065/$06.00/0

functional monomers have a profound effect on the condensation process and on the physical properties of the resulting condensation polymer. Little information exists on the detailed molecular structure of such polymers, in particular the extent to which these materials contain monofunctional or polyfunctional chains. This chapter was designed to provide that information.

CTB 2000X165 is a polybutadiene (PBD) synthesized free radically by using as initiator 4,4′-azobis[4-cyanopentanoic acid] (ADVA) Registry No. [2638-94-0] (*1*, *2*). If termination is via coupling, a carboxyl group should be on every chain end. However, termination by other than coupling of radicals or transfer reactions (e.g., to monomer, polymer, solvent, or impurities) would lead to monofunctional chains and/or to chains with more than two functional groups. Little agreement exists in the literature on the critically significant question of the functionality of such telechelic polymers. Karatavykh et al. (*1*) found little branching in PBD made with ADVA in acetone at 70 °C, and concluded that the role of transfer reactions in the polymerization was insignificant. Similarly, Ray (*3*) found that in the polymerization of styrene at 60 °C, combination was the major termination mode of the radicals. On the other hand, Valuev (*4*, *5*), examining copolymers of butadiene with acrylonitrile (AN) and with methacrylic acid, found an average functionality of two, yet found a large amount of trifunctional polymer. Finally, Reed (*6*, *7*) has examined carboxyl-terminated polymers from ADVA and found widely varying functionality, usually greater than two.

A careful fractionation of CTB 2000X165 was carried out, and each fraction was characterized to provide structure, molecular weight, molecular weight distribution, and functionality information. Number average molecular weights were obtained by both colligative and end-group counting methods [IR and NMR spectroscopies, and equivalent parts per hundred parts resin (EPHR)]. By comparing the data from these two methods, we assayed the distribution of carboxyl end groups and branches across the fractions. Although gel permeation chromatography (GPC) fractionation procedures (*8*) have been used by earlier workers, several of the end-group counting methods we report have not previously been used on telechelic polymers.

Experimental

Preparative Fractionation. GPC (Waters Associates AnaPrep) was used to fractionate CTB 2000X165 by molecular size. The conditions for the fractionation were as follows: Five 4 ft × 2.5 in. i.d. columns packed with Styragel of porosities 10^4 (1), 10^3 (1), 500 (2), and 100 (1) Å. The solvent was distilled toluene; the flow rate was 50 mL/min.; and the fraction size was 390 mL.

Eleven injections were made for a total throughput of 110 g of polymer. At the end of the run, each fraction was contained in 4.3 L of toluene. The toluene was removed by use of a rotary evaporator after addition of 0.5% Antioxidant 2246 (AO2246) (American Cyanamid) (based on polymer weight) to the solution. Antioxidant was required because the antioxidant originally present in the polymer would all be in the lowest molecular weight fraction. The fractions were taken down to about 150-mL volume, and final drying took place in a vacuum oven at 40 °C for ~ 96 h (to constant weight). The weight percent values were obtained by dividing the dry weight of each fraction by the weight of polymer charged.

Analytical GPC. Each of the fractions was analyzed for molecular size distribution (MWD) with a Waters Model 200 GPC modified for high-speed operation with microparticulate (10 μm) columns. The conditions for the analysis were as follows: The columns were 5-μm Styragel 30 cm × 7.8 mm i.d.; 10^2, 10^3, 10^4, 10^5, and 10^6 Å pore size (Waters Associates designation). The solvent was tetrahydrofuran (THF) (Fisher); the flow rate was ~2.0 mL/min; the sample size was 200 μL of 0.2% concentrate; and the detector was refractive index (Δn) at 4× attenuation.

The columns were calibrated by using narrow size distribution polystyrenes (PS). Because no universal calibration constants are available for these polymers, all GPC data were reported in terms of PS equivalent molecular weight.

Vapor Pressure Osmometry. The number average molecular weight ($\overline{M}_n$) of each fraction was determined with vapor pressure osmometry (VPO) (Corona/Wescan Model 232A molecular weight apparatus). The solvent used was ethyl acetate at 37 °C. The instrument was calibrated with benzil (MW 210.2) and sucrose octaacetate (MW 678.6). An instrument constant of 42495 was obtained ($K = MW \cdot \Delta R/C$ where C is in moles per liter). At least four concentrations were measured for each sample, and ΔR (change in resistance) was obtained. $\Delta R/C$ (where C is in grams per liter) was then plotted against C and (by using a least squares fit) extrapolated to zero concentration. The data were corrected for the amount of antioxidant in each fraction. $\overline{M}_n$ was then calculated from

$$\overline{M}_n = \frac{K}{\Delta R/C_{(C=0)}}$$

Functionality by Acid Group Titration. A sample (~1 g) was weighed into a 4-oz jar, and 50 mL of a toluene/2-propanol (60/40 by volume) mixture was added. The sample was mixed on a magnetic stirrer for ~20 min, or until dissolved. The mixture was then titrated with ~0.1 N KOH in ethanol (standardized) with a phenolphthalein indicator. The following equation was used to calculate acid content.

$$\text{COOH EPHR} = \frac{(\text{mL titrant})\ (N \text{ of titrant})}{(\text{grams sample}) \times 10}$$

The acid-equivalent weight (AEW) of each fraction was then calculated from AEW = 100/COOH EPHR.

Proton NMR Analysis of $\overline{M}_n$. Proton Fourier transform NMR (^{1}H FTNMR) spectra were obtained at ambient temperature at 200 MHz on a Bruker WH-200 spectrometer. Samples were dissolved in $CDCl_3$ with tetramethylsilane (TMS) as internal standard and were run in 5-mm o.d. NMR tubes. Typical conditions were 90° rf pulses of 4.6-μs duration; pulse repetition rate, 9.1 s; and line broadening, 0.4 Hz. The ^{1}H FTNMR spectra of the preparative GPC fractions were used to obtain AEW values. The triplet at about 2.5 ppm (Figure 1) is assigned to the methylene protons next to a COOH group on the basis of model compounds and expected chemical shifts. AEW values were calculated for comparison to other data. The calculation is based on the ratio of the peak area due to alkene protons (adjusted for 1,2-units) to the area due to the end groups. Hence, number of butadiene units = $(E + D/2)/2$, number of COOH = $C/2$, number of butadiene units/COOH = $(E + D/2)/C$, and AEW = (number of units/COOH) (54.0) + (126), where 54 and 126 are the molecular

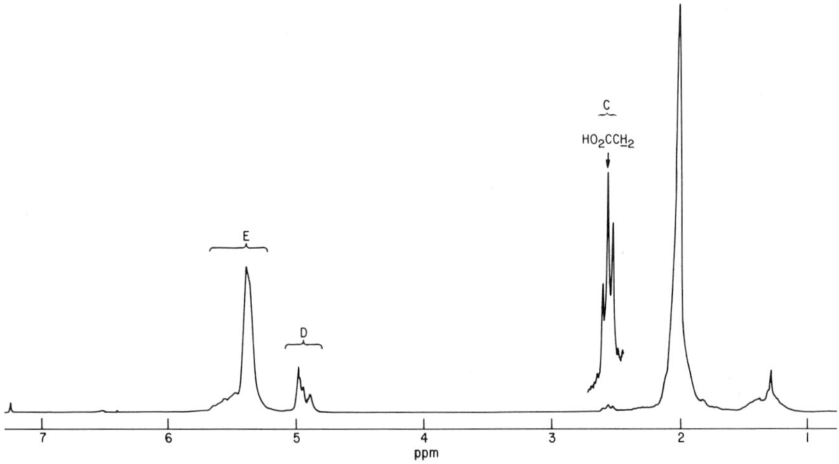

Figure 1. The 200-MHz ${}^1H$-NMR spectrum of fraction 5 of CTB 2000X165.

weight of a monomer unit and an end group, respectively. An adjustment was made for a small amount of antioxidant detected in Fractions 9–12.

Carbon-13 NMR Analysis of Microstructure. Carbon-13 FTNMR spectra were obtained at ambient temperature at 50.3 MHz on the Bruker WH-200 spectrometer. Samples dissolved in $CDCl_3$ with TMS were run in 20-mm o.d. tubes. Typical conditions were 90° rf pulses of 35-μs duration; 12,195-Hz spectral width accumulated in 32,000 points with quadrature detection; pulse repetition rate, 10.7 s; and line broadening, 0.5–2 Hz. Increasing the pulse repetition rate to 20 s did not noticeably change the spectrum, hence peak intensities are not attenuated by incomplete relaxation to equilibrium. For reasons of sensitivity, all of the ${}^{13}C$ spectra were acquired in standard fashion with nuclear Overhauser enhancement (NOE). To ensure that variable NOEs were not affecting our microstructural results, a ${}^{13}C$ spectrum of a fraction was acquired with NOE suppression (*9*). No significant change was observed relative to the results obtained with NOE.

Carbon-13 NMR spectra of PBDs containing *cis*-1,4-, *trans*-1,4-, and 1,2-units have been reported, and peak assignments made (*10, 11*). Using the published assignments, we obtained *cis*-, *trans*-, and 1,2- compositions for the preparative GPC fractions with ${}^{13}C$-NMR spectroscopy. A typical ${}^{13}C$ spectrum is shown in Figure 2. The percent vinyl was obtained from the ratio of the vinyl carbon areas ($X_1 + X_2$) to the area due to all olefinic backbone carbons ($X_1 + X_2 + Z$). For fractions 9 and 10, the area Z is corrected for the presence of A02246 by using the peak at about 31 ppm and the spectrum of A02246. The remaining non-1,2-unit content is divided between *cis*-1,4- and *trans*-1,4-units on the basis of the ratio of the respective peaks due to backbone CH_2 (*see* Figure 2).

Determination of AEW by IR Spectroscopy. AEW values were calculated for each fraction from the relationship AEW = 126.14/weight fraction of chain ends. The weight fraction of chain ends was obtained for each fraction by an absorbance subtraction method developed for this problem; the details will be published elsewhere (*12*). The method treats carboxyl-terminated polybutadiene (CTPB) as a mixture of PBD and of ends, and applies a binary mixture analysis (*13*). Traditional methods based on titration standards were used for confirma-

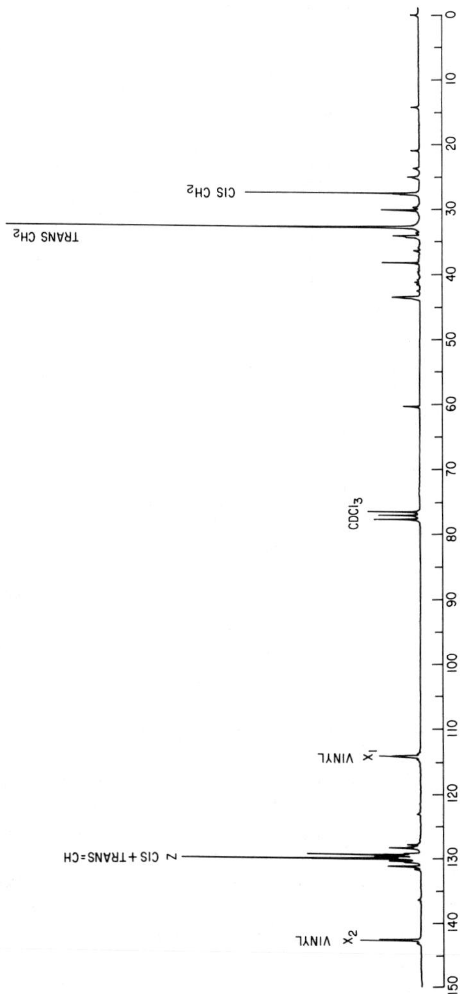

Figure 2. The 50.3-MHz ^{13}C-NMR spectrum of fraction 3 of CTB 2000X165.

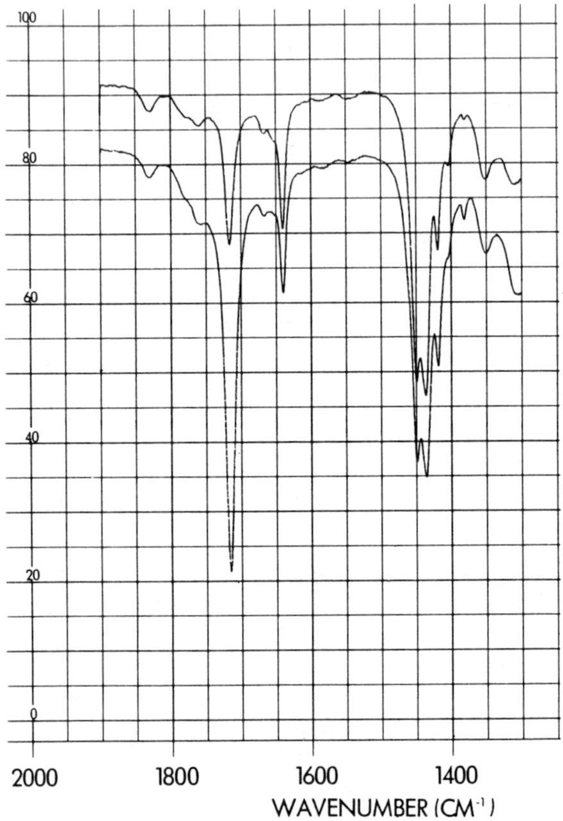

Figure 3. IR spectra of fraction 2 (upper curve) and fraction 7 (lower curve) of CTB 2000X165.

tion. Each fraction was dissolved in carbon tetrachloride (40 g/L), and a signal-averaged, solvent-compensated spectrum (0.2-mm cell) was recorded between 1900 and 1300 cm^{-1}. Measurements were made on a Perkin-Elmer 180 spectrometer at a spectral slit width of 1.8 cm^{-1}, a scan rate of 96 cm^{-1}/s, and a data interval of 0.5 cm^{-1}; two scans were averaged. The index bands for the analysis, which can be seen in the spectra of Figure 3, are the carboxyl dimer absorption at 1714 cm^{-1}, for the chain ends, and the vinyl double bond stretch at 1639 cm^{-1}, for the PBD. The analysis was made, two fractions at a time, by subtracting the spectra of the two fractions. First, the scaling factor (F_1) required to null the PBD in one subtraction was determined. Then the factor (F_2) required to null the carboxyl band in the inverse subtraction was determined. The weight fraction of chain ends in the two fractions, x and y, were calculated from these scaling factors.

$$x = \frac{1 - F_1}{1 - F_1F_2} \qquad y = \frac{(1 - F_1)F_2}{(1 - F_1F_2)} = F_2x$$

By taking the lowest molecular weight fraction (fraction 12) with each of the other fractions in turn, the weight fraction of chain ends was calculated for all three fractions except 1, 10, and 11.

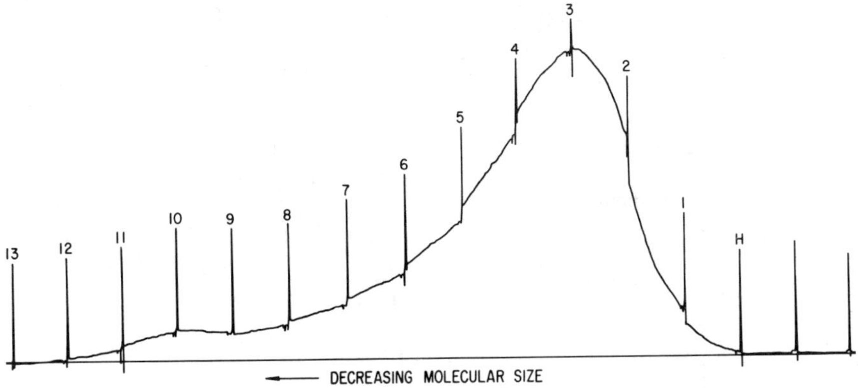

Figure 4. Anaprep preparative GPC size distribution profile.

The microstructure of the PBD portion of the whole polymer was determined by an in-house procedure that had been calibrated by ^{1}H-NMR measurements. The microstructures of the fractions were not determined directly. However, the fractions were judged to have the same microstructure because the *trans* and vinyl bands could be nulled out of any pair of spectra with the same scaling factor.

Results

Figure 4 shows a chromatogram taken during the preparative GPC fractionation. One of the objectives of this study was to prepare narrow fractions of CTPB suitable for calibrating the GPC analysis. The results of the fractionation, plotted in Figure 5, show that the efficiency of the fractionation is good, and the fractions are of suitably narrow distribution. The analytical GPC data (Table I) are plotted in Figure 6, in which the solid line is the least squares line calculated omitting fraction 1.

The $\overline{M}_n$ data from the colligative methods and the AEW data from the end-group counting methods are collected in Table II. In general, the molecular weight progresses from high to low values through the fractions. However, the molecular weight of fraction 1 is lower than that of fraction 2 with all methods except NMR spectroscopy and GPC. The $\overline{M}_n$ values obtained with the analytical GPC (calibrated with narrow size distribution PS fractions) are some 70% above the values obtained from VPO and are about equally above the values obtained from the various end-group counting methods as well. Thus, the need for recalibration is obvious. With the data plotted as the solid line in Figure 7 (omitting fraction 1) the relationship $\overline{M}_{n_{\text{actual}}} = 0.5697\ \overline{M}_{n_{\text{PS}}}$ is found. The standard error of the relationship is 525.

The microstructural results are given in Table III. Essentially no change is observed over the entire molecular weight range. Values determined by NMR and IR spectroscopy for the whole polymer are

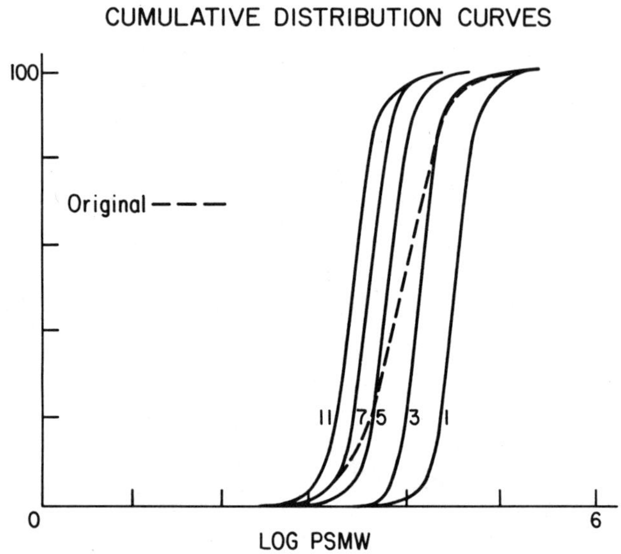

Figure 5. Fractionation efficiency for several GPC fractions from Figure 4.

in agreement. A full characterization of the sequence distribution of the *cis-*, *trans-*, and 1,2-units was not performed. Comparison of the ^{13}C spectra of the various fractions indicates no major change of sequence distribution with molecular weight. Comparison to spectra in the literature suggests a random distribution of the various units (*2*).

We did not detect any additional resonances in either the ^{1}H- or ^{13}C-NMR spectrum of fraction 1 that could be attributed to car-

Table I. Analytical GPC Results on CTB 2000X165

Fraction	$\overline{M}_w$	$\overline{M}_n$	*Peaks*
Original	14,900	9,000	12,700
1	38,500	28,300	30,300
2	23,100	19,100	18,900
3	15,100	12,600	12,700
4	9,620	8,300	8,080
5	8,110	5,960	7,230
6	6,560	4,920	5,790
7	5,720	4,290	4,910
8	5,600	4,070	4,650
9	4,440	3,470	3,730
10	3,870	3,070	3,170
11	3,000	2,760	2,840
12	2,670	2,470	2,690

NOTE: Excluding fraction 1, $\overline{M}_w = 1.1923\ \overline{M}_n + 284$; and $r^2 = 0.9947$.

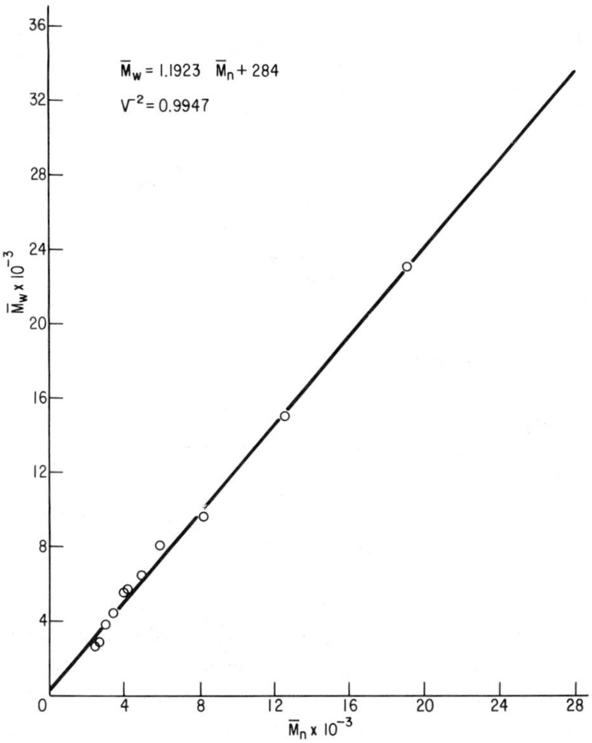

Figure 6. Molecular weight values for fractions of CTB 2000X165 obtained from analytical GPC.

Table II. $\bar{M}_n$ Values for CTB 2000X165 Fractions

Fraction	*Weight Percent*	*VPO*	*EPHR*[a]	*IR*[a]	*NMR*[a]	*GPC*[b]
1	3.47	7,178	3,437	4,885	7,100	28,300
2	14.51	10,365	5,406	5,592	6,250	19,100
3	25.35	8,365	4,762	3,983	4,410	12,600
4	18.76	4,941	2,591	2,880	3,335	8,300
5	11.97	3,821	2,020	2,305	2,485	5,960
6	7.96	3,256	1,642	1,805	1,975	4,920
7	5.42	2,588	1,376	1,511	1,625	4,290
8	3.90	2,266	1,134	1,468	1,530	4,070
9	3.04	1,534	1,035	1,312	1,285	3,470
10	2.61	833	—	—	1,415	3,070
11	1.64	1,560	903	—	1,235	2,760
12	1.21	1,389	769	983	1,050	2,470
Whole polymer		3,601	1,961	—	3,150	9,000

[a] Data given as acid equivalent weights.
[b] Based on calibration with PS.

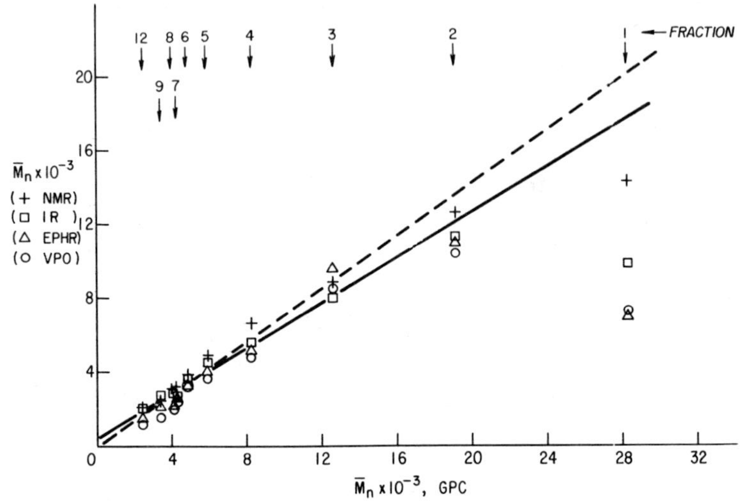

Figure 7. $\overline{M}_n$ values from end-group counting methods vs. $\overline{M}_n$ from analytical GPC.

boxyl-containing branches or any defect structures. However, the presence of such structures is indicated by the low AEW obtained for this fraction. Moreover, the carboxyl region of the IR spectrum could not be nulled out by the carboxyl region of any other fraction. An absorption always remained that overlapped the 1714-cm^{-1} acid dimer band. (An IR estimate of the molecular weight of fraction 1 was obtained from the linear plot of the IR absorbance ratio A_{1639}/A_{1714} versus AEW for the other fractions. This estimate is obviously low as a result of the overlap at 1714 cm^{-1}.) These observations suggest a unique structure in fraction 1.

Table III. Microstructure of HYCAR 2000X165 Fractions

Fraction	cis-*1,4-Units*	trans-*1,4-Units*	*1,2-Units*
Whole polymer[a]	23(22)	57(58)	20(20)
1	22	58	20
2	22	58	20
3[b]	21(21)	58(59)	20(20)
4	22	59	19
5	23	58	19
6	22	59	19
7	22	59	19
8	23	58	19
9[c]	21	58	21
10[c]	25	60	15

NOTE: Data given as percentages.
[a] Values in parentheses were obtained by IR spectroscopy.
[b] Values in parentheses were obtained with NOE suppression.
[c] Corrected for the presence of antioxidant.

Discussion

The agreement among the three end-group counting methods and VPO is quite gratifying. The data from each pair of methods were submitted to linear regression analysis. Data for fractions 10 and 11, as well as for fraction 1, were omitted from this analysis, because the IR spectra of fractions 10 and 11 showed large amounts of antioxidant, especially in 10. If all the 1.5 parts of AO2246 added initially had eluted with these two fractions, the antioxidant should constitute nearly 35% of the sum of the weights of fractions 10 and 11. Absorbance subtraction showed that fractions 9 and 12 also contained more AO2246 than the other fractions. Because this additional AO2246 was only tenths of a percent, data from fractions 9 and 12 were used in calculations. Regression lines were calculated for each pair of methods with first one method as the independent variable, then the other. Standard error of estimate values were obtained with each line, and the algebraic mean of the regression lines was taken as the least squares equation. Correlation coefficients were uniformly high, and precision was good in most instances.

In comparing the $\overline{M}_n$ values from the various methods ($\overline{M}_n$ values were calculated for the IR, NMR, and EPHR methods by assuming two carboxyls per chain), VPO values were consistently the lowest, NMR were consistently the highest, and EPHR and IR results were intermediate. The differences are not really large, however. The assumption of two carboxyls per chain can be tested with the slope of the plot of VPO (y) against any end-group method (x). The slopes of those plots are 1.87, 2.08, and 1.82 for the EPHR, IR, and NMR methods, respectively. However, the standard errors of the slopes are large enough so that these values are not statistically different, and the data can be pooled. The slope of the plot of VPO against the pooled end-group methods is 1.90. If our assumption (termination by coupling only) is correct, the slope should be 2. Thus, we have 1.9 functional groups per chain with a precision of about 0.1. The standard error of estimate of the pooled data is 502, which is in line with the data for the methods taken by pairs.

We attempted to account for the fate of the carboxyl end groups in the fractionation, as detailed in Table IV. About 15% of the carboxyl groups in the starting polymer were not accounted for by the measured fractions. Part of this loss may result from the necessity of estimating the EPHR for fraction 10, which contained much antioxidant. We calculated that the loss is equivalent to about 1% of the total weight as catalyst "cage" product. However, tests showed that, although the preliminary filtration step may have removed traces of (insoluble) cage product, 1% could not have been present. Thus the loss in carboxyl is probably a result of accumulated weighing and titration inaccuracies.

Fraction 1 is clearly unique. The anomaly is striking in Figure

Table IV. End-Group Accounting

Fraction	*Weight Percent*	*EPHR*[a]	*Equivalents*
1	3.47	0.0291	0.00101
2	14.51	0.0185	0.00268
3	25.35	0.0210	0.00532
4	18.76	0.0386	0.00724
5	11.97	0.0495	0.00593
6	7.96	0.0609	0.00485
7	5.42	0.0727	0.00394
8	3.90	0.0882	0.00344
9	3.04	0.0966	0.00294
10	2.61	(0.1)[b]	0.00261
11	1.64	0.1108	0.00182
12	1.21	0.1300	0.00157
	99.84		0.04335 (0.051 whole sample)
Lost	0.16	(4.78)	0.00765[c]

[a] Corrected for 0.5% AO2241.
[b] Estimated.
[c] 0.97% of total as cage product; cage product EPHR = 0.7928.

7. In this figure, the $\overline{M}_n$ values from NMR, IR, EPHR, and VPO measurements are plotted against the GPC $\overline{M}_n$ values (excluding fractions 10 and 11). The $\overline{M}_n$ values for the end-group counting methods are obtained by multiplying the AEW by 1.9. The solid line is the least squares line calculated, omitting fraction 1, by pooling the end-group counting and VPO data. The fact that all the end-group counting methods find far too many carboxyls suggests that fraction 1 is branched. However, the VPO result does not seem in accord with this; agreement among the methods is poor.

One sort of "branch" can be proposed to harmonize the data. Twenty-five years ago, Grassie and coworkers were studying the development of yellow color in polyacrylonitrile. They found (*14, 15*) that a potent catalyst for the color reaction was the imino anhydride, –C(O)OC(NH)–, that formed when a small amount of acrylic acid (2-propanoic acid) was copolymerized with the acrylonitrile and the acid hydroxyl was added across the C≡N bond. In some of our previous characterization work on RLPs, a cyclic imino anhydride was postulated as the starting point in heat aging changes (development of color, increased viscosity, and loss of carboxyl and nitrile groups) in butadiene–acrylonitrile copolymers (*16*). Some support for this structure was obtained by absorbance subtraction IR spectroscopy. In the present case, we suggest the possibility that fraction 1 consists of chains in which the carboxyl at the end of one chain has reacted with the nitrile at the end of another to give $(CN)C(CH_3)CH_2CH_2C(O)OC(NH)C(CH_3)(CH_2CH_2COOH)$. Such a reaction is really chain extension, rather than branching, because the branch is only three carbons long and the molecular weight is doubled.

With one such branch per molecule, NMR and IR spectroscopy should observe three carboxyls rather than two (except that absorption from the imino anhydride will complicate the IR spectra). But because an anhydride would be expected to hydrolyze easily, EPHR titration should observe four carboxyls. Similarly, VPO should measure two molecules per mole because interchange with the ethyl acetate solvent would split the anhydride. Using this logic and Figure 7, we find that the number of branches per molecule in fraction 1 is 0.6 from NMR, 1.3 from VPO, and 1.6 from EPHR data. As a result of the carboxyl overlap, only a maximum value, 1.5, can be calculated from IR data. These values are in reasonable agreement, especially considering that in the other fractions, VPO and EPHR consistently found more carboxyls than NMR and that all the methods have their smallest accuracy for fraction 1. We take an average value of 1.2 as the number of branches per chain in fraction 1.

Note the dashed line in Figure 7 that is the least squares line calculated omitting both fractions 1 and 2. Although all the points for fraction 2 fall below the line, the precision is such that no firm conclusions can be drawn about this fraction. But the data suggest that some branching may also exist in fraction 2. (The use of the dashed line only increases the calculated number of branches per chain in fraction 1 to ~1.6.)

Is there any experimental evidence for an imino anhydride structure in fraction 1? We have already noted that nothing was observed with ^{13}C-NMR spectroscopy that might relate to branches. This result may occur simply because the concentration of such structures would be not more than one-third that of the carboxyl groups, which is already quite low. However, carbonyl groups are among the strongest absorbers in IR spectroscopy and can be observed at very low concentrations. We noted previously that the carboxyl region in fraction 1 could not be nulled in absorbance subtraction by the carboxyl region of any other fraction. This result is shown in Figure 8, where the spectrum of fraction 2 is subtracted from fraction 1. Although the noise level is quite high because of the low concentration of functional groups, when the carboxyl dimer band is brought to zero at 1714 cm^{-1}, appreciable absorption remains. Fraction 1 contains some other carbonyl function with a maximum near 1740 cm^{-1} and possibly a second weaker band at ~1705 cm^{-1}. We have found no spectra of imino anhydride model compounds in the literature, but, by analogy with open chain anhydrides and imides (*17*), one might expect a pair of bands in this region. By itself, this evidence is weak, but it does support the proposed method by which "branching" could arise in fraction 1. Also, the overlap of this band makes the carboxyl dimer band too strong in the IR spectrum of fraction 1, and as a result, IR measures appreciably more carboxyl than NMR.

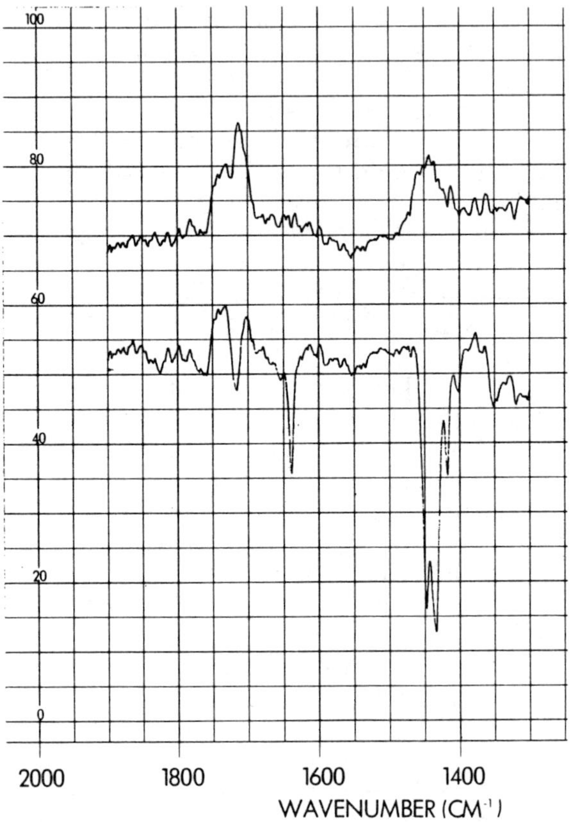

Figure 8. Difference in spectra obtained from IR absorbance subtraction: (fraction 1) − factor (fraction 2). Both spectra are Δ absorbance of 0.1 AUFS: upper curve, factor just sufficient to null the vinyl band at 1640 cm^{-1}; lower curve, factor large enough to bring the $(COOH)_2$ band at 1714 cm^{-1} to (slightly below) zero.

Finally, indirect evidence of the proposed structure is obtained by considering that fraction 1 did, in fact, elute first on the GPC (i.e., it was the highest molecular weight fraction). Because the mechanism we propose is really chain extension, which doubles the molecular weight with only a three carbon atom branch, such a result would be expected. The amount of a branched fraction such as fraction 1 that is found in a given lot of polymer will probably correlate with the heat history and time elapsed since manufacture of the given polymer. Thus, the observation of Valuev (*4*, *5*) of a polymer with an average functionality of about two but with a large fraction of trifunctional material is not necessarily inconsistent with this work.

In summary, CTB 2000X165 contains a small fraction (3.5%) with about three carboxyls per chain. The remaining polymer consists of about 90% difunctional chains and about 10% monofunctional chains.

The microstructure was uniform across the molecular weight distribution. The relationship of the actual $\overline{M}_n$ to that calculated from GPC (with PS calibration) is constant across the total MWD and allows a simple conversion.

Acknowledgments

The authors thank the following people for their assistance: W. R. Barr, for preparative GPC fractionation; J. E. Shockcor, for NMR spectroscopic measurements; R. J. Stevens, for EPHR measurements; and A. M. Fairlie, for statistical analysis support.

Literature Cited

1. Karatavykh, V. P.; Barantsevich, Ye. N.; Ivanchev, S. S., *Polym. Sci. USSR Engl. Transl.* **1976,** *18,* 991.
2. French, D. M., *Rubber Chem. Technol.* **1969,** *42,* 71.
3. Ray, G.; Palit, S. R. *Indian J. Chem.* **1971,** *9,* 1124.
4. Valuev, V. I.; Tsvetskovskü, I. B.; Trizna, N. N.; Schlyakhter, R. A. *Metody Anal. Kontrolya Kach. Prod. Khim. Prom-sti* **1978,** *3,* 24.
5. Valuev, V. I.; Shlyakhter, R. A.; Pavlova, I. L.; Rozinova, O. A. *Polymer Sci. USSR* **1973,** *15,* 3097.
6. Reed, S. F., Jr. *J. Polym. Sci. Part A-1* **1971,** *9,* 2147.
7. Ibid **1972,** *10,* 649.
8. Law, R. D. *J. Polym. Sci. Part A-1* **1971,** *9,* 589.
9. Freeman, R.; Hill, H. D. W.; Kaptein, R. *J. Magn. Reson.* **1972,** *7,* 327.
10. Elgert, K.-F.; Quack, G.; Stutzel, B. *Polymer* **1975,** *16,* 154.
11. Clague, A. D. H.; Van Broekhoven, J. A. M.; Blaauw, L. P. *Macromolecules* **1974,** *7,* 348.
12. Diem, H. E. unpublished results.
13. Diem, H. E.; Krimm, S. *Appl. Spectrosc.* **1981,** *35,* 421.
14. Grassie, N.; McNeill, I. C. *J. Polym. Sci.* **1959,** *34,* 211.
15. Grassie, N.; Hay, J. N. *J. Polym. Sci.* **1962,** *56,* 189.
16. Diem, H. E. private communication.
17. Bellamy, L. J. "Advances in IR Group Frequencies"; Methuen: London, 1968.

RECEIVED for review November 18, 1983. ACCEPTED April 16, 1984.

FRACTURE MECHANICS AND FAILURE MECHANISMS

7

Failure Behavior of Rubber-Toughened Epoxies in Bulk, Adhesive, and Composite Geometries

D. L. HUNSTON[1] and W. D. BASCOM[2,3]

[1]National Bureau of Standards, Polymers Division, Washington, DC 20234
[2]Naval Research Laboratory, Chemistry Division, Washington, DC 20375

Rubber-modified epoxies were first developed empirically in the 1960s to improve the poor crack-growth resistance of epoxies with a minimum sacrifice in other desirable properties. In 1971 a major effort was launched to study the failure behavior of these materials as structural adhesives. Since then, this program has been expanded to include numerous laboratories and reseachers, and the objectives have been broadened to include the failure behavior of bulk specimens and fiber-reinforced composites as well. We discuss the effects on failure behavior of rubber modification, test temperature, and loading rate for all three specimen types. For adhesive bonds, the effects of bond thickness are also discussed.

POLYMERS USED IN COMPOSITES AND STRUCTURAL ADHESIVES are generally glassy materials that are highly cross-linked to obtain a high modulus and minimize long-term creep. Unfortunately, high cross-linking can produce brittle polymers; thus, these materials must be toughened if they are to be used in structural applications. One of the most successful methods used to toughen cross-linked epoxy systems is the addition of an elastomer that phase separates during curing. The curing reactions are controlled so that the final material is a matrix of epoxy with small elastomer particles dispersed in and bonded to the matrix. The advantage of the elastomer-modified material is that the two-phase nature of the system makes possible toughening mechanisms that do not occur (or occur at a vastly diminished magnitude) in a single-phase material. These mechanisms can greatly increase

[3] Current address: Hercules, Inc., Magna, UT 84044.

0065-2393/84/0208-0083/$06.00/0

the material's resistance to crack growth. Moreover, because the matrix itself contains relatively little elastomer, the bulk properties, such as the moduli, are close to those of the unmodified epoxy. Thus, the crack-growth resistance is increased with a minimum sacrifice in other properties. For this reason, many commercial structural adhesives are based on a technology of this type. In addition, composite fabricators are increasingly investigating the use of these materials as matrix resins.

The first rubber-modified epoxies were developed in the mid 1960s by adhesive manufacturers who used primarily empirical methods. By the early 1970s, the first publications began to appear (*1–4*). These studies focused on the morphology and fracture of neat specimens (bulk sample of modified polymer to characterize the material itself) and clearly demonstrated the dramatic improvement in toughness that could be achieved with the proper addition of rubber. Because the major use of these materials at the time was in structural adhesives, researchers were naturally interested in extending such studies to adhesive bonds. An extension of this type, however, required the development of fracture mechanics techniques applicable to bonded geometries, and these techniques were not available. Coincidentally, a totally unrelated effort was addressing this problem during that same time period. This work was prompted by the suggestion from George Irwin, one of the pioneers in fracture mechanics, that fracture techniques might be applied to the study of adhesive bonds. By the early 1970s, the work of Corten, Mostovoy, Ripling, Patrick, and Bersch (*5–8*) had provided a scientific basis for such studies as well as several useful fracture specimen geometries.

On the basis of the work in these two areas, a program was initiated in 1971 at the Naval Research Laboratory to study the fracture behavior of elastomer-modified epoxies in adhesive bonds. The goal was to see how the dramatic improvement in fracture behavior that was observed for the rubber-modified epoxies in neat (i.e. bulk) specimen tests would be reflected in the properties of adhesive joints. Since 1971, this program has examined many of the variables that affect adhesive bond fracture. These studies have not only shed light on adhesive properties (*9–20*) but have also led to numerous questions about other aspects of the behavior of these materials. As a result, the program was extended to include investigations of the basic viscoelastic properties of such polymer systems and their fracture behavior in neat and composite geometries (*21–26*). To address these varied topics, the program has expanded to include many coauthors and numerous government, university, and industrial laboratories. To date a total of 23 coauthors at eight different laboratories have been involved. The breadth, length, and continuity of the program initiated at the Naval Research Laboratory is such that, after more than a decade of work, an overview of some of the most important results from this effort is now apropriate. This chapter provides such an overview.

Space limitations make it possible to highlight only a few selected achievements from the first decade of this program. Four topics have been chosen for discussion, and they will be presented in a more or less chronological order: mode-I adhesive bond fracture, including both toughness and bond thickness effects; mixed-mode adhesive fracture; temperature and loading rate effects for neat and adhesive specimens; and composite interlaminar fracture. The discussion of these topics is necessarily brief, but more detailed information can be found in the references cited. In addition, a number of important topics addressed in the research program but not covered in this chapter are discussed in the listed references: the effects of formulation on fracture behavior of unmodified and commercial resins (*9, 11, 13, 27*), the effects of concentration (*9, 22*) and formulation (*21, 22, 27*) on morphology and fracture of rubber-modified resins, fracture-surface topology (*11, 13, 14*) and effects of adherent surface roughness (*10, 16*) on mixed-mode failure, stress corrosion in adhesive bonding (*28, 29*), fracture-surface studies in bulk (*9, 12, 17, 23, 24*) and composite (*21*) geometries, and viscoelastic effects on the properties and fracture of rubber-modified epoxies (*17, 18, 22*).

Experimental

Details of the materials and test procedures have been published (*9–26*) and will only be briefly reviewed here. The experiments were performed with an unmodified epoxy and a series of rubber-modified epoxies made at different rubber concentrations. All materials were cured with piperidine (5 parts per hundred parts resin, phr) at 120 °C for 16 h. For the modified materials, the epoxy (diglycidyl ether of biphenol A, DGEBA) and rubber [carboxyl-terminated poly(butadiene–acrylonitrile)] were thoroughly mixed, and then the piperidine was added. For neat specimens, the liquid was poured into a metal mold coated with mold release and cured into plates. These plates were used to machine bars for dynamic mechanical analysis (DMA), compact tension specimens for mode-I loading, and center-crack plates for mixed-mode tests.

The adhesive bond specimens were prepared by fabricating both adherends from aluminum. The bonding surfaces were cleaned with standard techniques and then bonded together with polytetrafluoroethylene (PTFE) shims used to control the bond thickness. Mode-I tests were conducted with tapered double cantilever beam (TDCB) specimens; modified double cantilever beam and scarf joint specimens were used for mixed-mode tests. Composites were examined for interlaminar fracture with both width tapered and untapered double cantilever beam specimens. Samples were cut from plates supplied by manufacturers. Two types of plates were obtained: one with an unmodified epoxy resin matrix and another with rubber added to toughen the epoxy. Specimens contained either 12 plies of woven graphite-fiber reinforcement or 24 plies of unidirectional fibers (the fiber direction in the plies was down the length of the specimen).

To minimize the effects of differences in thermal history (physical aging), all specimens were given the same thermal pretreatment prior to testing (i.e., a very slow cool, >12 h, from above the glass transition temperature, T_g, to room temperature).

Mode-I fracture tests were conducted on neat specimens by inserting a sharp precrack with a knife edge and then loading to failure in a tensile test machine with crosshead speeds from 5 to 0.008 mm/s and temperatures from −60 to 60 °C. Standard ASTM equations were used to convert these results into fracture energies. Although the bending modulus is relatively insensitive to time and temperature over the range of conditions employed for the fracture tests, care was taken in the fracture energy calculation to use modulus values that were measured at approximately the same temperatures and time scales as the corresponding fracture experiments. Selected mixed-mode tests were performed with center-cracked plates in which the crack was at an angle to the applied tensile loading. Although the results were not analyzed in terms of fracture energies, the direction of crack initiation was carefully noted and compared to theoretical predictions.

Fracture experiments on composites and adhesive bond specimens were performed in a similar way. In the adhesive specimens, precracks were inserted by prying open the end of the specimen with a sharp wedge or by including a thin PTFE tab in the specimen during preparation. In addition to varying the temperature and crosshead speed, adhesive tests were conducted at a series of bond thicknesses ranging from 0.1 to 3 mm. A precrack was inserted in the composite samples by using a knife edge at one end of the specimen between the two center plies. The specimen was then loaded perpendicular to the precrack in a tensile machine. This arrangement forced the crack through the center of the specimen between adjacent fiber plies and caused a mode-I type failure. Although most composite tests were performed at 24 °C and a crosshead speed of 0.04 mm/s, some specimens were tested at other conditions to obtain preliminary information about rate and temperature effects. Standard equations were used to evaluate the fracture energies for both adhesive and composite test samples (*19, 21*).

Mode-I Adhesive Fracture

When brittle polymers are tested in adhesive bonds, the fracture energies obtained are similar to those found in tests on neat specimens of the same material (*6, 9, 30*). This result suggests that the failure mechanisms are the same in the two geometries and that the fracture energy obtained is a material property (i.e., it is independent of measurement geometry). By comparison, when the high fracture energy, rubber-modified materials were tested, results such as those shown in Table I were obtained (*9*). This table lists some typical data for neat and bonded specimens as a function of elastomer concentration. The results illustrate the high toughness of these materials in both geometries. This high resistance to crack growth was associated with a large crack-tip deformation zone in which cavitation of the rubber particles and large-scale yielding of the matrix resin were observed in the fracture surface of the region of crack-growth initiation (*1–4, 9–12, 17, 23, 24*). Detailed discussions of possible toughening mechanisms have been published (*1–4, 9, 12, 17, 23, 24*). In general, these papers argue that the large crack-tip deformation zone permits blunting of the crack tip, and this blunting decreases the local stress concentration at the tip of the crack. As a result, higher applied loads are required to cause failure. Such a picture is consistent with the observation that the magnitude of the fracture energy.

Table I. Bulk Versus Adhesive Fracture Energy

Concentration of Rubber (phr)	*Fracture Energy (kJ/m²)*	
	Bulk	*Adhesive*
0	0.11	0.13
5.5	1.1	2.3
11.7	2.7	3.6
18.5	3.5	3.5

correlates with the size of the deformation zone (9, *17*, *23*, *24*). The dimensions of this zone can be easily estimated because stress whitening is associated with the deformations.

The results in Table I also show that a tough polymer produces a tough adhesive. The correlation, however, is not as quantitative as with brittle materials; the initial experiments on adhesive bonds made with the modified epoxies exhibited unusually large scatter. After some study most of this scatter was found to be caused by the fact that an important parameter (bond thickness) was not controlled (9). As a result, a careful study was conducted to measure fracture energy as a function of bond thickness, h. The results (Figure 1) showed that the adhesive fracture energy was independent of bond thickness for brittle materials but exhibited a strong dependence for the modified epoxies (9). This significant result demonstrated that adhesive fracture energy for the modified materials was not a true materials property but rather a structural property because it depends on a structural parameter, namely, the bond thickness.

A possible explanation for this effect notes (9) that the size of the crack-tip deformation zone was comparable with the bond thickness.

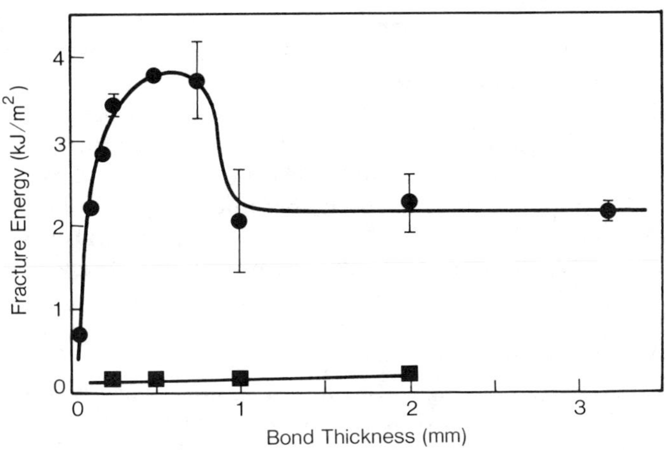

Figure 1. Adhesive fracture energy vs. bond thickness for the epoxy (■) and a modified epoxy containing 18.5 phr of rubber (●). Tests performed at 25 °C and a crosshead speed of 0.02 mm/s (12).

As a result, thin bonds tended to restrict the size of the zone and hence the toughness. This observation explained the decrease in fracture energy below the optimum bond thickness, h_m (i.e, that corresponding to the maximum fracture energy), but did not explain the behavior at larger thicknesses. The decrease in fracture energy for thicker bonds will be discussed later.

Another tenet of the proposed explanation was that the optimum bond thickness occurred when h was equal to the size of the deformation zone as exemplified by measurements on neat samples (9). Because this hypothesis could be tested if an estimate of the deformation zone could be made, a simple elastic–plastic model was suggested to represent a simplistic but perhaps useful first effort to estimate h_m (9). This model relates the fracture energy ($\mathcal{G}_{Ic}$), the radius of the deformation zone (r_c), and the yield stress (σ_y) and strain (ϵ_y). With this model, h_m could be estimated from measurements made on bulk samples (the plane-stress version (*31*) has been shown to give the best results).

$$h_m = 2r_c = \mathcal{G}_{Ic}/\pi\sigma_y\epsilon_y \tag{1}$$

With this procedure, excellent agreement was obtained between the predicted value of h_m and the data in Figure 1.

Because adhesive bond fracture is temperature dependent, the bond thickness experiment was repeated at a series of temperatures (*12*). The results, which are shown in Figure 2, demonstrated that the behavior at all temperatures was similar to that found in the room temperature experiments. In general, the temperature dependence of adhesive properties is complex and difficult to understand and predict. As seen in Figure 2, one reason for this difficulty is the bond thickness effect. Most previous studies have been performed at just one bond thickness, and consequently, only part of the total picture was determined. For the material used to generate the data in Figure 2, however, the temperature dependence can best be described by noting that an increase in temperature shifts h_m to larger values. If this shift could be predicted, a simple description for the temperature dependence would result. Initial efforts (*12*) to use the elastic–plastic model with temperature-dependent parameters gave a good quantitative prediction for h_m at all but the lowest temperatures. As will be discussed later, more recent treatments have produced good agreement at all conditions.

Mixed-Mode Loading

The next major area addressed in the program involved mixed-mode loading. This area is important because practical adhesive bonds are generally designed to minimize mode-I loads. With brittle materials, the fracture energy associated with mixed-mode loading was significantly higher than that found in mode-I tests (*32, 33*). This result

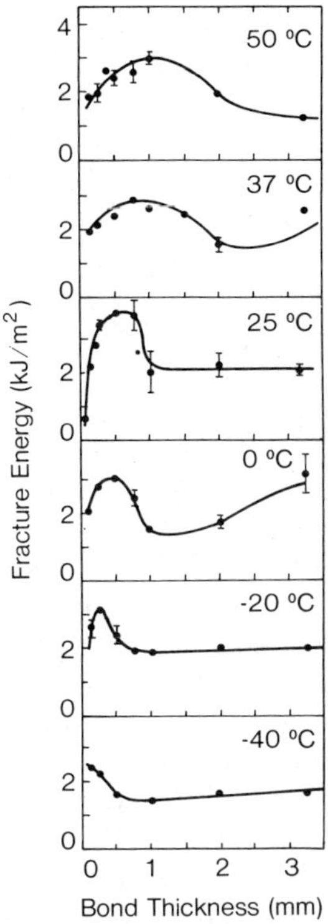

Figure 2. Adhesive fracture energy vs. bond thickness at a series of temperatures for a modified epoxy containing 18.5 phr of rubber tested at a crosshead speed of 0.02 mm/s (12).

reflects the fact that the crack would like to run in a direction perpendicular to the maximum resolved tensile stress at the crack tip (i.e., mode-I type failure). As a result, the crack will generally advance in this direction during loading until it reaches the adherend and stops. Substantially higher loads are required to cause the crack to change direction and advance along the interface because this direction is less favorable.

With tough materials (elastomer-modified epoxies) the situation is not as well understood, and many of the results are contradictory. Some experiments (*10, 11, 13–15*) have suggested that the mixed-mode fracture energy is lower than the mode-I value, but others have found exactly the reverse (*34*). A complete scientific explanation of this complex area will require much more detailed research. Nevertheless, some useful observations and conjectures can be offered on the basis of results now available.

The first question that might be asked is whether the tough materials behave in some unusual and unexpected manner in mixed-mode loading. To investigate this behavior in a qualitative way, plates with centrally located precracks were made from both modified and unmodified epoxies and loaded to failure in tension. The precracks were inserted at different angles to the applied load to obtain various types of mixed-mode loading. The failed specimens were then examined to determine the direction of crack initiation. The results showed that in all cases the initiation directions in the modified and unmodified materials were the same (*14, 15*). Moreover, regardless of the precrack angle, these directions were in good agreement with theoretical predictions based on initiation in a direction perpendicular to the maximum resolved tensile stress at the crack tip. In conclusion, therefore, at least in these tests, the modified epoxies exhibited no unexpected behavior.

Why then do the bonded specimens give conflicting results? Unfortunately, we have no completely satisfactory explanation at present, but some interesting observations can be made. Those tests that gave a high mixed-mode fracture generally involved a scrim cloth (a thin layer of cloth used in some adhesives), and the failure was center-of-bond along the scrim cloth. Moreover, failure often occurred when the load component perpendicular to the bond line reached the mode-I value. On the other hand, those tests that gave a lower mixed-mode fracture energy involved an adhesive without a scrim cloth. The crack was driven toward the adhesive–adherend interface and then propagated along the interface in the adhesive but very close to the boundary. These results may suggest that the scrim cloth provides either a change in the stress field or a preferred path (i.e., weaker) for the crack. This provision keeps the crack in the center of the bond despite the fact that the normal mixed-mode stress field would drive the crack toward the interface. Under these conditions, a reasonable failure criterion might predict that the initiation of rapid crack growth will occur when the load component normal to the scrim cloth plane reaches the mode-I failure load. For those tests where a scrim cloth is not present, the crack is driven toward the interface, as was seen in the unmodified epoxy experiments. If the fracture energy is lower than expected for these cases, it may be related to some type of interaction between the interface and the crack-tip deformation zone that restricts the toughness. More work is needed to clarify this situation. Examination of mode-I and mixed-mode fracture using the same adhesive with and without a scrim cloth would be of particular interest.

Temperature and Loading Rate Effects

The dramatic temperature dependence exhibited in adhesive bond tests on the rubber-modified epoxies suggests a dependence on loading rate. The literature, however, gives conflicting results for experiments designed to look at loading rate effects. This conflict is not surprising because experiments we conducted (*17, 22*) with the rubber-modified epoxy gave similar results—the fracture energy was found to increase, decrease, or remain unchanged depending on the temperature at which the tests were performed (Figure 3).

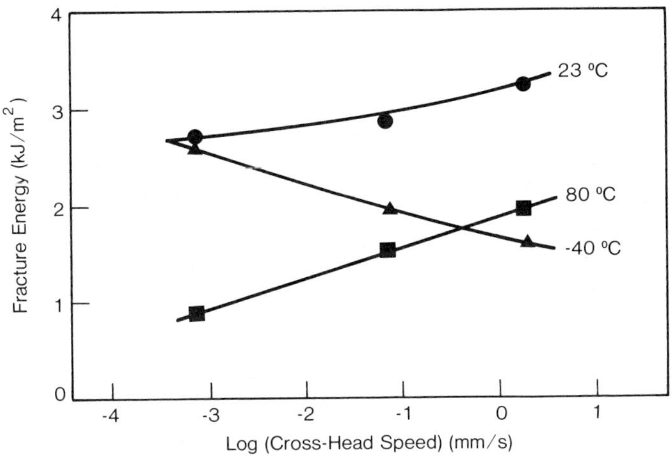

Figure 3. Adhesive fracture energy for a 0.25-mm bond tested at three temperatures as a function of crosshead speed. Adhesive is modified epoxy containing 18.5 phr of rubber.

To develop a better understanding of these results, temperature and loading rate studies were first conducted on neat specimens. Figure 4, for example, shows the data obtained for a rubber-modified epoxy measured at three crosshead speeds over a wide range of temperatures (22). These results indicated an inverse relationship between rate and temperature; that is, increasing the temperature produced a change similar to that obtained by decreasing the loading rate (crosshead speed). This observation suggested that the fracture

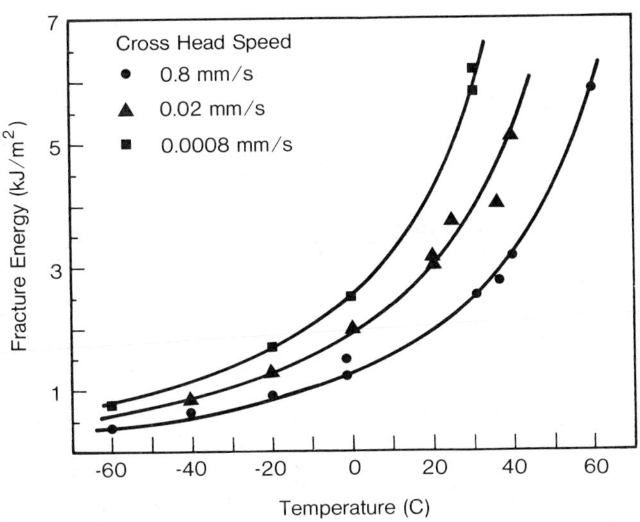

Figure 4. Bulk fracture energy of modified epoxy containing 18.5 phr of rubber tested at various temperatures and crosshead speeds.

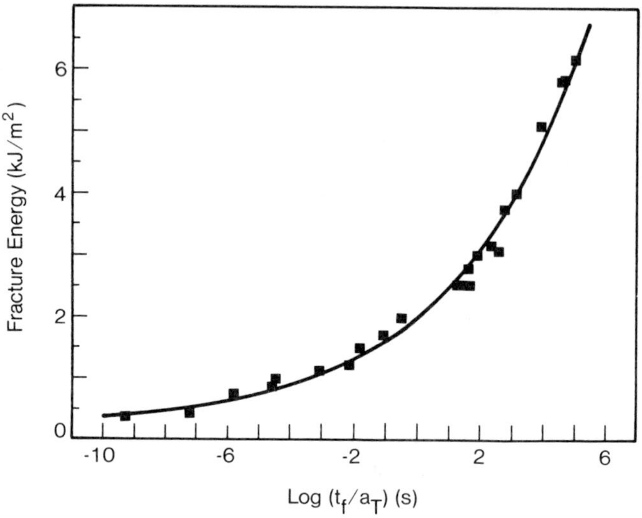

Figure 5. Bulk fracture energy of modified epoxy containing 18.5 phr of rubber plotted aginst time to failure normalized with a temperature shift factor. Curve represents best fit with Equation 2.

was dominated by viscoelasticity. Consequently, the data were analyzed to see if time–temperature superposition techniques could be used to reduce all of the data for a given polymer system to a single curve. The details of this analysis can be found in previous publications (*18, 20, 22*), but we concluded that at least one way to accomplish this superposition was by plotting the fracture energy against the time to failure, t_f, reduced with an appropriate temperature shift factor, a_T. Figure 5 shows a curve of this type for the data in Figure 4.

The results for the rubber-modified epoxy could be fitted to a number of empirical equations such as Equation 2 (*see* Figure 5),

$$\mathcal{G}_{\mathrm{Ic}} = A(t_f/a_T)^m + \mathcal{G}_{\mathrm{Icb}} \quad (2)$$

where $\mathcal{G}_{\mathrm{Icb}}$ is the minimum fracture energy at low temperatures and high crosshead speeds, and A and m are constants. The data for a_T could be fit to Equation 3.

$$a_T = e^{\Delta H/RT} \quad (3)$$

The parameter ΔH had the same value as that obtained by applying a similar superposition procedure to yield data. This superposition technique has now been applied to four different rubber-modified epoxy formulations with very successful results (*20*). Consequently, the data supported the assertion that the fracture behavior of neat specimens was dominated by viscoelastic effects.

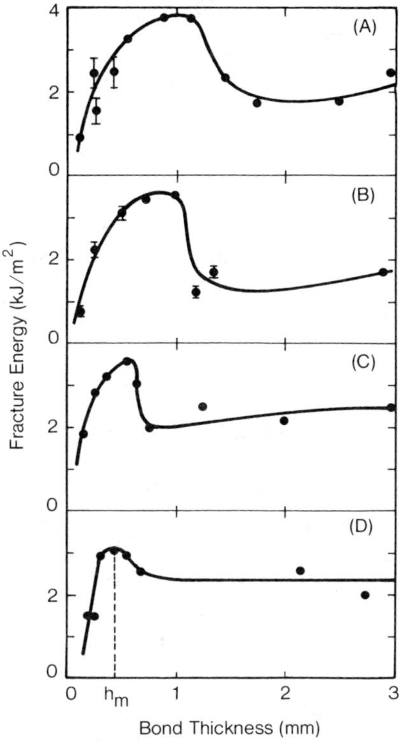

Figure 6. Adhesive fracture energy vs. bond thickness for a modified epoxy containing 15.0 phr of rubber tested at 22 °C and various crosshead speeds (m/s): A, 8.33 × 10^{-7}; B, 1.67 × 10^{-5}; C, 1.67 × 10^{-4}; and D, 8.33 × 10^{-4}. (Reproduced with permission from Ref. 31. Copyright 1981, Journal of Adhesion.*)*

These results suggested that the adhesive bond data might also reflect viscoelastic effects. If so, increasing the crosshead speed should produce the same trend as decreasing the temperature. On the basis of this idea, an examination of Figure 2 predicts that the change in fracture energy associated with an increase in crosshead speed would depend on the value of bond thickness compared to the optimum thickness, h_m. More specifically, $\mathcal{G}_{Ic}$ would be expected to increase, decrease, or remain constant for values of h that are less than, greater than, or about equal to h_m, respectively. A simple examination of this idea (*17, 22*) explained the results in Figure 3. More recently, Kinloch et al. (*31*) measured the bond thickness dependence of fracture energy at a series of crosshead speeds (Figure 6), and the results agree well with the proposed explanation.

This hypothesis also led to a possible revision in the model for predicting the optimum bond thickness. If viscoelastic effects are dominant, the quantities in Equation 2 will be loading rate and temperature dependent; consequently, appropriate values must be used

Table II. Deformation-Zone Size

Temperature (°C)	*Log Crosshead Speed (m/s)*	h_m *(mm)*	
		Measured	*Predicted*
20	−6.08	1.0	0.85
20	−4.78	0.8	0.70
20	−3.78	0.55	0.49
20	−3.08	0.4	0.43
37	−4.66	0.9	1.16
25	−4.66	0.6	0.57
0	−4.66	0.5	0.39
−20	−4.66	0.25	0.15

if good agreement is to be obtained. To examine this hypothesis, all of the quantities involved were measured as a function of time and temperature, and the time to failure in the fracture experiment was used to select the appropriate values in each case. When the quantities were substituted into Equation 2, values of h_m could be successfully predicted for adhesive bond tests performed at a wide variety of temperatures and crosshead speeds. Table II, for example, shows some typical results obtained in this way, and the agreement is excellent (*18*, *31*).

The final topic related to adhesive bond failure concerns the drop in $\mathcal{G}_{Ic}$ at high bond thicknesses. A possible explanation for this effect was suggested several years ago by S. S. Wang (University of Illinois) as part of his collaboration with this program (*18*, *20*, *22*, *35*). A detailed discussion of this hypothesis will be given in a future publication, but briefly it can be described in qualitative terms by asserting that two effects influence the size of the deformation zone and, therefore, the fracture energy. As the bond thickness decreases from large values where the fracture energy is similar to that obtained for neat specimens, the high stress field ahead of the crack tip extends farther down the bond line than expected because of the nature of the adhesive bond geometry (*35*). This extension causes the deformation zone to extend down the bond line and thus produces an increase in zone size. Consequently, the fracture energy is increased. Eventually, however, the bond thickness becomes small enough so that it begins to restrict the zone size in the thickness direction. Once this effect becomes predominant, the zone size decreases. As a result, the fracture energy decreases. Thus, the fracture energy is predicted to go through a maximum, as seen experimentally in Figures 1 and 2. Recently, high-speed motion pictures of adhesive bond fracture were made at various bond thicknesses. These movies show variations in zone size and shape that agree well with this proposed explanation for the bond thickness dependence.

Table III. Neat and Composite Fracture

Elastomer Concentration (phr)	Fracture Energy (kJ/m^2) Neat	Woven	Unidirectional
0	0.23	0.8	0.4
16.0	5.8	4.2	1.9

Interlaminar Fracture of Composites

The final area to be discussed concerns the interlaminar fracture of composites. A great deal of interest has focused on increasing the resistance of composites to delamination (i.e., the initiation and growth of cracks between the layers of fibers either in the polymer matrix or at the fiber–matrix interface). These cracks, which are often initiated by impact loading, can substantially reduce composite performance, particularly in compression loading (*36*). Moreover, such damage is generally difficult to detect and repair. To minimize this problem, development of resins with improved resistance to the initiation and growth of delaminations is desirable.

One suggestion is that composites fabricated with elastomer-modified epoxies would be superior to those made with normal epoxies because the resin itself is so much tougher. The contrasting argument is that on the basis of adhesive bond results, the small interlaminar spacing in a composite would be expected to restrict the crack-tip deformation zone. As a result the composite with the toughened matrix would be predicted to have about the same fracture energy as that for the composite with the brittle matrix.

To help resolve this difference of opinion, we conducted studies in which the double cantilever beam geometry was used to assess the resistance of selected composite formulations to growth of an interlaminar crack under mode-I type loading. The results (*21*) indicated that a 25-fold increase in the resin toughness produced a fourfold to eightfold improvement in the interlaminar fracture energy for woven-fiber reinforcement composites as measured in this test (Table III). When the specimens were examined with a scanning electron microscope, the crack-tip deformations in the toughened material were not restricted to the matrix layer between the plies on either side of the crack, but also extended to the matrix between the plies a number of layers above and below the crack plane. Thus, although the fibers seem to somewhat restrict the deformation-zone size, and hence reduce the improvement in toughness relative to the epoxy, the restrictions were much less than originally anticipated by the theory that asserted the zone would not extend beyond the first layer of fibers on either side of the crack (*21*).

The resin-rich areas at the crossover points in the weave were believed to be important to the toughening obtained with the modified system. Thus, the benefits of a tough resin would be much less in a unidirectional, fiber-reinforced composite. The results in Table III, however, show that adding rubber to the epoxy increased the interlaminar fracture energy of unidirectional composites by almost as much as it does in woven systems (*26*) (i.e., a 4.75-fold improvement in unidirectional materials versus a 5.25-fold improvement for woven materials). The major difference between the behaviors for the two types of materials was that the woven composites gave higher results regardless of the resin. Therefore, additional toughening mechanisms may exist in the woven systems. One possibility might be crack pinning at the crossover points in the weave.

Another interesting question with regard to the rubber-modified epoxy composites concerns the possibility of temperature and loading rate effects. With brittle epoxies, neither the neat resin fracture nor the composite interlaminar fracture was a strong function of these parameters as long as the conditions were well below the glass-to-rubber transition (*25*, *37*). With rubber-modified epoxies, however, the neat resin shows dramatic variations with temperature and loading rate. Consequently, we have important questions about the composite behavior. According to one hypothesis, because the fibers appear to restrict the deformation zone size, an unused potential may exist for deformation in the matrix that could be tapped if the temperature was decreased or the loading rate was increased. If so, the composite interlaminar fracture energy might show smaller variations as the experimental conditions are changed than the resin itself. Some preliminary tests (*25*) of this hypothesis were performed with woven-reinforcement composites; the results obtained are shown in Table IV. For the set of samples used in these tests, the unmodified epoxy composite gave a fracture energy of 0.8 kJ/m^2 so this particular formulation of rubber-modified epoxy composite yielded about a two-fold improvement in toughness. The results in Table IV indicate that this improvement was not lost as the temperature was decreased and/or the crosshead speed was increased. If anything, the toughness

Table IV. Fracture Energy at Various Crosshead Speeds

	Fracture Energy (kJ/m^2)		
Temperature (°C)	*0.0008*[a]	*0.04*[a]	*0.8*[a]
40	1.7	1.8	1.8
24	1.7	1.7	1.9
0	1.7	1.9	2.0
−25	1.9	2.1	2.4

[a] Crosshead speeds, mm/s.

increased slightly, although the experimental error was such that this trend could not be confirmed. Such results raise some interesting questions about the behavior of composites for different temperatures and loading histories. Obviously this area, particularly with regard to unidirectional lay-ups, represents an important area for future research.

Conclusions

The elastomer-modified epoxies represent a very useful and interesting class of materials. The two-phase nature of this system makes possible a number of superior performance characteristics, such as high fracture energy, that are difficult or impossible to achieve in similar single-phase materials. In return for these benefits, however, the mechanical properties and failure behavior can be more complex than for a single-phase material; this complexity may make design more difficult.

Recently, an effort has been made to develop a basic understanding of these complications in the elastomer-modified epoxies. For the most part, this understanding has been developed by first studying the behavior of neat specimens to determine the properties of the material itself and then investigating the behavior of adhesives and composites. These latter results can be understood by considering how the constraining effects of the adherends in the adhesive bond or the fibers in the composite modified and changed the basic behavior of the rubber-toughened epoxy itself. Through this approach, significant progress has been made in developing a basic understanding of the properties of these materials. Continued work along these lines will lead to a more complete knowledge of these effects.

List of Symbols

A	Constant in Equation 2
$\mathcal{G}_{Ic}$	Mode-I fracture energy, J/m^2
$\mathcal{G}_{Icb}$	Low temperature/high rate limit of $\mathcal{G}_{Ic}$, J/m^2
ΔH	Thermal activation parameter in Equation 3, kcal/mol
PTFE	Polytetrafluoroethylene
R	Gas constant, kcal/mol · T
T	Temperature, °C
T_g	Glass transition temperature, °C
TDCB	Tapered double cantilever beam
a_t	Shift fact
h	Bond thickness, m
h_m	Optimum bond thickness, m
phr	Parts by weight per hundred parts by weight resin
r_c	Radius of deformation zone at fracture load, m

t_f Time to failure, s
ϵ_y Yield strain
σ_y Yield stress, N/m^2

Literature Cited

1. Rowe, E. H.; Siebert, A. R.; Drake, D. S. *Mod. Plast.* **1970,** *47*, 110–117.
2. Riew, C. K.; Rowe, E. H.; Siebert, A. R. In "Toughness and Brittleness of Plastics"; Deanin, R. D.; Crugnola, A. M., Eds.; ADVANCES IN CHEMISTRY SERIES No. 154, American Chemical Society: Washington, DC, 1976; pp. 326–343.
3. Sultan, J. N.; Laible, R. C.; McGarry, F. J. *J. Appl. Polym. Sci.* **1971,** *16*, 127–136.
4. Sultan, J. N.; McGarry, F. J. *J. Polym. Eng. Sci.* **1973,** *13*, 29–34.
5. Ripling, E. J.; Mostovoy, S.; Patrick, R. L. *ASTM, Spec. Tech. Publ. 360*, **1963,** 5.
6. Mostovoy, S.; Ripling, E. J. *J. Appl. Polym. Sci.* **1971,** *15*, 641–659.
7. Ripling, E. J.; Mostovoy, S.; Corten, H. T. *J. Adhes.* **1971,** *3*, 107–123.
8. Mostovoy, S.; Ripling, E. J.; Bersch, C. F. *J. Adhes.* **1971,** *3*, 125–144.
9. Bascom, W. D.; Cottington, R. L.; Jones, R. L.; Peyser, P. *J. Appl. Polym. Sci.* **1975,** *19*, 2545–2562.
10. Bascom, W. D.; Timmons, C. O.; Jones, R. L. *J. Mater. Sci.* **1975,** *10*, 1037–1048.
11. Bascom, W. D.; Jones, R. L.; Timmons, C. O. In "Adhesion Science and Technology"; Lee, L. H., Ed.; Plenum: New York, 1975; Vol. 9B, pp. 501–510.
12. Bascom,W. D.; Cottington, R. L. *J. Adhes.* **1976,** *7*, 333–346.
13. Bascom, W. D.; Cottington, R. L.; Timmons, C. O. *J. Appl. Polym. Sci., Appl. Polym. Symp.* **1977,** *32*, 165–188.
14. Bascom, W. D.; Hunston, D. L.; Timmons, C. O. *Org. Coat. Plast. Chem.* **1978,** *38*, 179–184.
15. Bascom, W. D.; Hunston, D. L.; Timmons, C. O., unpublished results.
16. Bascom, W. D.; Oroshnik, J. *J. Mater. Sci.* **1978,** *13*, 1411–1418.
17. Hunston, D. L.; Bitner, J. L.; Rushford, J. L.; Oroshnik, J.; Rose, W. S. *J. Elastomers Plast.* **1980,** *12*, 133–149.
18. Hunston, D. L.; Rushford, J. L.; Wang, S. S.; Kinloch, A. J. *Proc. 37th Annu. Conf., Reinf. Plast./Compos. Inst. Soc. Plast. Ind.* 1982, Session 29-C, pp. 1–5.
19. Bascom, W. D.; Hunston, D. L. In "Adhesion 6"; Allen, K. W., Ed.; Applied Science: New Jersey, 1982; pp. 185–210.
20. Hunston, D. L.; Kinloch, A. J.; Shaw, S. J.; Wang, S. S. In "Adhesive Joints: their Formation, Characterization, and Testing"; Mittal, K., Ed.; Plenum: New York, 1984; pp. 789–807.
21. Bascom, W. D.; Bitner, J. L.; Moulton, R. J.; Siebert, A. R. *Composites* **1980,** *11*, 9–18.
22. Bitner, J. L.; Rushford, J. L.; Rose, W. S.; Hunston, D. L.; Riew, C. K. *J. Adhes.* **1981,** *13*, 3–28.
23. Kinloch, A. J.; Shaw, S. J.; Tod, D. A.; Hunston, D. L. *Polymer* **1983,** *24*, 1341–1354.
24. Kinloch, A. J.; Shaw, S. J.; Hunston, D. L. *Polymer* **1983,** *24*, 1355–1363.
25. Hunston, D. L.; Bascom, W. D. *Comp. Tech. Review* **1983,** 5(winter), 118–19.
26. Hunston, D. L.; Moulton, R. J.; Johnston, N. J., unpublished results.
27. Ting, R. Y.; Cottington, R. L. *J. Appl. Polym. Sci.* **1980,** *25*, 1815–1823.
28. Bascom, W. D.; Gadomski, S. T.; Henderson, C. M.; Jones, R. L. *J. Adhes.* **1977,** *8*, 213–233.
29. Bascom, W. D.; *Adhes. Age* **1979,** April, 29–35.
30. Gledhill, R. A.; Kinloch, A. J.; Yamini, S.; Young, R. J. *Polymer* **1978,** *19*, 574–581.
31. Kinloch, A. J.; Shaw, S. J. *J. Adhes.* **1981,** *12*, 59–77.

32. Trantina, G. G. *J. Composite Mater.* **1972,** *6,* 191–207.
33. Trantina, G. G. *J. Composite Mater.* **1972,** *6,* 371–385.
34. Mostovoy, S., personal communication.
35. Wang, S. S.; Mandell, J. F.; McGarry, F. J. *Int. J. Fract.* **1978,** *14,* 39–58.
36. Williams, J. G.; Rhodes, M. D. *NASA Tech. Memo. 83213,* October 1981.
37. Miller, A. G.; Hertzberg, P. E.; Ranatala, V. W. *Proc. 12th Nat. SAMPE Tech. Conf.* **1980,** 279–295.

RECEIVED for review December 23, 1983. ACCEPTED May 1, 1984.

1970 - 1983

8

Rubber-Toughened Polyimides

A. J. KINLOCH, S. J. SHAW, and D. A. TOD

Ministry of Defence (PE), Propellants, Explosives and Rocket Motor Establishment, Waltham Abbey, Essex, United Kingdom

The effects of adding up to 100 parts per hundred parts resin (phr) of a rubber to a bismaleimide resin on such properties as fracture energy, modulus, flexural strength, glass transition temperature, and decomposition temperature are described. The two-phase morphology resulting from this rubber addition is discussed in terms of data obtained from both scanning electron microscopy and dynamic mechanical analysis. The results obtained suggest that up to 50 phr may be incorporated to produce a 20-fold increase in fracture energy without major sacrifices in other important properties.

Polyimides are organic polymers that in general possess very high thermal and thermo-oxidative stabilities. They are used in a number of industrial applications, especially as adhesives and as matrices for fiber–composite materials.

Most polyimides may be classified either as condensation-type or addition-type polymers according to the type of polymerization reaction used to produce them from their constituent(s) (*1*). Condensation-type polyimides are generally prepared from dianhydrides and diamines via the formation of a soluble polyamic acid precursor. The formation of polyimides from the polyamic acid involves the evolution of volatiles, which can cause voids in the final product. Addition-type polyimide prepolymers are cured by an addition reaction that overcomes the problem caused by evolution of volatiles. Addition-type polyimides are prepared by several routes. One of the most common, and the one employed in this work, involves using the activated maleimide double bond.

However, one disadvantage of addition-type polyimides based upon bismaleimides is that they depend to a large degree upon a highly cross-linked structure for their high-temperature capability. This structure results in them being extremely brittle. Some of the more brittle polybismaleimides have been toughened by chain extension of the imide prepolymer molecules. The resulting polyimides

0065-2393/84/0208-101/$06.00/0
Published 1984 American Chemical Society

have more open, flexible molecular structures than their unmodified, brittle counterparts. However, the increase in toughness is usually offset by a major reduction in other desirable properties of the polyimides, such as glass transition temperature (T_g), thermal stability, and mechanical strength, because chain extension of the prepolymer reduces the density of intermolecular cross-links in the polyimide structure.

We describe some initial exploratory studies concerned with achieving a two-phase microstructure consisting of a dispersed rubbery phase embedded in and bonded to a matrix of polyimide. In brittle, cross-linked epoxy resins this microstructure increases the toughness without greatly reducing other important properties (*2–8*).

Experimental

Materials and Preparation. The basic bismaleimide resin employed was a commercially available material reported (*1*) to consist of three bismaleimide species (**I**) that form a eutectic mixture with a melting point of approximately 70 °C.

The rubber used was a carboxyl-terminated, random copolymer of butadiene and acrylonitrile that can be represented by HOOC-$[(CH_2CHCHCH_2)_x(CH_2CHCN)_y]_m$COOH.

The carboxyl content was 2.37 wt%, the acryonitrile content was 18 wt%, and the molar mass of the rubber was 3500 g/mol. Rubber modification was carried out by adding the liquid rubber to the molten bismaleimide at 120 °C. A prereaction period of 24 h at 120 °C was allowed to develop a copolymer from the constituent materials. Curing of the bismaleimide systems was achieved by heating for 2 h at 170 °C followed by 5 h at 210 °C.

A total of five bismaleimide–rubber formulations were studied. Formulations 1–5 contain 0, 10, 30, 50, and 100 parts per hundred parts resin (phr) of rubber, respectively.

Thermal Properties. The T_g values of the cured polymers were determined by using two techniques: thermomechanical analysis (TMA) and differential scanning calorimetry (DSC). The former was conducted with a thermomechanical analyzer (Stanton Redcroft Model 790) at a heating rate of 5 °C/min in static air. DSC was conducted with a differential calorimeter (Perkin-Elmer) at a heating rate of 20 °C/min in a nitrogen atmosphere.

The thermal stabilities of the cured formulations were determined by using two techniques. First, thermogravimetric analysis (TGA) was employed with a thermobalance (Linseis L81) in an atmosphere of helium. A heating rate of 5 °C/min was employed with sample weights of approximately 15 mg. Differential thermogravimetric analysis (DTA) was conducted simultaneously. Second, isothermal weight loss measurements were performed on formulations 1, 3, and 4 with a thermogravimetric analyzer (Perkin-Elmer). Samples were examined under an oxygen environment, at a gas flow rate of 40 mL/min. The temperature program consisted of an initial temperature of 250 °C, followed by heating at 200 °C/min to 330 °C, which was then maintained for 100 min.

Mechanical Properties. Sheets of cured material, 6 mm thick, were cast from the five formulations. The neat bismaleimide, or prereacted rubber–bismaleimide mixture, was degassed at 120 °C and then poured into a heated casting mold. The bismaleimide systems were cured by heating for 2 h at 170 °C followed by 5 h at 210 °C.

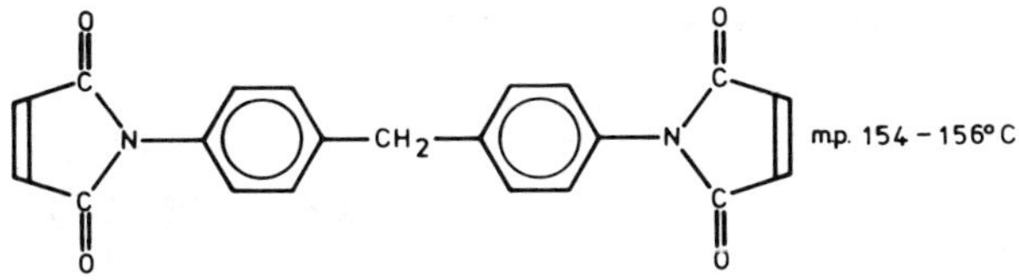

di (4-MALEIMIDOPHENYL) METHANE

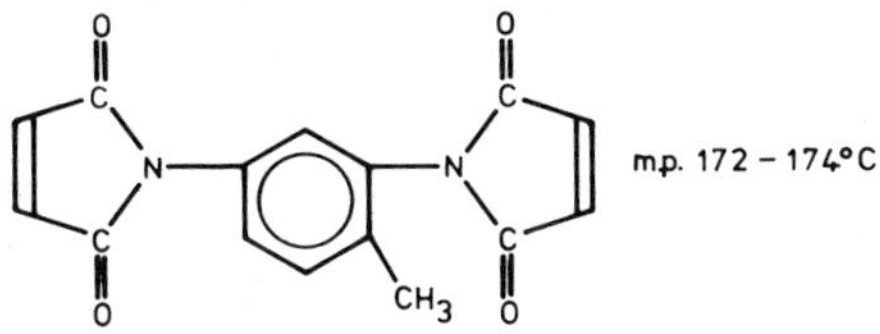

2, 4-BISMALEIMIDO-TOLUENE

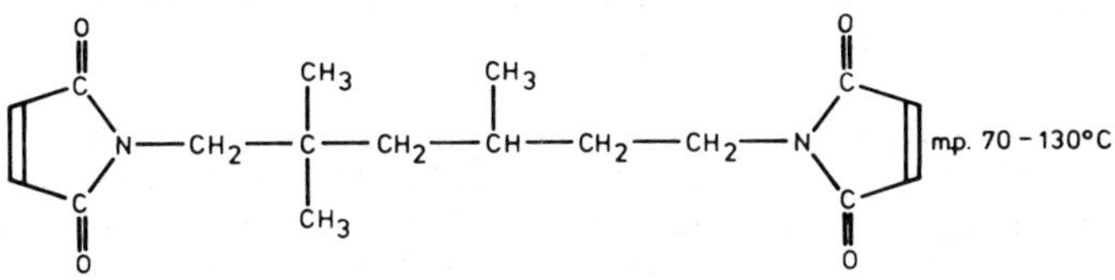

1, 6-BISMALEIMIDO-2,2,4-TRIMETHYLHEXANE

I

DMA experiments were conducted on rectangular strips measuring 85 × 10 × 6 mm by using a mechanical spectrometer (Rheometrics). The specimens were subjected to torsional sinusoidal oscillations at a frequency of 1 Hz. Values of storage shear modulus, G', and loss shear modulus, G'', were obtained, and values of the loss tangent, tan δ, were calculated from Equation 1. Measurements were taken at approximately 10 °C intervals between −160 and 340 °C, with a heating rate between test temperatures of approximately 5 °C/min.

$$\tan \delta = \frac{G''}{G'} \qquad (1)$$

Flexural experiments were conducted on rectangular strips measuring 110 × 20 × 6 mm by using a three-point-bend assembly attached to a mechanical testing machine (Instron). The experiments were conducted at a crosshead speed of 1 mm/min according to ASTM standard method D790-1971. Values of flexural modulus, E, and flexural strength and strain at failure were determined from the resultant load-displacement curves.

Fracture properties of the cured formulations were determined by using compact-tension specimens (3) (Figure 1), which were machined from the cast sheets. Prior to testing, a saw cut was made in the position indicated in Figure

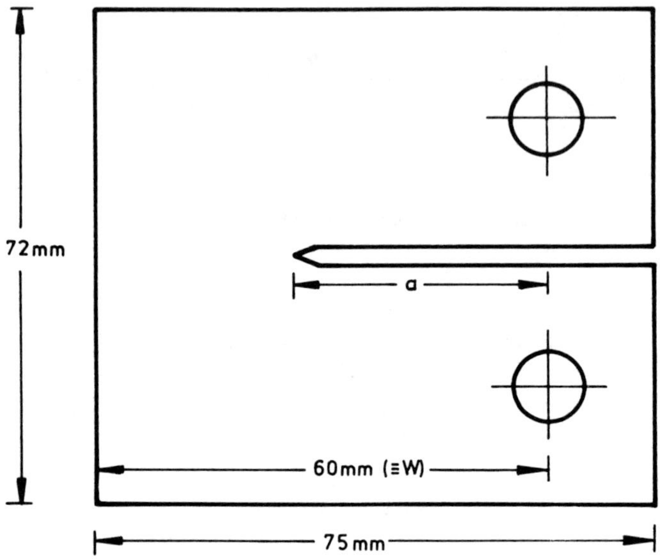

Figure 1. Compact tension specimen used for determining the stress-intensity factor, K_{Ic}*, and fracture energy,* $\mathcal{G}_{Ic}$*, of the polymers.*

1 of approximately the same length, but slightly shorter, than the final crack length required. A sharp crack was then introduced by carefully tapping a fresh razor blade into the base of the saw cut, which caused a sharp crack to grow for a short distance ahead of the blade. The specimen was then mounted in a mechanical testing machine (Instron) and was loaded at a constant crosshead rate of 1 mm/min. The result of the experiment was recorded as a load, P, versus displacement curve. The value of the stress-intensity factor, K_{Ic}, was calculated with Equation 2,

$$K_{Ic} = \frac{P_c}{B_w^{\frac{1}{2}}} Q \qquad (2)$$

where $Q = [29.6(a/w)^{1/2} - 185.5(a/w)^{3/2} + 655.7(a/w)^{5/2} - 1017(a/w)^{7/2} + 638.9(a/w)^{9/2}]$; P_c = load at crack initiation; B = thickness of specimen; w = width of specimen, as defined in Figure 1; and a = crack length.

The K_{Ic} values were converted to fracture energy, $\mathcal{G}_{Ic}$, by Equation 3

$$\mathcal{G}_{Ic} = \frac{K_{Ic}^2}{E}(1 - \nu) \qquad (3)$$

where ν is Poisson's ratio (taken to be 0.35).

Three specimens per test were examined for all the mechanical property studies.

Fractography. Fracture surfaces obtained from the fractured compact-tension specimens were mounted on metal stubs and coated with a thin evaporated layer of gold in order to improve conductivity and prevent charging. The specimens were then examined in a scanning electron microscope (SEM) (Cambridge Instruments) at an accelerating voltage of 20 kV.

Results and Discussion

Glass Transition Temperature, T_g. T_g values obtained from the TMA experiments are shown in Figure 2 plotted against rubber concentration. Somewhat surprisingly, rubber addition appears to increase the T_g by approximately 40 °C over a rubber concentration range of 10 to 100 phr. The exact reasons for this unusual effect are not currently understood, but a similar effect has been observed by St. Clair and St. Clair while working with silicone rubber modified polyimides based upon norbornene systems (as opposed to bismaleimides) (9). A similar trend is further shown in Figure 2 for two formulations that underwent a postcure for 16 h at 240 °C. However, as would be expected, they exhibit greater T_g values.

A typical result obtained from the DSC experiments is shown in Figure 3. The result is anomalous in that an endothermic transition, which is generally indicative of a glass transition, was not observed with any of the formulations. Instead, quite pronounced exothermic processes were observed, suggesting a postcure phenomenon. This hypothesis is supported by the fact that the two samples postcured for 16 h at 240 °C exhibited greatly reduced exotherms. The postcure exotherm would therefore be expected to swamp an

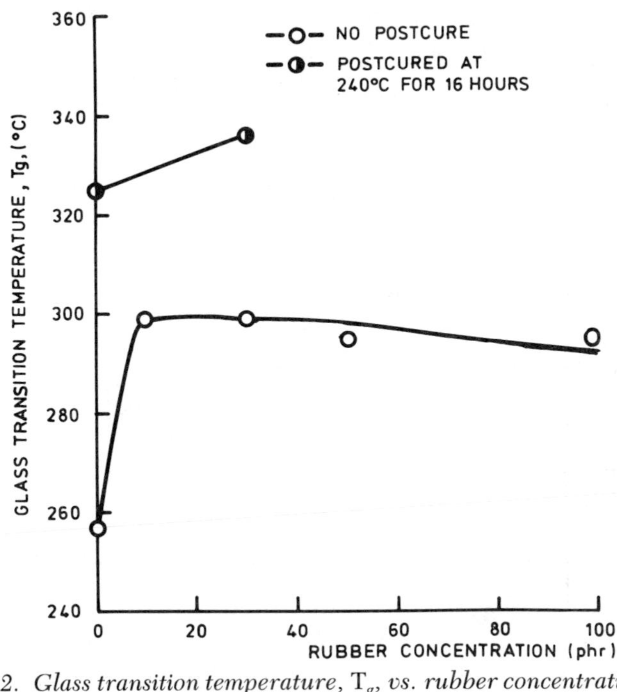

Figure 2. Glass transition temperature, T_g, *vs. rubber concentration from TMA.*

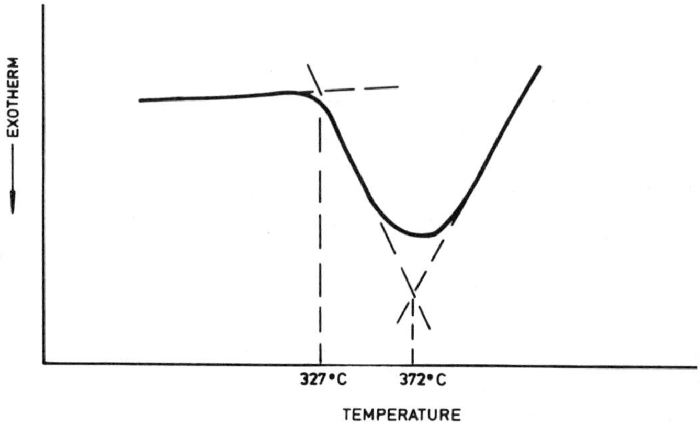

Figure 3. DSC scan obtained from 30-phr carboxyl-terminated butadiene–acrylonitrile (CTBN) composition (postcured for 16 h at 240 °C).

endotherm occurring as a result of a glass transition. Thus, the T_g would be approximately equivalent to the onset of the exotherm. In all cases, this temperature was within 5–10 °C of the T_g values obtained from TMA.

Deformation and Fracture Studies. Figures 4–6 show the results obtained from the flexural tests as a function of rubber concentration. As expected, increasing the rubber concentration produces a reduction in modulus from approximately 4.5 GPa for the unmod-

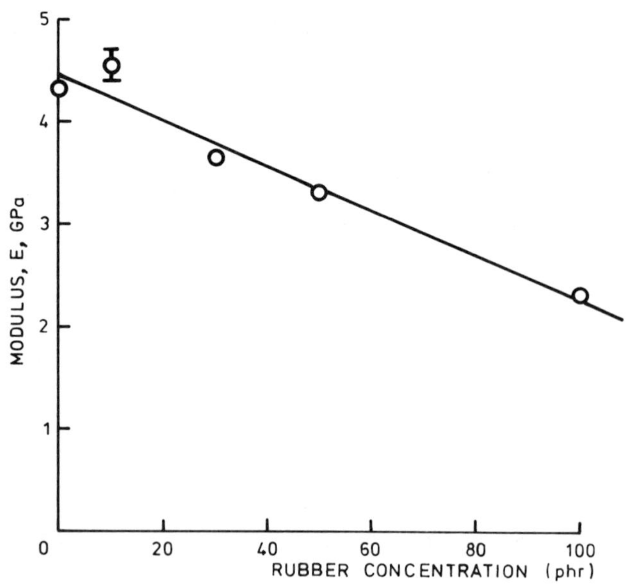

Figure 4. Flexural modulus vs. rubber concentration.

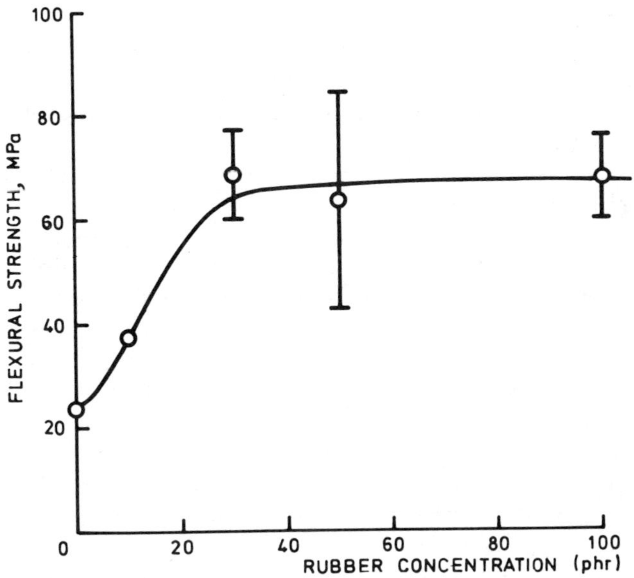

Figure 5. Flexural strength vs. rubber concentration.

ified polyimide to 2.3 GPa for a composition containing 100 phr of rubber. Although this reduction is quite severe, 2.3 GPa is only slightly lower than the value frequently observed for various rubber-modified epoxy resins (*3, 7*). However, the latter generally have T_g

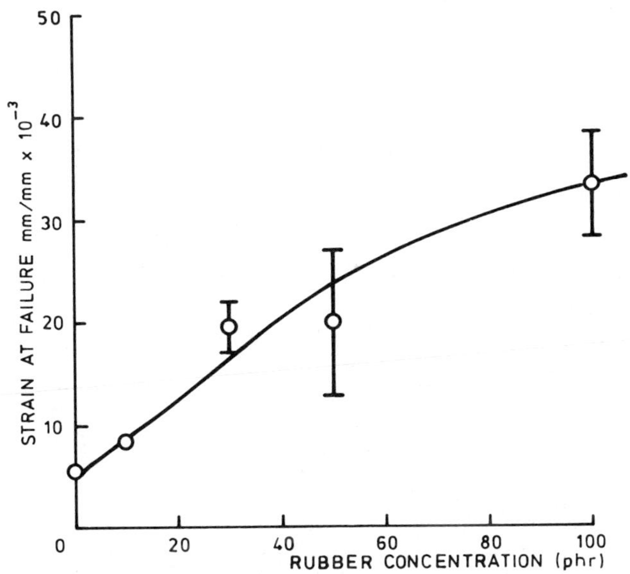

Figure 6. Flexural strain-at-failure vs. rubber concentration.

values of approximately 100 °C, which, as indicated in the previous section, is some 200 °C lower than for the rubber–polyimide formulations. Further increases in rubber concentration beyond that shown in Figure 4 have indicated a continuation in the linear trend leading to a modulus of approximately 0.3 GPa for a formulation containing 200 phr of rubber. Thus, rubber concentrations in excess of 100 phr would clearly not be feasible for structural applications.

As shown in Figure 5, increasing the rubber concentration increases flexural strength quite substantially up to 30 phr of rubber; further increases in rubber concentration have no effect. This trend is rather unusual in that the addition of rubbery materials to thermosetting polymers generally results in reductions in properties such as tensile and flexural strengths. In this case, the rubber has apparently reduced the flaw sensitivity and thus resulted in an increase in flexural strength.

The strain at failure increases with rubber concentration, as shown in Figure 6; the value for the 100-phr formulation is typical for unmodified epoxy resins.

In Figure 7, $\mathcal{G}_{Ic}$ is shown as a function of rubber concentration added to the polyimide. The fracture energy steadily increases as the rubber concentration increases. At a rubber concentration of 100 phr, the value of $\mathcal{G}_{Ic}$ is about 40 times greater than for the unmodified material.

Finally, for both the flexural and fracture experiments, the load-displacement curves were essentially linear. However, slight indications of nonlinearity just prior to failure were noted with the 50- and 100-phr rubber formulations.

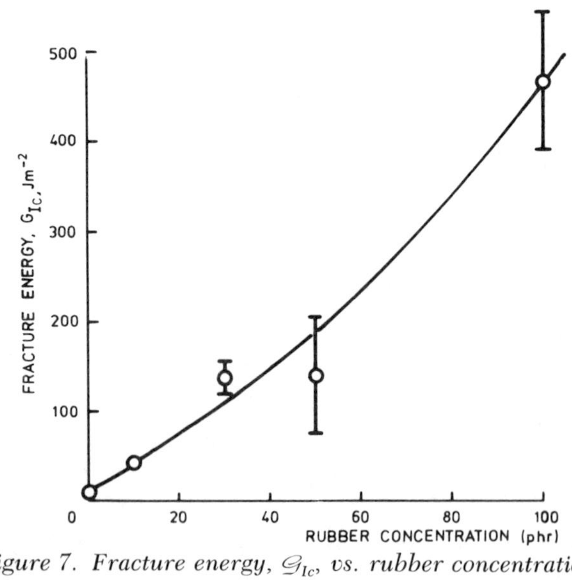

Figure 7. Fracture energy, $\mathcal{G}_{Ic}$, vs. rubber concentration.

Microstructure and Toughening Mechanisms. SEMs of fracture surfaces of compact-tension specimens of a polyimide containing various amounts of added rubber are shown in Figure 8. A two-phase microstructure is clearly visible with particles about 1–2.5 μm in diameter.

a, *b*, *c*

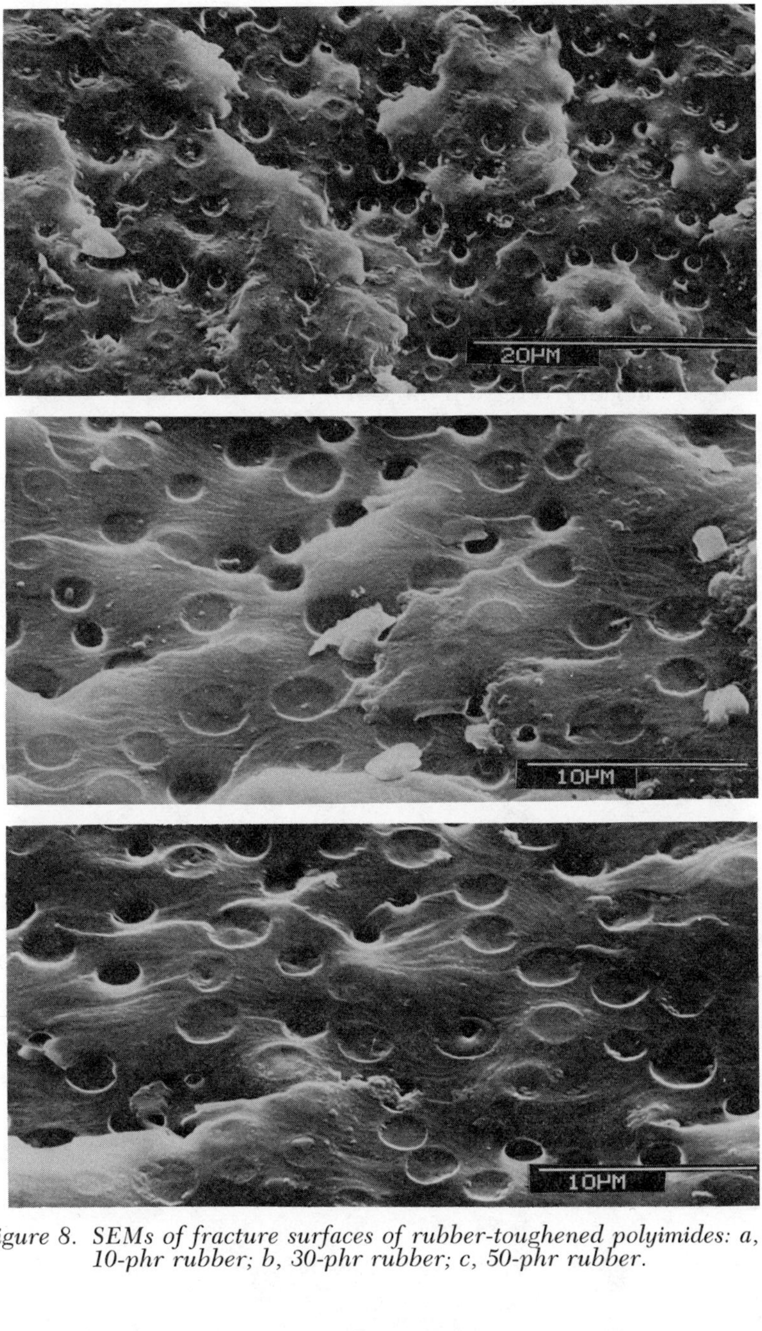

Figure 8. SEMs of fracture surfaces of rubber-toughened polyimides: a, 10-phr rubber; b, 30-phr rubber; c, 50-phr rubber.

DMA studies are extremely useful in interpreting the microstructure of two-phase materials. The values of G', G'', and tan δ for the rubber-modified polyimides are shown in Figures 9–11. The peak in tan δ at about −90 °C that occurs in all the materials could arise from a β-relaxation in the polyimide matrix. The α-relaxation (i.e., T_g) of the matrix is not observed within the temperature range of the mechanical spectrometer employed (i.e., >320 °C). However, the values of G'' do show evidence of an impending transition at 350 –400 °C. This value is somewhat higher than the T_g of the matrix measured with TMA or DSC. This result is to be expected and arises from differences in the types of measurement, timescales that the tests employed, definitions of the transition temperature used, and thermal histories of the specimens.

The T_g of the rubber employed in this work is about −45 °C. Because no apparent transition occurs in this temperature region, no significant concentration of unreacted rubber should exist within the cured systems. Because the rubber would not be expected to react with itself, the second-phase particles must result from the formation of a copolymer derived from the rubber and bismaleimide molecules. The remaining peak in tan δ at about 150 °C may be ascribed to the α-relaxation of this copolymer, and this assignment is reinforced by the decrease in T_g and the increase in peak height as the added rubber concentration increases. Clearly, this result would imply that the second-phase particles would be rigid at room temperature, where the fracture tests were conducted. However, improvements

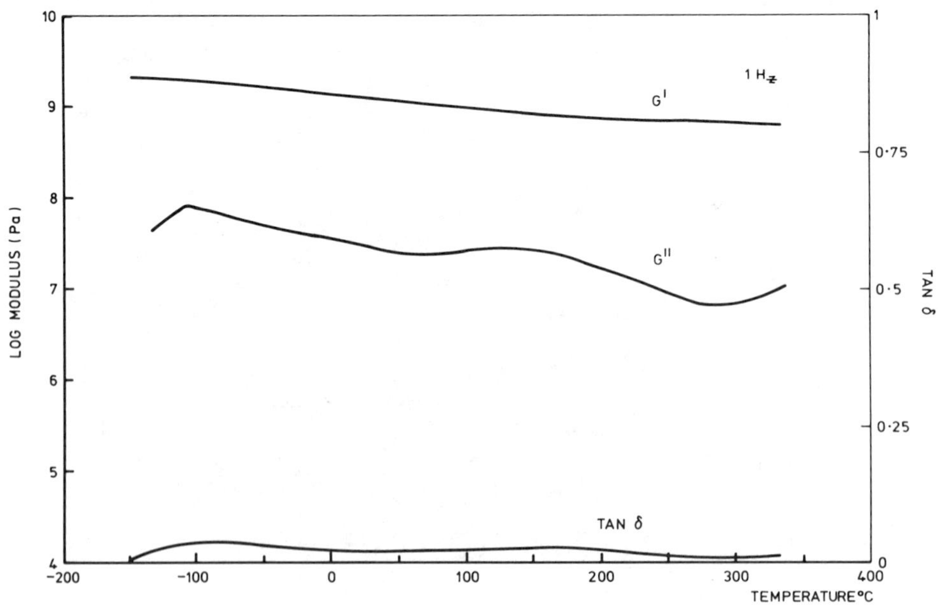

Figure 9. DMA results for 30-phr rubber formulation.

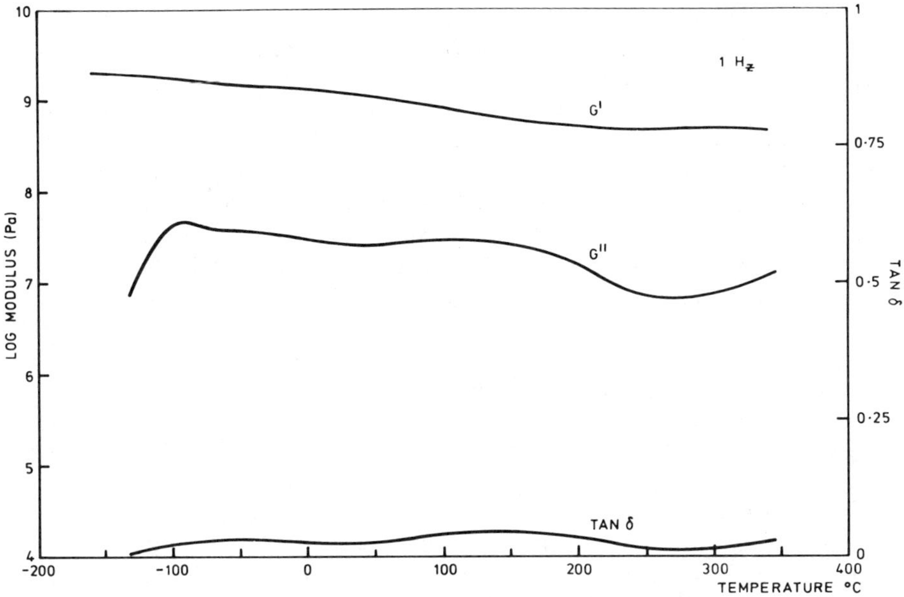

Figure 10. DMA results for 50-phr rubber formulation.

in toughness could still be expected because investigations with rubber-modified epoxies have shown that substantial degrees of toughening can still be imparted by polymeric particles at temperatures below their glass transition (*6*).

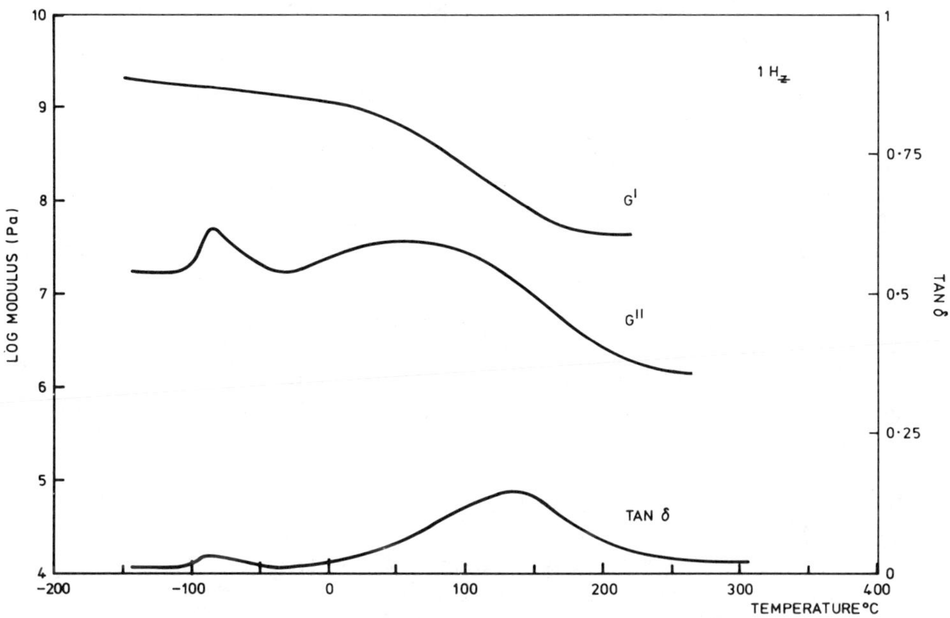

Figure 11. DMA results for 100-phr rubber formulation.

The interpretation of the DMA spectra is made difficult by the absence of the spectrum for the unmodified material. This absence arises from the difficulty of working with this extremely brittle material, but experiments to obtain this important data are continuing. Nevertheless, the relatively small effect that the rubber addition up to 50 phr has had on the mechanical properties at high temperatures may be clearly seen by the relatively small change in G' values over the temperature range −150 to +320 °C.

Finally, the micrographs shown in Figure 8 enable some tentative conclusions to be drawn concerning the mechanisms whereby the presence of a two-phase microstructure increases the toughness. Under an applied stress, the particles will cause stress concentrations, probably at their equators. In the vicinity of the crack tip, these concentrations will augment the crack-tip stress and act as multiple initiation points for localized shear yielding in the matrix. No signs of stress whitening accompany the fracture process. Thus, unlike the rubber-toughened epoxies (*6, 7*), microvoid formation in the rubber particles (which can lower the degree of local constraint and allow further shear yielding to occur) does not appear to be present. The lower toughness of the rubber-toughened polyimides compared to the rubber-toughened epoxies (45–500 and 2000 J/m^2, respectively) might be a reflection of this feature. However, another aspect that undoubtedly influences this comparison is that the polybismaleimide matrices would be expected to possess significantly higher yield stresses, because of higher cross-link densities and more rigid main backbone chain. A high yield stress will obviously inhibit and reduce the extent of localized plastic shear yielding in the matrix around the particles.

Thermal Stability. The influence of rubber concentration on thermal stability is indicated in Figure 12. The decomposition temperature, T_D, refers to the temperature at which 10% weight loss was experienced by the samples during the thermogravimetric experiments. The T_D for the unmodified material (462 °C) compares favorably with that obtained by Knight (*1*) for the same material in a nitrogen atmosphere and indicates the good inherent thermal stability of the base resin (*1*). Although the addition of rubber results in a reduction in T_D, the decrease is relatively minor. For example, a 50-phr addition of rubber produces only a 33 °C reduction in T_D.

Results obtained from the isothermal weight loss experiments are shown in Figure 13. All three samples show a 3.5–5% reduction in weight after 100 min at 330 °C. The unmodified material exhibits somewhat greater thermal stability than the rubber-modified formulations, although the differences between the neat polybismaleimide and the rubber-modified polymers are relatively small.

Work is in progress to ascertain the long-term high-temperature stability of these formulations. Although only limited data are avail-

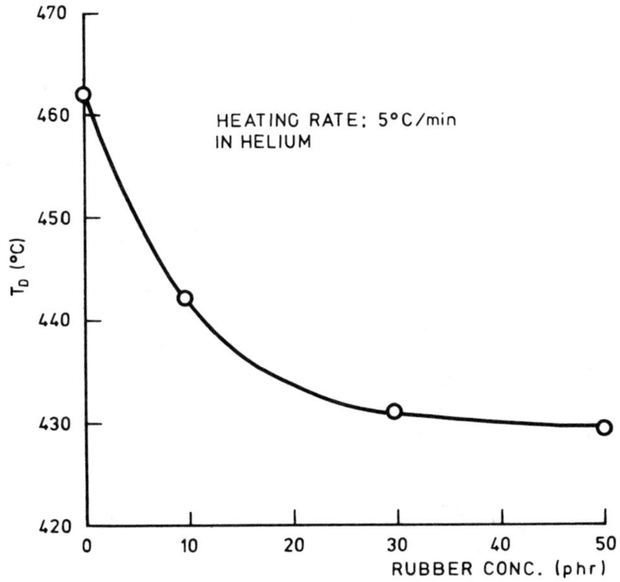

Figure 12. Decomposition temperature (10% weight loss) vs. rubber concentration.

able, the initial results show no significant reduction in fracture energy after 500 h at 170 °C. This result confirms the good thermal stability indicated in Figures 12 and 13.

Conclusions

The addition of up to 100 phr of rubber increases the fracture energy from 11 to 470 J/m^2. This increase is accompanied by improvements in other mechanical properties such as flexural strength and strain-at-failure. The modulus shows a reduction from 4.3 to 2.3 GPa at 100 phr of rubber; this latter figure compares favorably with modulus values generally obtained for some epoxy structural adhesives.

Rather surprisingly, the T_g passes through a very broad maximum with rubber concentration and reaches a peak at approximately 300 °C over a rubber concentration range of approximately 10 to at least 100 phr. Postcuring for an additional 16 h at 240 °C results in an increased T_g of up to 335 °C. Addition of rubber up to 50 phr appears to have only a minor effect on thermal stability as determined from TGA and long-term hot aging.

Thus the results obtained so far suggest that up to 50-phr rubber may be added to the bismaleimide resin employed to produce significant improvements in fracture energy and strain-to-failure without major sacrifices in other properties.

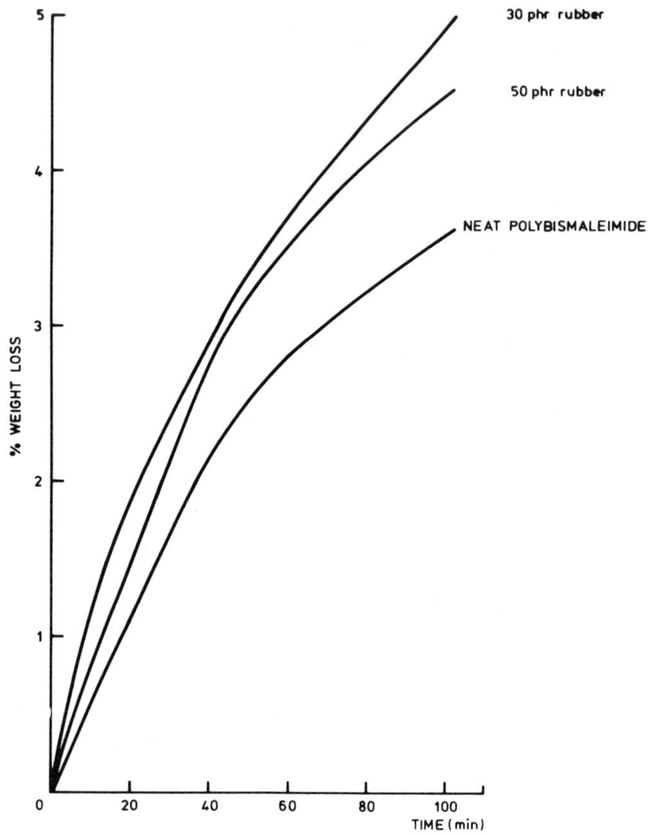

Figure 13. Weight loss vs. time at 330 °C.

Acknowledgments

The authors would like to thank Mr. McKee for experimental assistance.

List of Symbols

- a Crack length
- w Specimen width
- B Specimen thickness
- E Flexural modulus
- $\mathcal{G}_{Ic}$ Fracture energy
- G' Storage shear modulus
- G'' Loss shear modulus
- K_{Ic} Stress-intensity factor for crack initiation
- P_c Load at crack initiation
- Q Geometry factor
- T_g Glass transition temperature

T_D Thermal decomposition temperature
δ Phase angle
ν Poisson's ratio

Literature Cited

1. Knight, G. J. In "Developments in Reinforced Plastics—1"; Pritchard, A., Ed.; Applied Science: London, 1979; p. 145.
2. Drake, R.; Siebert, A. *SAMPE Q.* **1975,** *6* (4), 11.
3. Kinloch, A. J; Young, R. J. "Fracture Behaviour of Polymers"; Applied Science: London, 1983.
4. Bucknall, C. B.; Yoshii, T. *Br. Polym. J.* **1978,** *10,* 53.
5. Sultan, J. N.; McGarry, F. J. *Polym. Eng. Sci.* **1973,** *13,* 29.
6. Kinloch, A. J.; Shaw, S. J.; Tod, D. A.; Hunston, D. L. *Polymer* **1983,** *24,* 1341.
7. Kinloch, A. J.: Shaw, S. J.; Hunston, D. L. *Polymer* **1983,** *24,* 1355.
8. Maxwell, D; Young, R. J.; Kinloch A. J. *J. Mater. Sci. Lett.* **1984,** *3,* 9.
9. St. Clair, A. K.; St. Clair, T. L. *Int. J. Adhes.* **1981,** *2,* 249.

RECEIVED for review November 18, 1983. ACCEPTED March 14, 1984.

9

Effects of Monotonic and Cyclic Loading on Some Rubber-Modified Epoxies

D. N. SHAH, G. ATTALLA[1], J. A. MANSON, G. M. CONNELLY, and R. W. HERTZBERG

Department of Chemistry, Department of Metallurgy and Materials Engineering, and Materials Research Center, Lehigh University, Bethlehem, PA 18015

The effect of two carboxyl-terminated nitrile rubbers on the fatigue crack propagation (FCP) behavior of two amine-cured bisphenol-A-type epoxies (glass transition temperatures, T_g, ≅ *80–90 °C) was studied at sinusoidal frequencies of 10 or 15 Hz. Measurements of tensile response and Izod impact strength were made, as well as dynamic mechanical spectra. In the rubber-modified specimens, the* T_g *of the matrix phase was reduced; this result indicates some solubility of rubber in the epoxy. In addition, because of thermal stresses the* T_g *values of the rubbery phase were lower than those of the neat rubber. Ductility under monotonic loading was increased by rubber modification, as was the maximum attainable value for the range in stress intensity factor,* ΔK. *However, the FCP rates at a given* ΔK *were not very different from the control sample. Therefore, the gain in energy dissipation that was due to plastic deformation in the matrix was offset by some deleterious factor such as a decrease in modulus.*

ENGINEERING PLASTICS AND COMPOSITES are frequently subjected to cyclic loads during service, and such fatigue is inherently more damaging than the corresponding monotonic loading. For these reasons, an understanding of fatigue behavior is extremely important in many components, from molded articles to adhesive joints. Moreover, fabricated articles may contain defects that may develop into significant flaws that grow until catastrophic failure occurs. Because such growth often constitutes an important, and sometimes dominant, component

[1] Visiting scientist, on leave from the Istituto G. Donegani, Montedison Group, 28100 Novara, Italy.

0065-2393/84/0208-0117/$06.00/0

of the overall fatigue life, a program was initiated to study the effects of polymer properties and loading conditions on the kinetics and energetics of fatigue crack propagation (FCP) in a variety of plastics and related multicomponent systems (for a review, *see* Reference 1).

The addition of rubber to brittle or notch-sensitive polymers, such as polystyrene (PS) (*2*, *3*), poly(vinyl chloride) (PVC) (*4*), and nylon 66 (*5*), increases the FCP resistance as well as the general toughness exhibited during static loading. As is the case with static (monotonic) deformation and fracture of such polymer blends, the second phase appears to stimulate shear yielding or crazing in the matrix. Thus, the driving force required to maintain a given crack-growth rate is increased. At the same time, if the modulus is significantly decreased as a result of the presence of the second phase or hysteretic heating, the benefit of modification may be more limited in magnitude.

Increasing attention has been given to the modification of brittle thermoset resins such as phenolics and epoxies by the incorporation of rubbery components to improve toughness. Indeed, even with quite highly cross-linked resins, significant improvements in static toughness can often be achieved. For example, rubber-toughened epoxies now find many applications for both structural and nonstructural purposes (*6*, *7*). Considerable attention has also been given to research on relationships between structure, composition, morphology, and properties (*8–11*). Evidently, cavitation around the rubber particles and shear yielding in the matrix are important mechanisms for energy dissipation (*8–14*); microvoiding and tearing of the rubber particles may also occur (*14–16*).

Although most studies have involved monotonic properties such as impact strength or fracture toughness, little work has been reported on FCP in neat epoxies (*17–19*). We have found only one study of FCP in a rubber-modified epoxy (present in an adhesive joint) (*20*). In the one study of FCP in well-characterized neat epoxies (*18*, *19*), FCP rates and values of fracture toughness were directly and inversely proportional to the cross-link density and the glass transition temperature, T_g, respectively.

In view of the success in increasing static toughness by the addition of an appropriate rubbery phase, we decided to study the role of composition, state of cure, and test frequency on the rates and micromechanisms of fatigue crack propagation in such systems. For our studies, amine-cured bisphenol-A-based epoxy resins with fairly low T_g values (~80–90 °C) were selected for modification (*21*); studies with higher-T_g resins, described elsewhere (*22*), are in progress. This chapter presents and discusses characterization and results for two rubber-modified epoxies and their controls in comparison with those reported separately (*18*, *19*).

Experimental

The epoxy used was a low molecular weight liquid diglycidyl ether of bisphenol A (DER 331, Dow Chemical Company); the curing agent (in stoichiometric proportion) was an aliphatic diamine (Jeffamine D-230, Jefferson Chemical Company). The carboxyl-terminated acrylonitrile–butadiene (CTBN) rubbers were used in the form of epoxy adducts (Spencer Kellogg Division, Textron, Inc.) Kelpoxy G272-100 (series 1) and Kelpoxy 293-100 (series 2). The rubbery components contained 26 and 18% acrylonitrile (AN) in series 1 and 2, respectively, and the overall rubber content was 40% in each case. The formulation was adjusted to provide 10-wt% rubber in the final resin. After mixing the desired proportions of prepolymer epoxy and modifier (heated to 62 °C for 2 min), the solution was degassed, mixed with the curing agent, and cast in a steel mold. The cure cycle was as follows: 24 h at room temperature, 1 h at 66 °C, 2 h at 93 °C, and 4 h at 150 °C, followed by cooling to room temperature under ambient conditions.

The extent of cure was determined with a differential scanning calorimeter (DSC) (Perkin-Elmer model DSC-1B) at a scan rate of 40 °C/min. Because no curing peaks were observed in the DSC scans, reaction was judged to be essentially complete. Dynamic mechanical spectra were obtained at 110 Hz with an Autovibron unit, model DDV-IIIC; T_g values were determined from the positions of the peaks in the tan δ curves. Ultimate mechanical properties were determined with dog-bone-type specimens machined from the cast materials. Tensile properties were obtained with an Instron tester at a crosshead speed of 0.4 mm/s; notched impact tests were also made. Values of K_{Ic}, the plane-strain fracture toughness measured in the tensile-opening mode, were obtained for several specimens by means of a three-point bend test (*10*) with the Instron tester. Quasi-static values of fracture toughness, K_c', were also estimated by dividing values of ΔK_{max} (the maximum ΔK noted prior to fatigue fracture) by 0.9 [i.e., by $(1 - R)$, where R = the ratio of minimum to maximum load] (*18*).

Fatigue tests were made under ambient conditions following standard procedures (*18*) on notched wedge-opening-load specimens (6.3 cm × 4.8 cm × 6 mm) (*23*) with an electrohydraulic closed-loop test machine at a sinusoidal frequency of 10 or 15 Hz and a ratio of minimum to maximum load of 0.1. Experience has shown that FCP curves are usually reproducible to within ±20%. Values of crack-growth rate per cycle, da/dN, were correlated with ΔK, the stress intensity factor (a measure of driving force at the crack tip); ΔK is given by $Y\Delta\sigma \cdot a^{\frac{1}{2}}$, where Y is a geometrical factor, $\Delta\sigma$ is the range in applied stress, and a is the crack length. Although the control was tested at 10 Hz, experiments with series 1 at 15 Hz revealed no significant difference between curves obtained at the two frequencies. Scanning electron micrographs (SEMs) of fracture surfaces were made on Au/Pd-coated specimens with a scanning electron microscope (ETEC Autoscan). For transmission electron microscopy (TEM), which was performed with an electron microscope (Philips model 400), sections of the resins were ultramicrotomed and then stained with osmium tetroxide to reveal the rubbery phase. Crack-tip temperatures were measured for the series 1 material with an IR microscope (Barnes model RM-2B).

Results and Discussion

Morphology. The TEMs shown in Figure 1 reveal bimodal and unimodal distributions of rubbery-phase particles in series 1 and series 2 resins, respectively. Average particle diameters and rubbery-phase volume fractions (RPVFs) were estimated by direct measure-

a

b

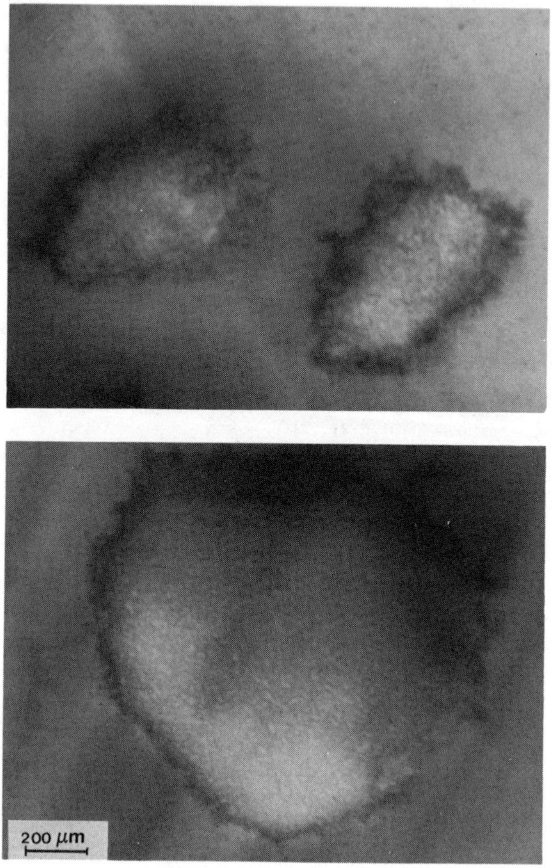

Figure 1. TEMS of rubber-modified epoxies stained with OsO_4: (a) series 1, (b) series 2.

ments of particle areas seen in the micrographs; we assumed that these areas were approximately equal to volume fractions (*see* Table I).

A discrete rubbery phase that sometimes has a multi-modal size distribution is commonly observed in epoxies modified with CTBNs containing either 18 or 26% AN (*8*, *24–29*). The micrographs of

Table I. Characteristics of Rubbery Phases

		Large Particles		*Small Particles*	
Resin	*AN (%)*	*Average Size (µm)*	*RPVF*	*Average Size (nm)*	*RPVF*
Series 1	26	0.4	0.19	10–20	0.06
Series 2	18	0.9	0.19	—	~0

Figure 1 especially resemble those of Reference 26 in several respects: size and shape, the presence of a large volume of unstained epoxy domains within the rubbery phase, and the concentration of rubber in the interfacial region. Especially for series 1, the average sizes of the large particles also resemble those reported by Sohn et al. (*27*) for epoxies containing CTBN components with the same compositions as used in this study. The tendency of the higher AN-content CTBN adduct (series 1) to yield smaller particles is consistent with the findings of Sohn et al. (*27*), who attributed the effect to an increase in the solubility parameter (and thus an increase in miscibility with the epoxy) with an increase in AN content in the CTBN.

The observation of epoxy domains in the rubbery phase (Figure 1) is quite consistent with other experience with rubber-modified glassy polymers including epoxies (*9, 25, 26*). Also, the CTBN component has been prereacted with epoxy to form, in effect, an ABA block molecule (30/40/30 by weight). Thus, in a particle very much larger than the ABA molecules, at least 60 wt% of the particle must consist of epoxy. The resolution of the microscopy does not permit detailed analysis of the interfacial region; X-ray analysis would be required (*28*).

The very small rubbery particles are approximately 10–20 nm in size and the extended chain-length and random-coil dimension of the rubbery component of a typical CTBN elastomer are approximately 20 and 2 nm, respectively. Even though the estimates of particle sizes are subject to considerable error, clearly only relatively few CTBN sequences are probably involved, perhaps on the order of 10^3. In any case, phase separation of rubber on a scale of ~10–20 nm may contribute to a degree of miscibility with the epoxy matrix (*30*); the effect of this miscibility on dynamic spectra is discussed in a later section.

Although the distribution of masses cannot be calculated with certainty, an estimation may be made by taking into account the densities of the epoxy and the CTBN [1.20 and 0.95 g/cm^3, respectively (*27*)] and assuming the additivity of volumes. If we use the stated composition of the prereacted CTBN 60-wt% epoxy) and the estimated value of RPVF (19%), a maximum value for the volume fraction (v_f) of rubber in the composite particles would be 0.46. This value is obtained if we neglect the possibility of an anomaly in rubber concentration at the interface and assume that no epoxy from the matrix is present in the rubbery particles. Because the overall value of v_f (rubber) is 0.12, the overall values of v_f (rubber) in the composite phase and in the epoxy matrix are estimated to be ~0.09 and 0.03, respectively (the latter figure constitutes a lower bound). If we assume an error in the estimation of the RPVF of ±20% (probably a minimum range), then the estimated v_f for rubber in the epoxy phase would lie between 0.02 and 0.05. If, as shown in Figure 1, the concentration of rubber in the interfacial region is higher than in the interior, these numbers would be reduced—for example, by about 30% in the limiting case of pure rubber. On the other hand, if epoxy from the matrix is present in the rubbery particles, the numbers would be increased. In any case, these estimates may be compared with the admittedly crude estimate of 0.06 for the v_f of small rubber particles dispersed in the epoxy. Thus, in view of the assumptions and uncertainties involved, the volume balance appears to be plausible. In comparison, an estimation of a maximum value of 0.05–0.09 for the v_f of rubber in the matrix was obtained from analysis of dynamic mechanical spectra. Again, the values estimated are in reasonable agreement.

Dynamic Mechanical Behavior. Dynamic mechanical spectra are presented in Figure 2, and a summary of pertinent experimental and derived data is shown in Table II. For the sake of convenience, the spectra for the series 1 and 2 specimens have been vertically

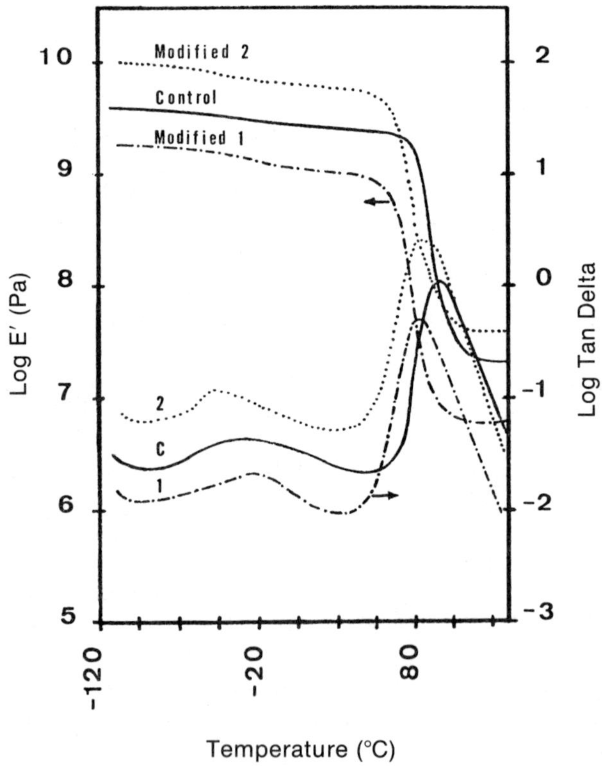

Figure 2. Dynamic mechanical spectra (at 110 Hz) for rubber-modified epoxies (series 1 and 2) and control. For convenience, the curves for series 1 and 2 are displaced above and below the curve for the control (see text).

Table II. Dynamic Mechanical Parameters for Control and Rubber-Modified Epoxies

Specimen	*AN in CTBN (%)*	ET_g[a] *(°C)*	RT_g[b] *(°C)*	ET_β[c] *(°C)*	E_r[d] *(MPa)*	M_c[e]
Control	—	94	—	−29	20.0	590
Series 1	26	82	−24	−30	12.5	920
Series 2	18	84	−50	−19	15.5	740
CTBN	26	—	−30[f]	—	—	—
CTBN	18	—	−45[f]	—	—	—

NOTE: Dynamic mechanical parameters at 110 Hz.
[a] T_g of epoxy phase. [b]T_g of rubber phase. [c]T_g of epoxy phase β relaxation. [d]Rubbery modulus. [e]Average molecular weight between cross-links, estimated by the approximation $E_r/3 = G = dRT/M_c$. [f]Data from torsional braid analysis (*27*).

displaced above and below the spectrum for the control, whose curves correspond to the axes as labelled. (The shift corresponded to ~0.35 on the log E' axis.)

Each spectrum shows three transitions: one for the glass transition of the epoxy matrix (ET_g), one for the first secondary (β) relaxation of the epoxy matrix (ET_β), and one for the rubbery component (RT_g). Although not seen as sharply as in torsional braid studies (*26*), the last peak (RT_g) is superimposed on the broad and slightly asymmetrical β-peak of the epoxy, which appears to reach a maximum height near −30 °C in each case. (The peak breadth precludes a more precise assignment.)

The T_g for the epoxy phase is depressed by 10–12 °C, presumably a result, at least in part, of the presence of dissolved or finely dispersed rubber. However, each value of RT_g is lower than the T_g values of the pure CTBNs after correction for differences in test frequency: −25 and −10 °C for CTBN containing 18 and 26% AN, respectively. This difference undoubtedly is due to unrelaxed thermal stresses resulting from differences in the coefficients of expansion on cooling from the curing temperature (*26, 27, 31*).

We had hoped that the phase compositions could be estimated from the shifts in ET_g and RT_g from the values for the pure components. Such a calculation would involve allowing for the effect of thermal stresses. Unfortunately, such effects appear not to be straightforward (*see*, for example, the values of T_g quoted in Reference 27 for a pure rubber and for the same rubber dispersed in an uncured and a cured epoxy). Also, the equation used (*31, 32*) for the calculation is exceptionally sensitive to any deviation of the Poisson's ratio (ν) of the rubber from precisely 0.5—the equation contains a term $(1 - 2\nu)$. Because of these uncertainties, the estimation of phase composition was restricted to the epoxy phase.

Equation 1 was used to estimate the weight percent rubber in the epoxy phase;

$$\frac{1}{T_g} = \frac{w_E}{T_{gE}} + \frac{(1 - w_E)}{T_{gR}} \tag{1}$$

T_g is the observed T_g, w_E is the weight fraction of epoxy, and T_{gE} and T_{gR} are the T_g values for pure epoxy and pure rubber, respectively. The values of T_{gE} and T_{gR} were corrected for the difference in frequency between the torsional braid technique used in Reference 7 and our tests by adding 15 °C; the prediction would vary by only 1% if 20 °C were used. In this way, a value of 7 wt% (10 vol%) was predicted for the concentration of rubber in the epoxy matrix (v_f = 0.05 to 0.09 from TEM analysis).

At the same time, variations in the cross-link density of the epoxy phase may also contribute to the lowering of T_{gE}. Thus, nominal values of M_c, the average molecular weight between cross-links, were significantly higher for the modified resins than for the control. Although the presence of rubber would in itself increase M_c by diluting the cross-linked network, the increase in M_c appears greater than expected if M_c were increased in proportion to the volume fraction of rubber.

Equation 1 was also used to check the sensitivity of predictions with respect to two factors: (1) the effect of using the experimental value of RT_g in the estimation of rubber content in the matrix; and (2) the effect of AN content in the CTBN on the observed RT_g. Sensitivity was negligible in both cases. Thus, the rubber content was calculated to be 7 wt%. At this rubber content, the predicted difference in T_g between series 1 and 2 was only 3 °C, within the reproducibility range for T_g from dynamic spectra. Of course, if the cross-link density is lower in the modified resins, then the true value of rubber content would be somewhat less than 7 wt%.

Static Mechanical Properties. Results of the tensile and fracture toughness tests are given in Table III. All the control and modified specimens yielded before failure; the modified specimens exhibited a drop in stress after yielding. This drop in stress is charac-

Table III. Tensile and Fracture Toughness Behavior of Control and Rubber-Modified Epoxies

Sample	σ_y *(MPa)*	E *(GPa)*	ϵ_b *(%)*	IS *(J/m)*	K_c' *(MPa · m^½)*
Control	65	4.5	15	64	— (0.89)[a]
Modified-1	52	4.2	21	91	— (1.67)[a]
Modified-2	54	—	26	36	2.2[b] (2.00)[a]

Note: Unless noted otherwise, the results given are the average of five tests.
[a] Values in parentheses are for K_c', estimated by dividing ΔK_{max} by 0.9 (*18*).
[b] Average of six tests.

teristic of many rubber-modified polymers (*9*). As expected, the modified specimens exhibited lower tensile moduli (E) and yield strengths (σ_y), but higher elongations to break (ϵ_b), than the control. However, although the impact strength (*IS*) of modified series 1 was significantly greater than that of the control, the impact strength of modified series 2 was significantly lower.

Interestingly, the IS results for series 2 are quite different from those reported for other rubber-toughened systems (*24*) in which the reported preferred AN content for toughening was between about 13 and 18%. The morphology may be the cause of such a difference; for example, the somewhat greater particle size in the series 2 specimens may be beyond the optimum value. Also, the T_g of the rubber phase was significantly lower in series 1 than in series 2 (−50 °C and −24 °C, respectively). However, because of the relatively low impact strengths and the consequent likelihood of significant error in the measurements, the value of impact strength for series 2 may be anomalous, especially in light of the more reasonable correlation between values of K_{Ic} and ϵ_B (Table III).

In contrast to the impact strength, the general toughness (as implied by the areas under the stress–strain curves) and, at least for series 2, the fracture toughness K_c' were higher in both modified series than in the control. A lack of correlation between impact strength and K_c' was reported earlier for several epoxies (*18*). Moderate whitening was also noted on and below fracture surfaces of the modified, but not the control, specimens. All these results are consistent with a cross-linked network capable of relatively facile segmental mobility in comparison to the higher cross-link density resins studied earlier (*18, 19*). At the same time, the discrete rubbery phase has induced a measure of energy dissipation in addition to that characteristic of the control, at least as manifested in static (or quasi-static) fracture toughness and tensile tests.

Fatigue Crack Propagation. The FCP curves for the control and modified resins are shown in Figure 3. In accordance with the Paris equation, $da/dN = A\Delta K^n$, where A and n reflect material and test parameters (*33*), the response of the control was quite linear on the log–log plot. As mentioned earlier, tests showed no significant effect of frequency on going from 10 to 15 Hz, in agreement with earlier results with epoxies (*34*). ΔK_{max}, the maximum value of ΔK attained at fracture, was ~0.8 MPa · $m^{\frac{1}{2}}$, a value between the values for a 1:1 and a 2:1 methylene dianiline (MDA)–Epon 828 resin [~0.6 and 1 MPa · $m^{\frac{1}{2}}$, respectively (*18*)]; however, the slope was lower for the latter two resins. In fact, the FCP response generally resembled that of a nonstoichiometric MDA/Epon 828 resin except for a slightly lower value of ΔK_{max}. This result is not surprising in view of the T_g for the control, 94 °C compared to 180 °C and 102 °C for the 1:1 and a 2:1 MDA/Epon 828 resin, respectively (*18, 19*). The T_g is an effec-

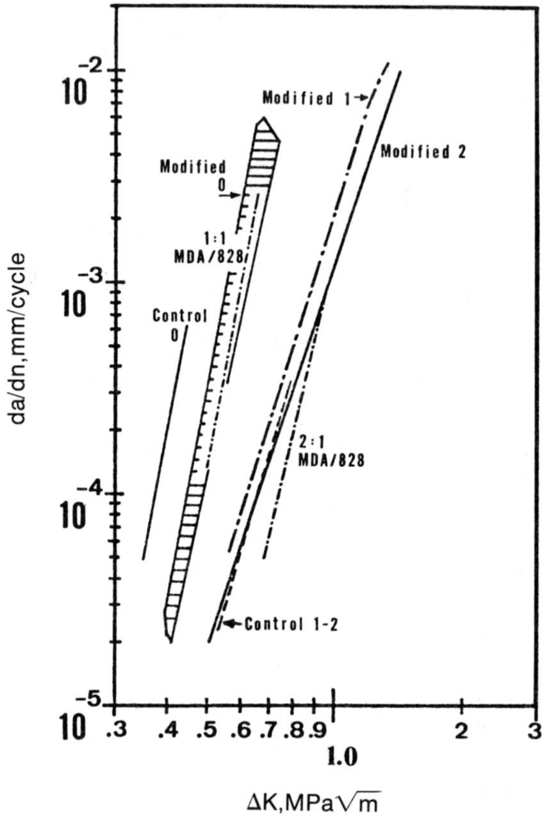

Figure 3. Fatigue crack propagation plots for series 1 and 2 control and rubber-modified epoxies. Curves also shown for two unmodified high-T_g *epoxies* (18) *and for another control and rubber-modified system, series 0* (21).

tive measure of cross-link density, and, in general, the lower the cross-link density, the lower the FCP rate (*18, 19*). However, at a given value of M_c, the T_g values of the polymers of this study are about 10 °C lower than those of the MDA-cured resins.

The FCP curves for series 1 and 2 also followed the Paris equation quite well, at least over much of the range in ΔK. However, at a ΔK value of ~1 MPa $m^{\frac{1}{2}}$ a slight downward curvature was clearly evident with series 1. An explanation for the curvature can be inferred from an analysis of the fracture surfaces and will be discussed in the section on fractography.

Curiously, the response of both series 1 and 2 closely followed the response of the control. Although the control was tested at 10 Hz rather than 15 Hz, tests showed no significant effect of frequency in this range, in agreement with earlier results with epoxies (*34*). Further measurements will be required to verify whether the apparently slightly greater FCP resistance (implicit in the lower values of da/dN at medium to high values of ΔK) of series 2 is real. In the modified resins, however, stable crack growth was maintained to significantly higher values of ΔK, 1.5 MPa $m^{\frac{1}{2}}$. This result is consistent with a higher level of static toughness (Table II); ΔK_{max} often closely approximates the fracture toughness (*1*, *18*). At the same time, extension of stable crack growth beyond the value of ΔK_{max} for the control involved only ~7000 load cycles (series 1), that is, about 7% more cycles in the propagation state than was observed for the control. As such, the addition of rubber to the epoxy improves the fracture toughness, but does not contribute much to an increase in fatigue life.

These specimens should also be considered in the perspective of noncross-linked polymers. The FCP response of the modified resins resembles that of a medium molecular weight poly(vinyl chloride) (PVC) (*4*) or a high molecular weight poly(methyl methacrylate) (PMMA) (*35*). Thus the modified resins fall in a middle range of FCP behavior when generally compared with thermoplastics (*1*). This ranking also holds for the control, with respect to FCP rates as a function of ΔK, although the value of ΔK_{max} is lower than for thermoplastics with similar rate behavior.

However, in contrast to the epoxies of this study, all other rubber-modified systems studied to date have exhibited lower FCP rates at a given value of ΔK than the unmodified resins, if the test frequency was low enough to preclude large-scale hysteretic heating. These systems include PVC (*4*), nylon 66 (*5*), a rubber–PS interpenetrating network (*3*, *36*), and a very brittle epoxy (*21*). For the brittle epoxy, tensile and impact properties were made worse by the rubber, presumably because of inappropriate formulation, and FCP rates were decreased to the general range characteristic of the 1:1 MDA–epoxy system (*see* Figure 3).

The fracture-surface morphologies are described below and compared with fracture surfaces from monotonically loaded specimens; the paradoxical behavior with respect to the lack of toughening in series 1 and 2 is also discussed later.

Fractography

On visual examination, all the fracture surfaces of the control were flat and glasslike. Although the fatigue-fracture surfaces of the rubber-modified materials were also smooth, they (but not the control) exhibited whitening (Figure 4) at high values of ΔK, ~1.5 and ~1.4 MPa $m^{\frac{1}{2}}$ for series 1 and 2, respectively. (The occasional markings seen across the specimen thickness are artifacts resulting from stopping the test). In series 1, but not series 2, a small clear zone also developed at a ΔK of 1.1 MPa $m^{\frac{1}{2}}$. Although the clear zone did not exhibit the slant-type fracture characteristic of classical shear lips,

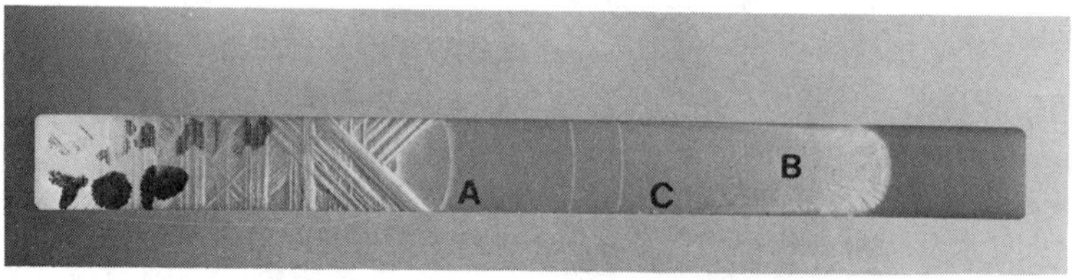

Figure 4. Low-magnification photograph of entire fatigue-fracture surface for a rubber-modified epoxy, series 1: A, initial notch; B, onset of whitening; and C, onset of shear zone.

we suggest that occurrence of the zone reflects a higher component of shear in the stress state at the surface than in the interior. The intensity of whitening in both series 1 and 2 and the size of the shear zone in series 1 increased progressively as ΔK (and the crack length) increased until the critical value of ΔK for catastrophic fracture was reached. Except for these differences in whitening and in the presence of the shear zone in series 1, the general aspect of the two fracture surfaces was similar.

Note that whitening on deformation normally reflects some kind of plastic deformation, typically involving microcavitation (often as crazing) in the matrix or rubbery phase of a toughened polymer (*9*, *10*). The associated dissipation of energy tends to stabilize crack growth to a higher level of ΔK than is possible for a control resin. Also, the intensity of whitening and the size of the shear zone tend to increase with increasing ΔK. Thus, the occurrence of these phenomena provides a plausible explanation for the downward curvature in the FCP plot for series 1 resin (Figure 3); furthermore, the lack of downward curvature in series 2 may well reflect the lack of a visible shear zone. Ordinarily, FCP plots tend to curve up as the onset of fast fracture is approached.

In any case, the occurrence of a shear zone is consistent with the observation that the deformation of epoxies is probably dominated by a shear response rather than crazing (*9*, *10*); indeed, no evidence for crazing was found in any specimen.

In Figure 5, SEMs of fatigue-fracture surfaces of the two rubber-modified epoxies are shown. The pictures were taken at different values of ΔK; for each material ΔK increases from top to bottom. The micrographs in the middle were taken in the whitened area just before catastrophic failure, and those at the bottom were taken in the fast-fracture region. Apart from the sizes of the particles (larger in series 2), the morphologies of the two fracture surfaces are very similar. Also, in each case, regardless of the magnitude of ΔK, fracture surfaces in the region of stable, fatigue crack growth appear cavitated. This observation has already been reported by several authors (*12*, *13*), and an explanation has been proposed (*12*): the rubber, in a state of negative pressure because of its higher thermal expansion coeffi-

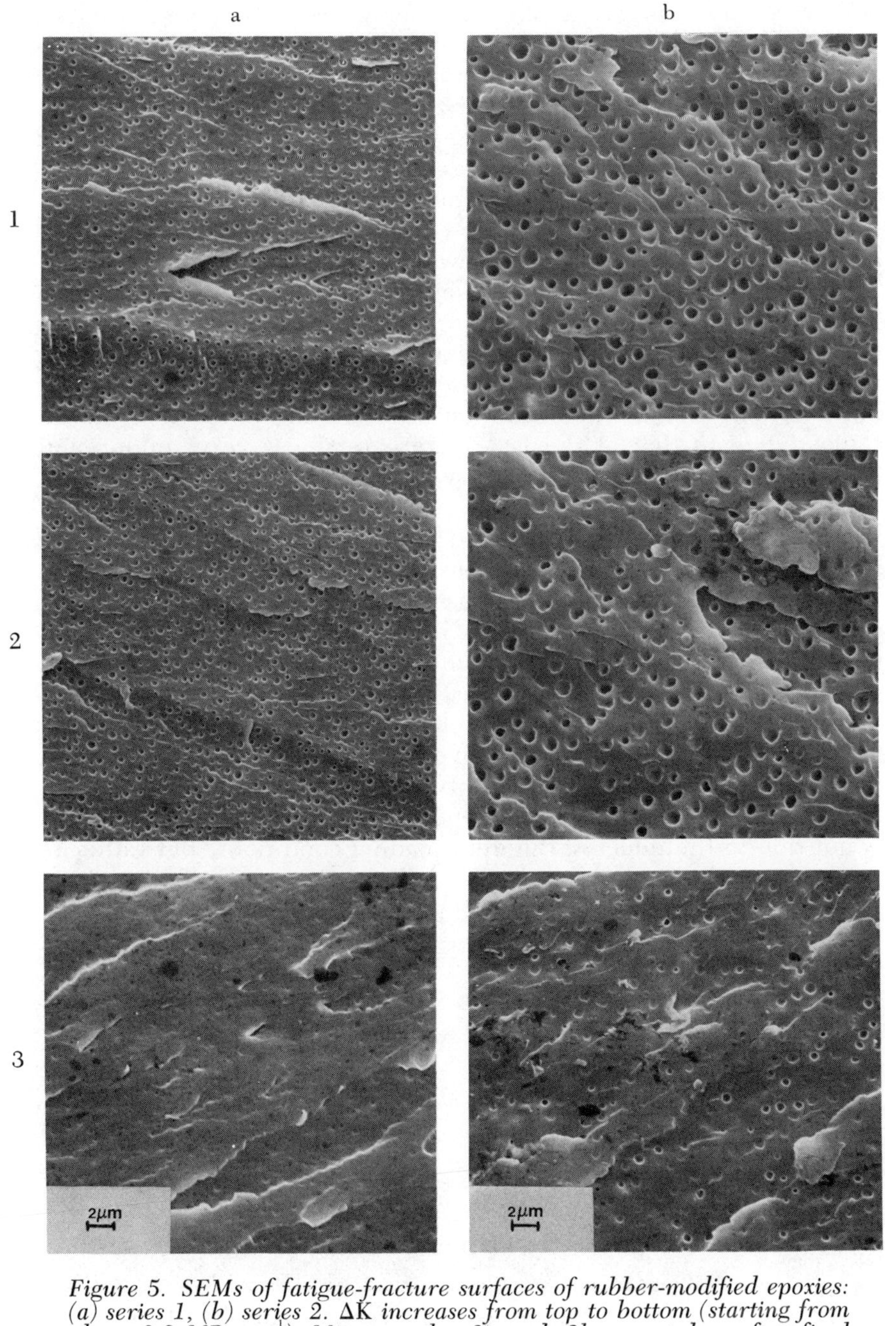

Figure 5. SEMs of fatigue-fracture surfaces of rubber-modified epoxies: (a) series 1, (b) series 2. ΔK increases from top to bottom (starting from about 0.9 MPa · m$^{\frac{1}{2}}$). Micrographs 3a and 3b were taken after final fast fracture had begun.

cient, retracts while the fracture proceeds. Indeed Pearson and Yee (*13*) detected the presence of the rubber at the bottom of the cavities from the backscattering of the electrons after staining the fracture surface with OsO_4. At high magnification, the morphology in the high-*K* whitened area is essentially the same as in the low-*K* region (micrographs not shown). Thus, the cause of the whitening during stable crack growth could not be deduced from the micrographs.

The morphology of the fast-fracture region differs considerably from that of the fatigue area; the cavities are fewer and shallower, and a number of bumps and hollows appear on the fracture surface. Throughout the fracture surface several extended steps, parallel to the direction of propagation of fracture, are present. Interestingly, as others have observed for nonfatigued specimens (*12–14*), the steps do not transect the rubber particles, but rather proceed along the particle boundaries. The transition from the morphology of the fatigue region to that of the fast-fracture region is very sudden and exactly corresponds to the last curved marking in the whitened area (Figure 4).

Examination of the fracture surfaces of specimens broken under monotonic loading conditions (tensile, impact, and fracture-toughness tests) cast some light on the nature of the transition to fast-crack propagation. In Figures 6 and 7, SEMs are shown for fracture surfaces obtained from the tensile and fracture-toughness tests. The morphologies of the slow- and fast-fracture regions in these specimens resemble very much those characteristic of slow-fatigue crack growth and fast fracture, respectively, in the fatigue specimens. In view of the presence of the same type of transition both in fatigued and monotonically loaded specimens, we speculate that the transition is not due to the effect of different loading conditions, but rather to

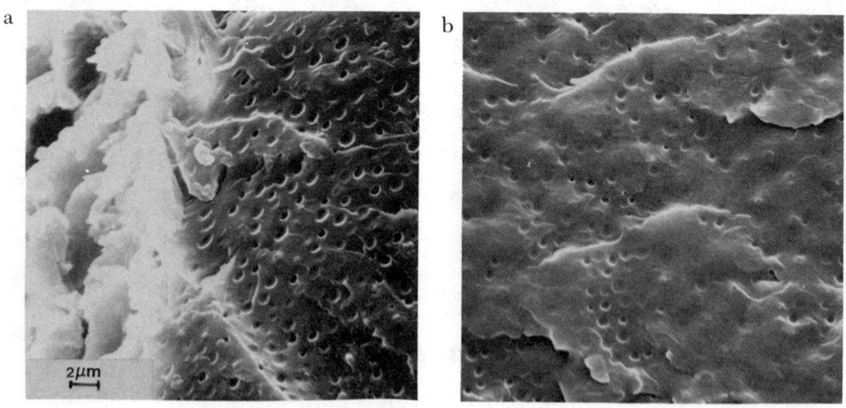

Figure 6. SEMs of fracture surfaces of a rubber-modified epoxy (series 2) broken during an impact test. Micrographs a and b correspond to the early and late stages of fracture (i.e., slow and fast rates), respectively.

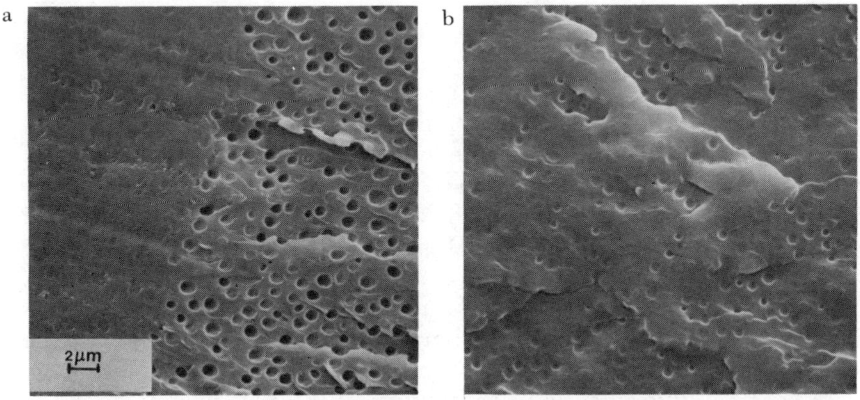

Figure 7. SEMs of fracture surfaces of a rubber-modified epoxy, series 2, broken during a fracture toughness test. Micrographs a and b correspond to the early and late stages of fracture (i.e., slow and fast rates), respectively.

an effect of crack-propagation rate, and we hypothesize the following mechanism. When the crack propagates at low speed, the rubber particles are subjected to a tearing process (*16*). The large amount of strain energy accumulated causes the rubber to snap back into the cavities after rupture and partial debonding from the matrix. At a high crack-growth rate, however, the stiffness of the rubber increases so that the tearing process tends to be hindered; at the same time, debonding from the matrix is less frequent. Thus, many rubber particles break with much less deformation, remain well bonded to the matrix, and produce shallow cavities and bumps on the fracture surface.

The Toughening Paradox. The fatigue results obtained in this study raise a serious question: our apparently well-made rubber-modified plastics exhibit yielding and toughening under monotonic loading but do not manifest much, if any, toughening under tensile fatigue loading, at least with respect to crack-propagation rates. So far, toughening in fatigue has been noted only for an epoxy that had otherwise unacceptable properties (e.g., poor tensile and impact response) (*21*). To be sure, various measurements of toughness do not necessarily correlate well in many polymers, including rubber-modified epoxies (*35–37*).

Why does the rubber not decrease FCP rates significantly (at a given ΔK) even though it improves the static behavior? As mentioned previously, the incorporation of a discrete rubbery phase in PVC, PS, and nylon 66 improves both impact strength and FCP resistance (*2–5*). Although the assumption of generally similar micromechanisms of failure under static and fatigue loading appears to be valid with most polymers (*1*), this assumption is clearly not true for the epoxies of this study. Several possible factors, not necessarily mutually exclusive, may be considered.

Of course one major difference between the modified thermoplastics and epoxies exists: cross-linking. The capacity for plastic deformation in the latter tends to be lower than in the former, although the epoxies discussed herein behave rather like a medium molecular weight PVC, PMMA, or amine-rich, high-T_g epoxy in terms of FCP resistance (*4, 5, 18*) (Figure 4). Hence, the thermoset nature per se of the epoxies concerned cannot be responsible for the paradox. Moreover, with at least some matrices (*36, 38*), FCP resistance is improved at a lower level of rubber content than is required for the significant improvement of impact strength.

Another difference is in the nature of the plastic zone ahead of the crack tip. Whereas the other toughened materials studied exhibited visible and relatively large plastic zones, the modified epoxies do not. However, because this behavior seems to exist in both static and fatigue failure, this difference is unlikely to provide an explanation.

A third factor may be considered: the contribution of rubber deformation. The fractographs in Figures 6 and 7 certainly imply increased energy dissipation resulting from the hole formation during both monotonic and cyclic loading. However, the series 1 and 2 resins behave as though this energy sink is balanced by other factors such as a diminished contribution of rubber deformation or a lower effective modulus than is the case with monotonic loading. Although the role of energy dissipation as a result of rubber deformation is somewhat controversial (*10, 16*), one would expect that more energy would be dissipated at high rather than low crack-growth rates (i.e., during monotonic rather than cyclic failure) (*38*). Indeed, at least with series 1, the impact strength is increased by rubber modification, even though the fractographs of Figures 6 and 7 apparently indicate that the deformation of many particles is restrained dissipation. Thus, the toughening effect of rubber on an inherently ductile high molecular weight PVC is much less than for a less ductile, lower molecular weight resin (*4*).

Finally, possible effects of modulus cannot be discounted. For example, the large-scale hysteretic heating in toughened nylon 6 that occurs when the test frequency is increased to 15 Hz causes a significant decrease in modulus and a relative increase in FCP rate (*1*). Even though the extent of hysteretic heating at the crack tip appears to be nearly negligible in the epoxies (a maximum temperature of 3 °C), the modulus in bulk is reduced somewhat by the presence of the rubber; the *effective* modulus in the crack region may be reduced still more. If the growing crack sees the holes as voids, then the latter phenomenon may well be important because of the decreased load-bearing ability. At the same time, any debonding that detaches one hemisphere of a rubber particle prior to crack advance may be expected to minimize energy dissipation because of tearing of the rubber. Thus, at the slow crack-growth rates in fatigue, the voiding may occur sufficiently earlier than the crack so that a weakening effect becomes manifest. Indeed, if the modulus were reduced in proportion to the volume fraction of rubber, normalization of the FCP data with respect to E (i.e., by plotting vs. $\Delta K/E$) would yield curves showing FCP rates of series 1 and 2 to be one-fifth of the control's rate. In this regard, the proportion of holes in the specimens of Reference 21 that showed improved FCP resistance is much lower than in series 1 and 2; instead, many of the rubber particles were fractured at their boundaries or transected. Conceivably, then, the rubber particles that did *not* form holes were able to induce toughening in fatigue. Thus, any benefit derived from rubber-induced plastic deformation during fatigue crack growth is probably offset by debonding and a lower effective modulus. Such an adverse effect of debonding has also been noted by Sayre et al. (*40*) for epoxies containing a cross-linked rubbery phase.

Although these speculations seem reasonable, clearly much more research on the detailed micromechanisms and energetics of failure in fatigue will be required to unequivocally resolve the question of paradoxical behavior. Also, the effect of rubber on very brittle, high-T_g epoxies remains to be seen.

List of Symbols

a	Crack length, mm
A	Pre-exponential constant in Paris equation
CTBN	Carboxyl-terminated butadiene–nitrile copolymer rubber
d	Density, g/cm^3
da/dN	Fatigue crack-propagation rate, mm/cycle
E_r	Tensile modulus in the rubbery state, GPa
ET_g	Glass transition temperature of epoxy phase, °C
FCP	Fatigue crack propagation
IS	Impact strength, J/m^2
K, ΔK	Stress intensity factor and range in stress intensity factor, respectively, MPa · m$^{\frac{1}{2}}$
ΔK_{max}	Maximum value of K noted before fast fracture, MPa · m$^{\frac{1}{2}}$
K_c'	Quasi-static fracture toughness estimated from ΔK_{max}, MPa · m$^{\frac{1}{2}}$
K_{Ic}	Plane-strain fracture toughness in tensile opening mode, MPa · m$^{\frac{1}{2}}$
M_c	Average molecular weight between cross-links
MDA	Methylene dianiline
n	Exponent in Paris equation
N	Number of cycles

R	Ratio of minimum to maximum load; also gas constant, 8.31 J/mol/degree
RPVF	Rubbery-phase volume fraction
RT_g	Glass transition temperature of rubbery phase, °C
SEM	Scanning electron microscopy (or micrograph)
TEM	Transmission electron microscopy (or micrograph)
T_g	Glass transition temperature, °C
T_{gE}, T_{gR}	Glass transition temperature of pure epoxy and rubber, respectively, °C
v_f	Volume fraction
w_e	Weight fraction of epoxy
ϵ_b	Elongation to break, %
γ	Geometrical factor
σ, $\Delta\sigma$	Stress and range in stress, respectively, MPa
σ_y	Tensile-yield stress, MPa
ν	Poisson's ratio

Acknowledgments

We wish to acknowledge financial support through the Polymers Program of the National Science Foundation, Grant No. DMR 8106489. We are also grateful to J. C. Michel and J. S. Crompton of Lehigh University for help with the FCP tests and for the TEM, respectively. Finally, we are indebted to R. S. Drake, C. M. Lizak, and D. R. Egan of The BFGoodrich Company for their advice on the formulation and curing of epoxies.

Literature Cited

1. Hertzberg, R. W.; Manson, J. A. "Fatigue in Engineering Plastics"; Academic: New York, 1980.
2. Hertzberg, R. W.; Manson, J. A.; Wu, W. C. ASTM STP 536, p. 391 (1973).
3. Qureshi, S.; Manson, J. A.; Sperling, L. H. In "Polymer Applications of Renewable Resource Materials"; Carraher, C. E., Jr.; Sperling, L. H., Eds.; Plenum: New York, 1983; p. 249.
4. Skibo, M. D.; Manson, J. A.; Hertzberg, R. W.; Collins, E. A. In "Durability of Macromolecular Materials"; Eby, R. K., Ed.; ACS SYMPOSIUM SERIES No. 95; American Chemical Society: Washington, D.C., 1979. p. 311.
5. Skibo, M. D.; Hertzberg, R. W.; Manson, J. A. Proc. Conf. Deformat. Fract. Plast. and Rubber Inst. London, 1979; p. 41.
6. Drake, R. S.; Siebert, A. R. *Org. Coat. & Appl. Sci. Proc.* 1983, *48*, 491.
7. Drake, R. S. presented at the 186th Natl. Meet. Am. Chem. Soc. Polym. Mater. Sci. Eng., Washington, D.C., August 1983.
8. Siebert, A. R. this volume, Ch. 13.
9. Bucknall, C. B. "Toughened Plastics"; Applied Science: London, 1977.
10. Kinloch, A. J.; Young, R. J. "Fracture Behavior of Polymers"; Applied Science: London and New York, 1983.
11. Riew, C. K.; Rowe, E. H.; Siebert, A. R. In "Toughness and Brittleness of Plastics"; Deanin, R. D.; Crugnola, A. M. Eds.; ADVANCES IN CHEMISTRY SERIES No. 154; American Chemical Society; Washington, D.C., 1976; p. 326.
12. Hunston, D. L.; Bascom, W. D. this volume, Ch. 7.
13. Pearson, R. A.; Yee, A. F. presented at the 186th Natl. Meet. Am. Chem. Soc. Polym. Mater. Sci. Eng., Washington, D.C., August 1983.

14. Kinloch, A. J.; Shaw, S. J. this volume, Ch. 8.
15. Hussain, A. presented at the 186th Natl. Meet. Am. Chem. Soc. Polym. Mater. Sci. Eng., Washington, D.C., August 1983.
16. Kunz, S. C.; Sayre, J. A. this volume, Ch. 15.
17. Sutton, S. A. *Eng. Fract. Mech.* 1974, *6*, 587.
18. Kim, S. L.; Skibo, M. D.; Manson, J. A.; Hertzberg, R. W.; Janiszewski, J. *Polym. Eng. Sci.* 1978, *18(14)*, 1093.
19. Manson, J. A.; Sperling, L. H.; Kim, S. L. "Influence of Cross-linking on the Mechanical Properties of High-T_g Polymers"; *Technical Report AFML-TR-77-109*, July 1977.
20. Mostovoy, S.; Ripling, E. J. *Polym. Sci. Technol.* 1975, *9B*, 513.
21. Shah, D.; Report, M. S. Lehigh University, 1983.
22. Gillham, J. K.; Chan, L. C. this volume, Ch. 16.
23. Saxena, A.; Hudak, S. J. *Int. J. Frac.* 1978, *14*, 453.
24. Rowe, E. H.; Siebert, A. R.; Drake, R. S. *Mod. Plast.* 1970, *47*, 110.
25. Bucknall, C. B.; Yoshii, T. *Br. Polym. J.* 1978, *10*, 3.
26. Manzione, L. T.; Gillham, J. K.; McPherson, C. A. *J. Appl. Polym. Sci.* 1981, *26*, 889, 907.
27. Sohn, J. E.; Emerson, J. A.; Chen, J.-K.; Siegel, A. F.; Koberstein, J. T. presented at the 186th Natl. Meet. Am. Chem. Soc. Polym. Mater. Sci. Eng., Washington, D.C., August 1983.
28. Sayre, J. A.; Assink, R. A.; Lagasse, R. R. *Polymer* 1981, *22*, 87.
29. Sultan, J. N.; McGarry, F. J. *Polym. Eng. Sci.* 1973, *13*, 29.
30. Olabisi, O.; Robeson, L. M.; Shaw, M. T. "Polymer–Polymer Miscibility"; Academic: New York, 1979.
31. Bates, F. S.; Cohen, R. E.; Argon, A. S. *Macromolecules* 1983, *16*, 1108.
32. Beck, R. S.; Gratch, S.; Newman, S.; Rusch, K. C. *Polym Lett.* 1968, *6*, 707.
33. Paris, P. C., Proc. 10th Sagamore Res. Conf. 1964, p. 107.
34. Skibo, M. D. unpublished data, Lehigh University, 1977.
35. Kim, S. L.; Skibo, M. D.; Manson, J. A.; Hertzberg, R. W. *Polym. Eng. Sci.* 1977, *17*, 194.
36. Qureshi, S. Ph.D. Thesis, Lehigh University, 1982.
37. Daly, J.; Pethrick, R. A.; Fuller, P.; Cunliffe, A. V.; Datta, P. K. *Polymer* 1981, *22*, 32.
38. Michel, J. C.; Manson, J. A.; Hertzberg, R. W.; Liégeois, J. M. In "Toughened Plastics"; Plastics and Rubber Institute: London, 1981; p. 15.1.
39. Greensmith, H. W. *J. Polym. Sci.* 1956, *21*, 175.
40. Sayre, J. A.; Kunz, S. C.; Assink, R. A. this volume, Ch. 15.

RECEIVED for review November 18, 1983. ACCEPTED May 10, 1984.

10

Siloxane Modifiers for Epoxy Resins

E. M. YORKGITIS, C. TRAN, N. S. EISS, JR., T. Y. HU, I. YILGOR, G. L. WILKES, and J. E. McGRATH

Polymer Materials and Interfaces Laboratory, Virginia Polytechnic Institute and State University, Blacksburg, VA 24061

Functionally terminated polydimethylsiloxane and its statistical copolymers with either methyltrifluoropropylsiloxane or diphenylsiloxane have been used to chemically modify bisphenol A based epoxy resins. Increasing the percentage of methyltrifluoropropyl (TFP) or diphenyl units relative to the dimethylsiloxane content of the oligomers enhanced their compatibility with the epoxy resin. This enhancement produced smaller rubber particle sizes and altered particle morphology. Improved fracture toughness relative to the cycloaliphatic diamine-cured control resin was achieved in resins modified with polysiloxane copolymers containing 40% or more TFP content or 20% diphenylsiloxane. Comparisons in these and other aspects are made with similarly prepared amino-terminated butadiene–acrylonitrile modified and carboxyl-terminated butadiene–acrylonitrile modified epoxies. Wear rate was quite dramatically reduced with some of the modifiers. The enhanced behavior may be related to a selective migration of the siloxane segment of the copolymer to the air–polymer interface. Wear results are discussed in terms of the fatigue-wear theory.

EPOXY RESINS ARE A VERY IMPORTANT CLASS of thermosetting polymers that often exhibit high tensile strength and modulus, excellent chemical and corrosion resistance, and good dimensional stability (*1–3*). However, these highly cross-linked systems are usually brittle and consequently have limited utility in applications requiring high impact and fracture strengths. Such applications include reinforced plastics, matrix resins for composites, and coatings.

The incorporation of elastomeric modifiers into the glassy epoxy matrix has in recent years served as a successful means of enhancing the fracture toughness and impact resistance of epoxy networks. For example, liquid butadiene–acrylonitrile copolymers with both carboxyl and amine end groups have been widely used as epoxy modifiers (*4–11*). During the curing process, the elastomeric component

0065-2393/84/0208-0137/$07.25/0

separates within the reacting medium to form a second dispersed phase. Through various proposed mechanisms, these rubbery domains alleviate crack propagation under low loads. The fracture energy of the base epoxy can often be significantly improved with the addition of as little as 5% of an elastomeric modifier (*12, 13*). Furthermore, this improvement in fracture resistance can be achieved with little sacrifice in bulk properties.

In this chapter, we investigate the use of liquid functional polysiloxanes as new modifiers for epoxy networks. Polydimethylsiloxanes exhibit important characteristics such as low glass transition temperature (T_g) (−120 °C), high flexibility, good weatherability, and good thermal and oxidative stability (*14*). In addition, because of their low surface energy and nonpolar structure, polysiloxanes tend to migrate to the air–polymer interface and provide a very hydrophobic surface for the substrate. Therefore, the presence of siloxane should strongly influence the friction and wear behavior of these modified resins.

Initial results on the synthesis and characterization of the siloxane-modified epoxy networks have already been presented (*15*) and will not be repeated here in detail. This chapter extends those studies to polysiloxane copolymers containing dimethyl groups as well as trifluoropropyl and diphenyl groups. This chapter is meant to be an overview of the following topics: chemistry and synthesis, mechanical properties, fracture toughness, morphology, and friction and wear characteristics.

The glassy network system under investigation was obtained from the curing reaction of a bisphenol A diglycidyl ether (DGEBA) based epoxy resin (Epon 828) with a cycloaliphatic diamine, bis(4-aminocyclohexyl)methane (PACM-20). Several of the siloxane oligomers were synthesized in our laboratory for this study (Structures **I-III**). The present investigation is restricted to the use of the 2-aminoethylpiperazine (AEP) end group.

The solubility parameter of polydimethylsiloxane, about 7.5, is much lower than that of bisphenol A epoxy resin, approximately 9.4. The solubility parameter of the oligomer, along with its molecular weight, is considered to be a key factor governing its phase separation from the epoxy resin. Ideally, one would like to control the solubility parameter in such a way that the siloxane oligomer will be miscible with the epoxy resin in the initial stages of cure but will phase separate during the later stages. Copolymerizing dimethylsiloxane with the more polar trifluoropropylsiloxane or the partially aromatic diphenylsiloxane should raise the solubility parameters closer to that of the base epoxy resin.

The primary objective of rubber modification is the improvement of fracture properties with a minimal decrease in stiffness and mechanical strength. As previously mentioned, this improvement has been achieved in part with the use of the commercially available ATBN and CTBN (amino-terminated and carboxyl-terminated bu-

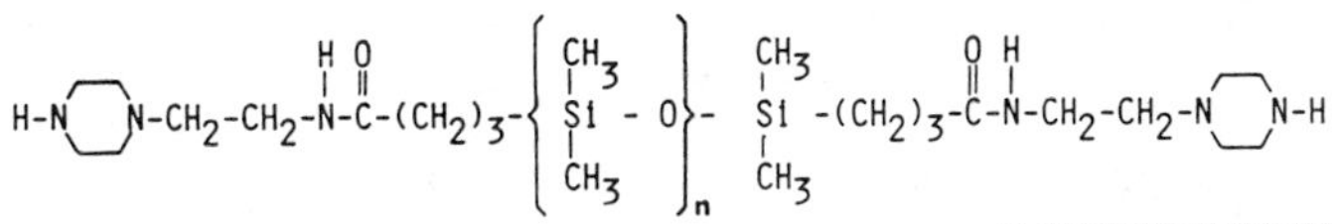

2-AMINOETHYLPIPERAZINE

I

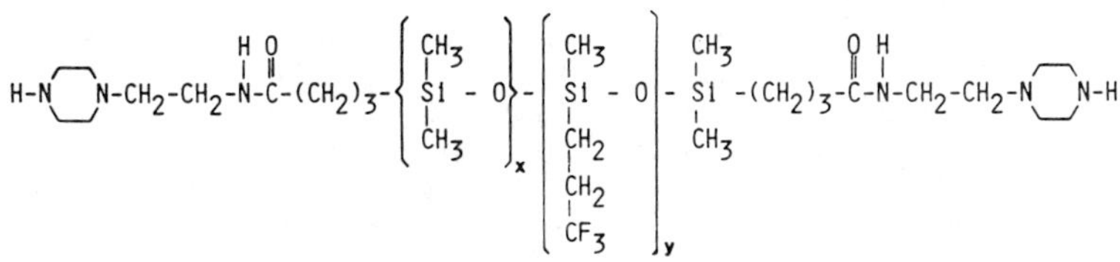

II

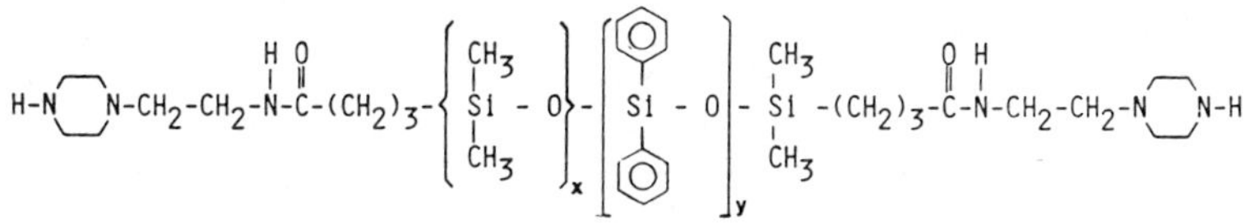

III

tadiene–acrylonitrile, respectively) copolymers manufactured by The BFGoodrich Company. However, the relatively high glass transition temperatures of acrylonitrile–butadiene copolymers limit their low temperature flexibility; their highly unsaturated structure makes them unsuitable for use at elevated temperatures (*16*, *17*). Siloxanes possess lower T_g values than conventional elastomers along with very good thermal stability.

Because a large proportion of the work on rubber-modified epoxies has dealt with the ATBN and CTBN materials, we also prepared networks containing these oligomers and studied them along with the new siloxane oligomers. For each modified resin, the quantities of flexural modulus, Young's modulus, and K_{Ic} (plane-strain fracture toughness) were determined. The fracture-surface morphology was examined as well.

The low surface energy of chemically bound siloxane promotes a nonfugitive "slippery surface." This unique circumstance has led to our interest in the friction and wear characteristics of siloxane-modified epoxies. Friction and wear are known to be functions of surface energy, toughness, and fatigue (*18*). Fatigue itself is thought to be a reflection of various factors including contact stress and elastic modulus. Tests designed to study the overall friction and wear phenomena were performed by using three variables: the normal load between two sliding surfaces, the percent rubber in the epoxy, and the percent of trifluoropropyl content in the siloxane modifier.

Experimental

Materials. The number average molecular weight of the epoxy resin, Epon 828 (Shell Chemical Company), is about 380 g/mol. The curing agent was bis(4-aminocyclohexyl)methane (PACM-20, Du Pont). Siloxane oligomers of controlled molecular weights were synthesized in our laboratory with varying weight percents of dimethyl-, trifluoropropyl (TFP)-, and diphenylsiloxane units. Dimethylsiloxane tetramer was obtained from Petrarch Systems, Inc. The TFP methyl cyclic trimer and the diphenylsiloxane tetramer were provided by the Silicone Division of the General Electric Company. The actual weight percentage of a specific kind of siloxane unit in an oligomer can be calculated in relation to either the entire oligomer, including its end groups, or its siloxane units alone. With the same oligomer, the percentage found by including the end groups will be smaller than that found by the second method. This effect is more pronounced with low molecular weights because the end groups comprise a more appreciable fraction of the total oligomeric weight.

The oligomers were prepared via equilibration reactions with base catalysts such as ammonium tetramethylsiloxanolate or potassium hydroxide (*19*). TFP–dimethyl copolymers contained 0, 20, 40, 70, or 85 wt% TFP based on the total oligomeric weight. We have also prepared dimethylsiloxane–diphenylsiloxane copolymers containing 20 and 40% diphenyl oligomer. The oligomers were characterized by proton NMR and FTIR techniques. Number average molecular weights of the oligomers were determined by end-group titration. All siloxane oligomers were viscous liquids at room temperature.

Siloxane-modified networks were prepared for mechanical and wear testing via two steps. First, a linear precursor was formed by reacting Epon 828 with the siloxane oligomer at 65 °C for 1 h; a vacuum was applied during this reaction to remove trapped air bubbles. In previous kinetic studies with DSC (*20*), 1 h was more than adequate for all of the AEP end groups to completely react with the epoxy resin. Next, the curing agent was added, and the mixture was rapidly stirred for 5 min under vacuum at 50 °C. The piperazine-terminated siloxane oligomers are much more reactive with the epoxy resin than PACM-20 (*15*).

The epoxy/siloxane/PACM-20 mixture was poured into a hot, room-temperature vulcanizing (RTV) silicone mold of the precise shapes to be used for either mechanical or wear tests. The mixture was cured at 160 °C for 2.5 h. In early studies, the T_g of the Epon 828–PACM-20 network was 150 °C (*21*). Consequently, the curing time and temperature were chosen to provide enough mobility for the chain ends to react and form the cross-linked network. The concentration of the siloxane modifiers was 5, 10, or 15% by weight.

The ATBN and CTBN rubbers (The BFGoodrich Co.) were of low (10%) and high (17–18%) acrylonitrile (AN) content. The reported molecular weights are about 3700. The ATBN oligomers are produced from CTBN oligomers of the same AN content and should therefore have just slightly higher molecular weights. Considerable disparity exists between ATBN and CTBN *titrated* molecular weights because of the excess AEP remaining in the ATBN after production. We have used the titrated molecular weight to determine the appropriate amount of curing agent for complete network formation. All networks, regardless of the modifier, were prepared identically. Hence, all modified resins were subjected to the same criteria concerning the ratio of rubber to resin to curing agent and were cured according to the same schedule.

The T_g values of samples were determined with a differential scanning calorimeter, a Perkin-Elmer DSC-II or DSC-IV, at heating rates of 10 or 20 deg/min, respectively.

Sample Designation. Each oligomer or modified resin is described by two or three numbers, respectively (*see* Box). As an example, a sample containing 10% by weight of a 20% TFPsiloxane rubber of molecular weight 2330 g/mol will be designated 10-2330-20F. The oligomer would be labelled 2330-20F. Note that the differences between specific types of siloxane copolymers are given by changes in the third number of the sample code, and differences between the end groups of the butadiene oligomers are given in the first number of the code.

Explanation of Sample Codes

The numbers given are the weight percent of rubber–the molecular weight of modifier–the weight percent of comonomer, in that order.

Examples:

Pure polydimethylsiloxane	= 5–2190–0
Trifluoropropylmethylsiloxane	= 10–2070–40F
Diphenylsiloxane	= 5–2250–20D
CTBN	= 10C–3880–17AN
ATBN	= 15A–2560–10AN

Mechanical Properties. Tensile moduli were measured from standard dog-bone samples (2.0 mm thick, 4.7 mm wide, and 22.0-mm gauge length) in a Model 1122 Instron at a crosshead speed (CHS) of 1.0 mm/min. Flexural modulus and fracture toughness were determined by using a three-point bend geometry. The testing apparatus consisted of two aluminum–steel pieces attached to the Instron, which was fitted with a tensile load cell. This device effectively performs an inverted three-point bend; the two side bars remain stationary above the sample as the central bar below the sample moves upward. Flexural samples measured approximately 52.0 × 1.7 × 13.1 mm. They were tested with a 2.54-cm span (distance between the two side bars) at a CHS of 1.0 mm/min. Dimensions of fracture toughness samples were approximately 3.2 × 6.4 × 38.1 mm. A thin saw notch less than 1 mm deep was cut into the center of the 3.2-mm side. Into this notch was placed a sharp, one-sided razor blade that was tapped lightly to make a short "precrack." The CHS was 0.5 mm/min. After fracture, the precrack lengths were measured with vernier calipers and a magnifying glass and were calculated as the average of three values (side, center, and side.) All modulus- and fracture-toughness testing was done at ambient temperatures. (*See* Reference 22 for a more detailed description of this procedure.)

Scanning Electron Microscopy. Fracture surfaces of cold-snapped and fracture-toughness samples were examined (22) in a scanning electron microscope (SEM) (ISI, Inc., Super III-A). Cold-snap samples were fractured after being submerged in liquid nitrogen for 10 min. Following fracture-toughness testing, the precrack front regions of such samples were examined in the SEM.

Friction and Wear. Polysiloxane-modified networks were cast into plates 3 to 4 mm thick and 50 mm square. After casting, the plates were immersed in 2-propanol for 20 min to remove any unreacted surface species or adsorbed substances. The plates were then stored in a desiccator overnight before testing.

The friction and wear testing was performed on a pin-on-disk machine (23) in which a stationary chromium steel sphere (pin) (3.2-mm diameter) was held in contact with the rotating epoxy plate (disk) as shown in Figure 1. Three normal loads were used (2, 5, and 10 N), and the sliding speed was constant at 0.63 m/s. The experiments were performed in laboratory air at 21–23 °C. The frictional force encountered during testing deflected a cantilever beam holding the pin, and the deflection was detected by a proximity detector. When wear occurred, the sphere generated a groove in the epoxy disk. Wear was estimated by measuring the cross-section profile of the groove with a surface profile meter. *See* Figure 1.

After wear began, the test was run for 14 kilocycles (kc), and the cross-section profile was measured every 2 kc. The wear was proportional to the number of cycles as shown in Figure 1. The slope of the wear versus cycles

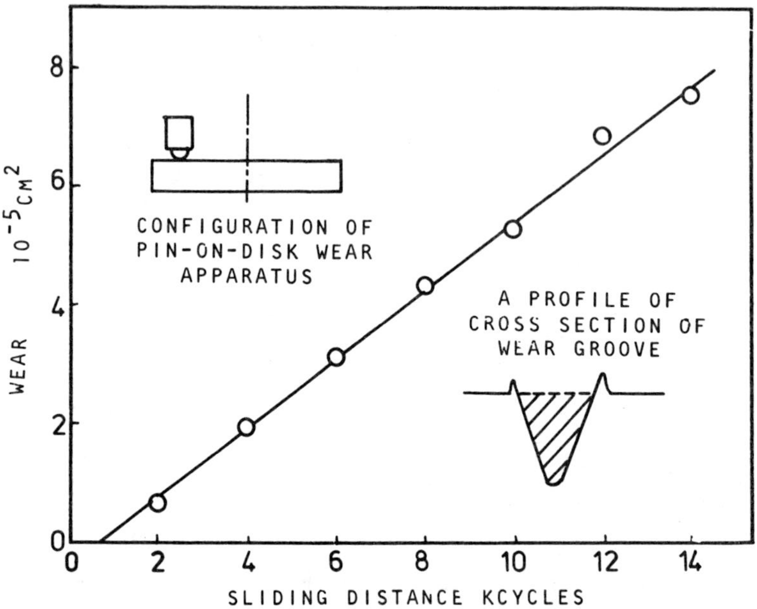

Figure 1. Wear (cross-section area) as a function of the number of cycles of disk rotation.

curve, as calculated by linear regression, is the wear rate. Ninety-five percent confidence limits of the wear rate were also calculated.

Results and Discussion

The curing reaction of the siloxane–epoxy system is depicted in Scheme 1. In the first step, the piperazine-terminated siloxane oligomer was reacted with the epoxy resin to make a linear soluble precursor with epoxide end groups. Then the siloxane–epoxy adduct (plus the excess epoxy) was cured with PACM-20 to yield the modified cross-linked network. Liquid siloxane oligomers of number average molecular weights from 1000 to 5000 g/mol have been used for epoxy modification, but only a few of these have been studied in depth. Table I shows the T_g values for pertinent siloxane oligomers. As expected, the T_g values of the oligomers increase with TFP content. The T_g values of the siloxane-modified networks are shown in Table II. For the samples modified predominantly with dimethyl units, the T_g values remained essentially unchanged with respect to the control sample. This finding indirectly implies that phase separation has occurred in these siloxane-modified epoxy networks. On the other hand, the T_g depression seen for the networks modified with certain percentages of some of the elastomers clearly indicates partial miscibility between the siloxane and epoxy phases.

Rubber–Epoxy Compatibility: Effect of TFP Content. Pure polydimethylsiloxane possesses no polar groups across its silicon–

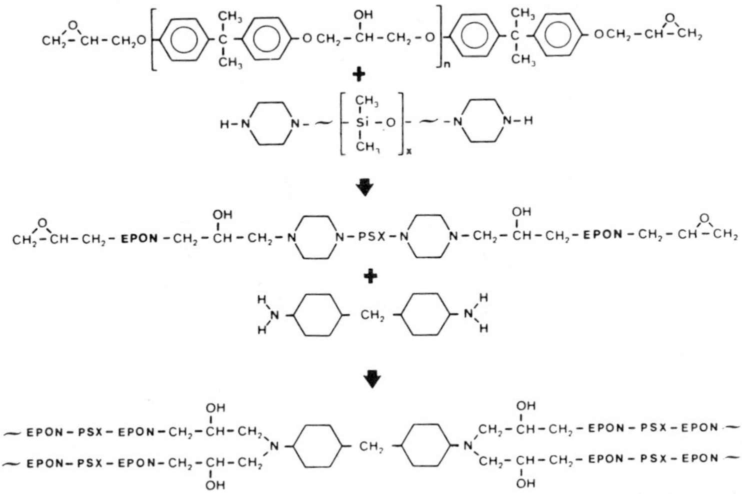

Scheme 1

oxygen chain. However, if one of the two methyls of a monomer unit is replaced with a more electronegative TFP group, this latter group can couple with the remaining electron-releasing methyl group to create a dipole across the siloxane chain. The epoxy resin has a greater affinity for this new polar elastomer, and the attraction between the two grows as TFP content increases. The effect of this situation is twofold. First, as TFP content is increased, additional siloxane will remain dissolved in the epoxy matrix after cure is complete. Second (again as TFP content increases), siloxane that does not actually dissolve in the matrix will precipitate out later in the curing reaction. Because the mobility of the network system diminishes continuously with cure time, at the point where incompatibility

Table I. Glass Transition Temperatures of Dimethylsiloxane–Methyltrifluoropropylsiloxane Oligomers

End Group	$\overline{M}_n$	*TFP*[a] *(wt%)*	T_g *(°C)*
AEP[b]	2190	—	−126
AEP	2330	25	−116
AEP	2070	50	−91
AEP	2730	75	−76
AEP	3130	100	−58
AEP	1500	100	−45
PIP[c]	1230	100	−48

NOTE: DSC data, 10 K/min.
[a] Weight percent trifluoropropylmethyl siloxane without end groups.
[b] Aminoethylpiperazine-capped polysiloxane.
[c] Piperazine-capped epoxy polysiloxane.

Table II. Glass Transition Temperatures of Siloxane-Modified Networks

$\overline{M}_n$ (g/mol)	TFP[a] (wt %)	Siloxane (wt %)			
		0	5	10	15
2190	0	150	158	153	152
2330	25	150	157	147	151
2070	50	150	147	136	150
2730	75	150	156	156	140
3130	85	150	150	153	147
1500	70	150	151	145	140

NOTE: DSC data, 10 K/min.
[a] Weight percent trifluoropropylmethylsiloxane without end groups.

is reached, the elastomeric segments of each network chain can coalesce with only so many other such segments. The net result is a particle size that is inversely proportional to TFP content. This same argument explains the effect of an increase in AN content on the particle size of ATBN and CTBN in cured epoxy resins.

Figure 2 shows representative cold-snap fracture surfaces for the control and four TFPsiloxane-modified epoxies at 15% rubber content. Particle size decreases as TFP content increases, from about 50 to less than 5 μm. Not only the size but the makeup of the particles has changed. At low TFP content, the rubber domains are not simply large but appear to contain both epoxy and elastomer. In studies of CTBN-modified epoxies, other workers have found evidence for mixed resin–rubber domains by viewing stained thin sections with the transmission electron microscope (*24*). At 70% TFP content, the siloxane oligomers form small particles that appear to be uniform in composition.

In order to at least qualitatively determine the distribution of siloxane in these materials, we have examined the fast-crack regions of 10–2330–20F and 10–2190–0 fracture-toughness samples. These relatively smooth surfaces contain few fracture artifacts and so are reasonably true cross sections of the samples. We noted an apparently even distribution of the elastomeric phase from one outside edge to the other.

Although molecular weight effects have not been emphasized in this research, we do note a primary effect of increasing molecular weight. Doubling the molecular weight of the high percentage TFP-siloxane modifier (Figure 3) introduces a few large particles while maintaining the small particles found alone at the 1500 molecular weight. Compare the upper and lower micrographs in Figure 3.

Flexural Modulus. The brittleness of epoxy networks is one of their most undesirable properties, but this drawback can be alle-

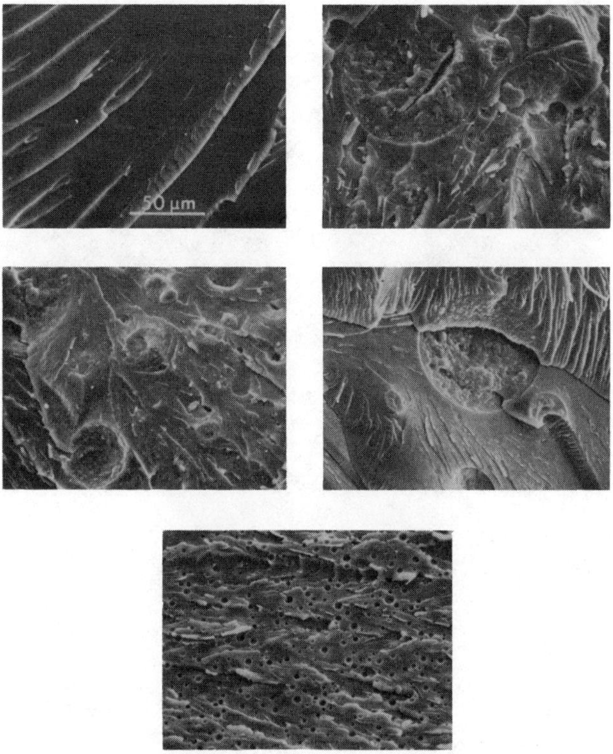

Figure 2. SEMs of cold-snap fracture surfaces for unmodified resin control (top left) and four siloxane-modified resins: 2190–0 (top right), 2330–20F (middle left), 2070–40F (middle right), and 1500–70F (bottom). Magnification is 345×.

viated by rubber modification. However, maintaining their favorably high moduli and overall good mechanical strength is important. Figure 4 illustrates that siloxane and TFPsiloxane modification only slightly influenced the flexural modulus of the unmodified epoxy. As expected, the flexural modulus decreases as rubber content increases. The rate at which it decreases depends at least partly on the TFP content of the rubber.

Compatibility effects can account for changes in morphology, but modulus effects are not so easily explained. As shown in Table I, the T_g of the TFP oligomers increases dramatically with TFP content, from −126 °C for the 2190–0 oligomers to −45 °C for 1500–70F. Because flexural testing is done at room temperature, well above these transition temperatures, the difference in T_g between the various rubbers does not bear heavily on the modulus.

A plausible explanation of the flexural modulus results combines particle size, compatibility, and interfacial adhesion. Siloxane–siloxane contacts consist primarily of van der Waals forces, whereas the

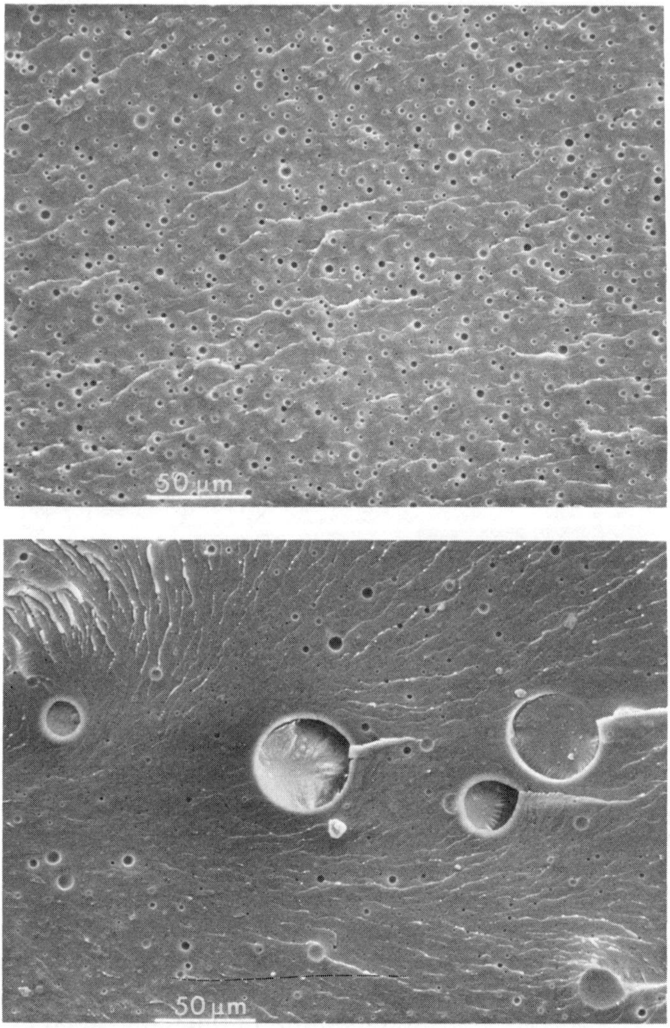

Figure 3. SEMs of cold-snap fracture surfaces for 15–1500–70F (top) and 15–3130–85F (bottom). Magnification is 255×.

forces across the siloxane–epoxy interface also include inherently stronger covalent bonds and possible dipole–dipole interactions. As particles become smaller and the surface-to-volume ratio increases, the resin–rubber interface may strongly contribute to the modulus in addition to the rubber. If such is the case, the effect of siloxane on flexural modulus will be less detrimental as TFP content is increased. In addition, the available volume fraction of the phase-separated elastomer will be effectively lowered if a significant amount of the elastomer is actually miscible with the matrix. Studies with

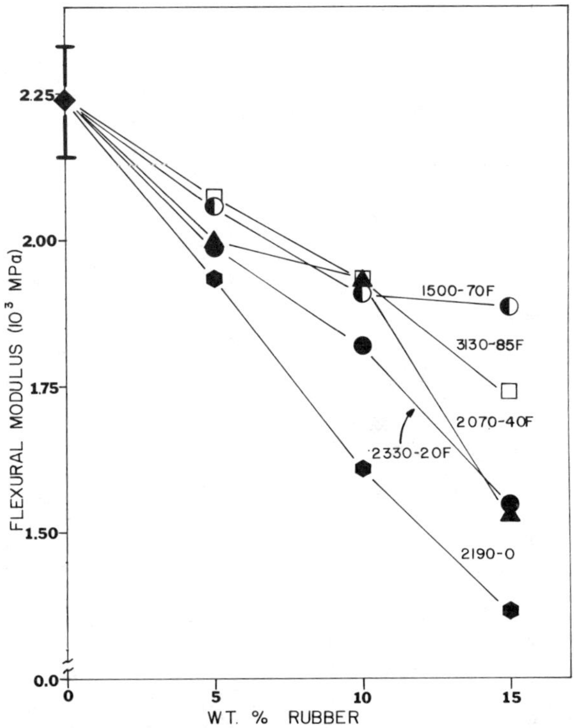

Figure 4. Changes in flexural modulus with rubber content and TFP percentage in rubber. Error bars on data point for control are typical for all data points. Crosshead speed is 1.0 mm/min.

CTBN-modified epoxies (*24*) have shown that dissolved rubber is not as harmful to modulus as phase-separated rubber.

Figure 5 gives flexural modulus data for the CTBN- and ATBN-modified epoxies. Again modulus drops as the rubber content increases. Within the percentage of error, the effect of increasing AN content appears to parallel that of increasing TFP content in the siloxanes. When the results of Figure 5 are compared with those of Figure 4 (both figures are drawn on an identical scale), the values for the TFPsiloxane-modified resins are generally higher than those for the butadiene-modified resins. This effect is rather subtle, but it is reinforced by Young's modulus data (Table III).

These differences may be at least partially explained by differential scanning calorimetry (DSC) data such as that in Figure 6. The T_g regions of the ATBN- and CTBN-modified epoxies are generally broader and have a lower midpoint than those of the control and the two siloxane-modified resins. Therefore, the butadiene oligomers may be relatively more miscible with the epoxy and may soften it. We recognize, however, that this observation does not completely

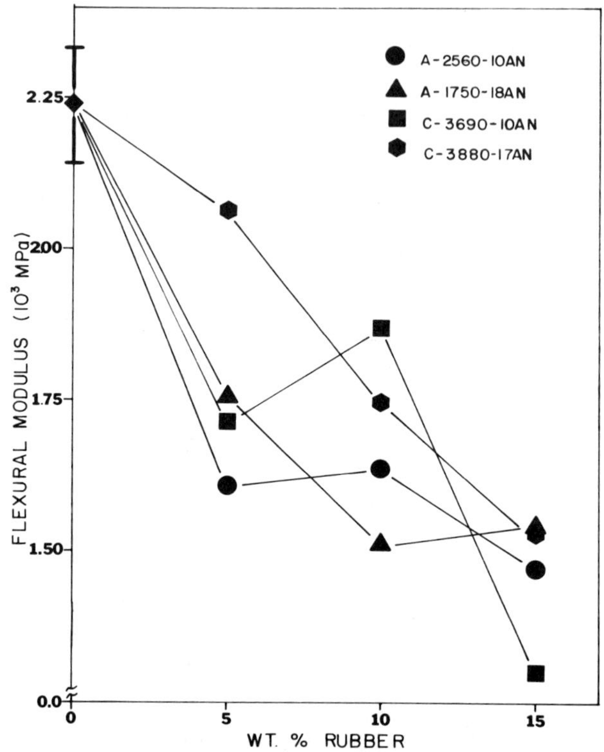

Figure 5. Changes in flexural modulus with rubber content and AN percentage in ATBN and CTBN rubbers. Error bars on data point for control are typical for all data points. Crosshead speed is 1.0 mm/min.

coincide with previous statements concerning the effect of dissolved rubber on modulus.

Fracture Toughness. Fracture toughness was monitored through K_{Ic}, plane-strain fracture toughness. K_{Ic} values for at least five samples of each material were calculated according to Equation 1 (25),

$$K_{Ic} = \frac{6\,P}{B\,W^{\frac{1}{2}}} f\left(\frac{a}{W}\right) \qquad (1)$$

where P is the critical load, B is the sample's width, W is its thickness, and a is the length of the precrack. In this study, P is the load at break. Letting $R = a/W$, the geometry factor $y = f(a/W)$ can be given as in Equation 2.

$$Y = 1.93\,R^{1/2} - 3.07\,R^{3/2} + 14.53\,R^{5/2} - 25.11\,R^{7/2} + 25.80\,R^{9/2} \qquad (2)$$

Table III. Selected Young's Moduli

Sample	E*(10³ MPa)*
Control	1.11 ± 0.17
15–2190–0	0.92 ± 0.09
15–2070–40F	0.95 ± 0.07
15C–3690–10AN	0.88 ± 0.02
15C–3880–17AN	0.86 ± 0.04
15A–2560–10AN	0.68 ± 0.06
15A–1750–18AN	0.77 ± 0.05

NOTE: Crosshead speed = 1.0 mm/min.

The criteria of ASTM E399 were followed as closely as possible. The only criterion that could not be consistently satisfied was a straight precrack front, ±5%.

When K_{Ic} results are presented graphically, the error bars given for the control are typical of all those data points that do not have their own error bars. For samples containing 15% rubber, error was sometimes greater than 10%, and individual error bars are provided and labelled with the corresponding symbol. Such a large deviation results from the clear violation of the homogeneity criterion of linear elastic fracture mechanics at 15% of certain oligomers.

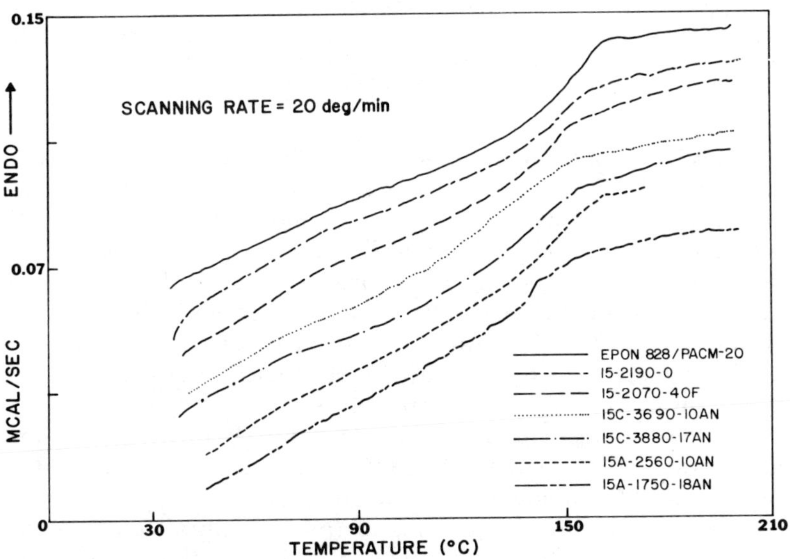

Figure 6. Normalized DSC traces of control and resin modified with 15% of one of six different oligomers. Six of the curves have been shifted vertically.

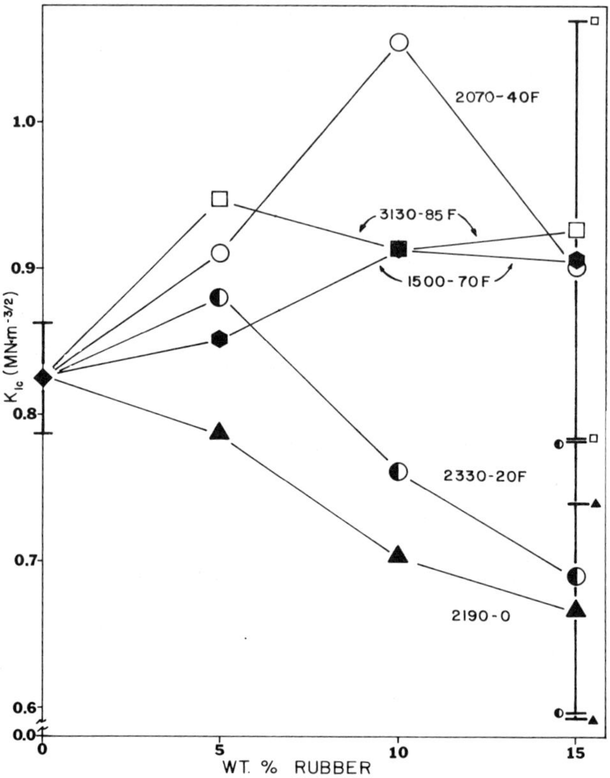

Figure 7. Changes in K_{Ic} *with rubber content and TFP percentage in rubber.* See *text for explanation of error bar usage. Crosshead speed is 0.5 mm/min.*

Figure 7 shows K_{Ic} results on the TFPsiloxanes. Modification with pure PDMS lowers K_{Ic} almost linearly with rubber content. After an initial increase at 5% rubber, modification with the 2330–20F oligomer lowers K_{Ic} in a linear fashion. As TFP content is raised to 40% and above however, K_{Ic} improves considerably and reaches a high value in the 10–2070–40F material. Because the K_{Ic} value is quantitatively derived from processes occurring in the precrack front, these fronts were examined with SEM. Figure 8 shows two of the outstanding morphological features observed.

The first of these features is illustrated by Figure 8a, an example of a precrack front of the 10–2330–20F material. This material showed a K_{Ic} value below that of the control. Particularly evident are the large heterogeneous particles of (presumably) both epoxy and elastomer. These particles, rather than absorbing the energy of fracture, tear at some small angle to the direction of crack propagation. Kunz and coworkers have postulated (*26*) a toughening mechanism involving particle stretching and failure by tearing. The tearing we

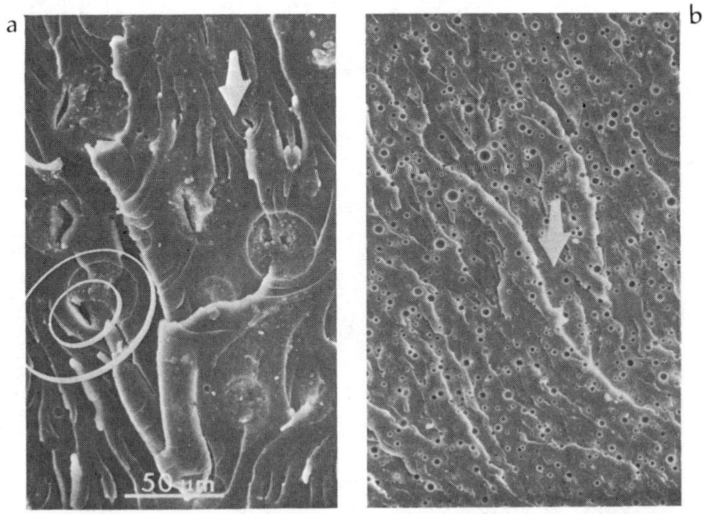

Figure 8. SEMs of precrack fronts of fracture-toughness specimens: left, 10–2330–20F and right, 15–1500–70F. Ellipses highlight torn rubber domain. Arrows indicate direction of crack propagation. Magnification is 255×.

observed is not entirely unexpected in light of this theory and the fact that many siloxane elastomers have relatively poor tear resistance at room temperature (*14*, *27*). This poor tear resistance may be related to the large difference (approximately 145 °C) between room temperature and T_g (*28*). However, the extent to which tearing occurs in this case and the associated K_{Ic} values suggest that the tearing is not a significant part of an energy absorptive process.

The second morphological feature we wish to point out is pictured in Figure 8b. This micrograph exemplifies the K_{Ic} fracture surface of the 15–1500–70F material that gave a K_{Ic} value above that of the control. The holes that cover the surface are examples of what has commonly been called the *dilatation effect* (*29*, *30*). In this effect, rubber particles expand in mutually perpendicular directions under the application of triaxial stresses.

Superimposing K_{Ic} data for the 2070–40F oligomer on data from the ATBN and CTBN oligomers in Figure 9, we find that this particular siloxane oligomer is quite competitive with the butadiene oligomers at both 5 and 10% rubber content. SEM micrographs reveal that, although some mixed resin–rubber particles are found in this material, we also find many small and presumably homogeneous rubber domains. (Data given in Figure 9 for the 2250–20D oligomer will be discussed later.)

For the butadiene oligomers, at high AN content, the CTBN-modified resins have higher K_{Ic} values than the ATBN-modified resins (*see* Figure 9). However, at low AN content, the K_{Ic} values of

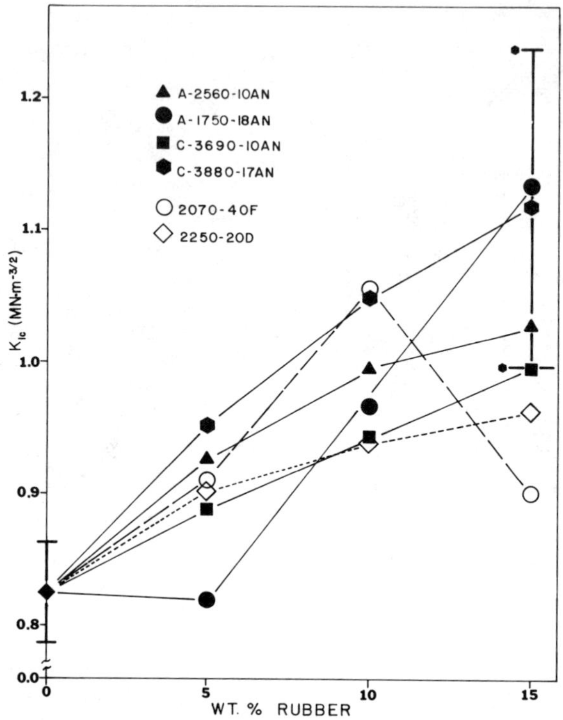

Figure 9. Changes in K_{Ic} *with rubber content for resin modified with ATBN, CTBN, 2070–40F, and 2250–20D. See text for explanation of error bar usage. Crosshead speed is 0.5 mm/min.*

the ATBN-modified systems are higher than those of the CTBN-modified resins. Close examination of the ATBN and CTBN domains revealed a somewhat unexpected difference in particle morphologies that may well explain this reversal in K_{Ic} values. As an example, in Figure 10 we show low-magnification micrographs of resins modified with 10% of each of the four butadiene rubbers. Although three of the samples show exclusively small-particle morphology, the 10C–3690–10AN sample contains large resin–rubber particles that coexist with small particles. The lower micrographs in Figure 10 clarify the nature of these small particles. For the CTBNs of either AN content, these small particles have a homogeneous texture. However, although ATBN incorporation never gives rise to *large* nonhomogeneous particles, the small particles appear to be resin–rubber particles themselves. This difference in morphology probably follows from the excess AEP, the amino versus the carboxyl end groups, or a combination of the two. The diameters of the small particles for either elastomer are roughly equivalent regardless of their morphological texture.

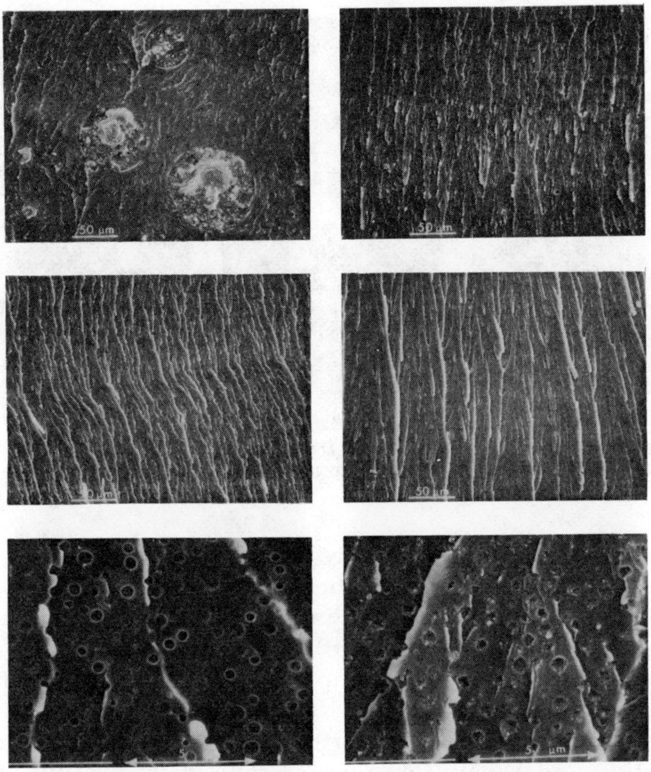

Figure 10. Upper four photomicrographs are low magnification views of K_{Ic} fracture surfaces of resin modified with 10% of each butadiene oligomer: 10C–3690–10AN (top left), 10A–2560–10AN (top right), 10C–3880–17AN (middle left), and 10A–1750–18AN (middle right). Lower two micrographs magnify the small particles of the two high AN oligomers and eludicate the differences in small particle makeup between the ATBN- and CTBN-modified resins. Magnifications of upper four are 345×, of lower two, 11,200×.

Figure 9 also presents K_{Ic} results for a siloxane modifier containing 20% diphenylsiloxane; the K_{Ic} values are among the best seen for any of the siloxane modifiers thus far. The micrographs in Figure 11 exhibit the small-particle morphology of these materials. Such a morphology was anticipated. The more hindered, bulky, and partly aromatic diphenylsiloxanes have higher T_g values (−88 °C for 2250–20D) than the corresponding dimethylsiloxane–TFPsiloxanes. Thus, during cure, phase separation is somewhat more difficult. No tearing of the 2250–20D domains was observed, in contrast to the epoxies modified with 2330–20F. Diphenylsiloxane also effectively raises the solubility parameter closer to the value for the resin. These characteristics appear to encourage good modifying properties. Other work with a 40% diphenyl oligomer also shows improvement in K_{Ic} along

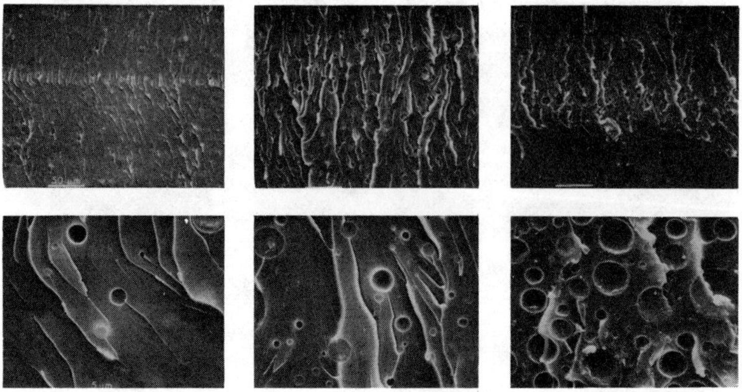

Figure 11. SEMs of precrack fronts of resin modified with 2250–20D oligomer. Rubber contents increase from 5 to 10 to 15 wt% from right to left. Magnification of upper row is 345×, of lower row, 3450×.

with smaller particle sizes. Future work in this area will especially focus on these copolymeric modifiers.

The fracture-toughness results we present are derived from a fairly small three-point bend (3PB) specimen. In limited work with a proportionately larger 3PB sample ($B \sim 6.4$ mm) and a standard size compact tension (CT) sample ($B \sim 6.5$ mm, $W \sim 27.2$ mm), we have found comparable results for the Epon 828–PACM-20 control.

Friction and Wear Behavior. Let us now turn our attention to investigations of the friction and wear behavior of these materials. In Figure 12, the wear rates are plotted versus the percent rubber for the three normal loads of 2, 5, and 10 N and TFP percentages of 0, 20, 30, and 70. To determine the main effect of the percent rubber on the wear rate, the wear rates at different loads and percent TFP were averaged at each percent rubber. (*See* Figure 13.) In an analogous manner, the main effect of percent TFP was calculated (Figure 14). The interaction between percent rubber and percent TFP was obtained by averaging the wear rates at the different loads for each combination of percent rubber and percent TFP. (*See* Figure 15.)

The coefficient of friction for the unmodified epoxy did not vary significantly with normal load. For the polysiloxane-modified epoxies, increasing the load from 2 to 10 N decreased the coefficient of friction between 22 and 46%. Although the coefficient of friction did not change for increases in percent rubber and percent TFP in the rubber at the 2- and 5-N loads, a significant change was found at the 10-N load. (*See* Figure 16.)

The interpretation of the friction and wear data first required the identification of the predominant wear mechanism. The three most prevalent mechanisms are adhesive transfer, abrasive cutting,

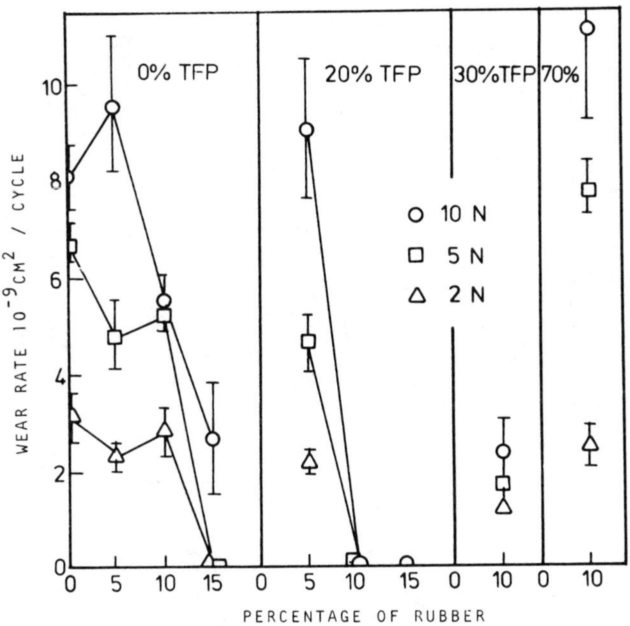

Figure. 12. Wear rate as a function of percent polysiloxane rubber, percent TFP in the rubber, and normal load. Indicated bars are 95% confidence limits.

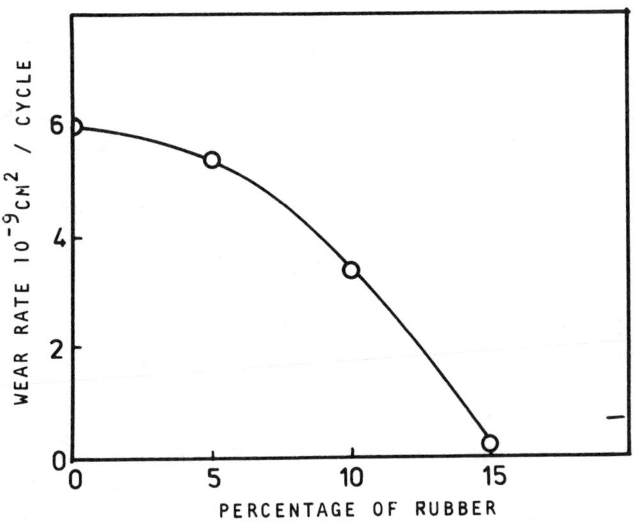

Figure 13. Main effect of percent rubber on wear rate.

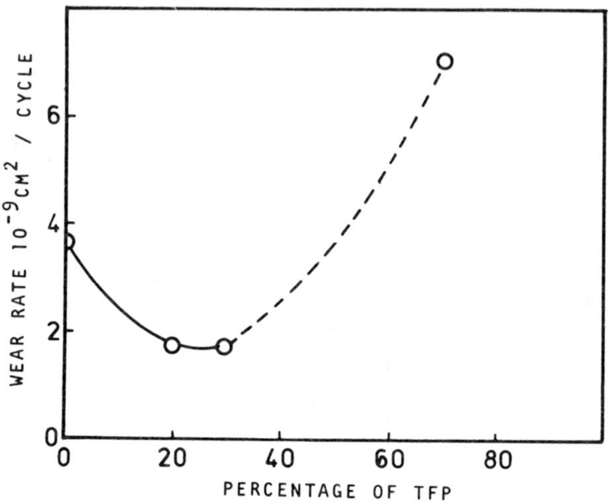

Figure 14. Main effect of TFP percentage on wear rate.

and fatigue. Of these three, only fatigue is consistent with our observation that an initiation time was required before wear occurred. That is, in most cases, wear did not occur immediately as the testing began. In a few instances, no wear was detected even after 30 kc of disk rotation. For a given sample, the number of cycles needed to initiate wear generally increased as the normal load decreased.

Fatigue wear has been shown to be proportional to the stress in the polymer raised to a power t, where t varies from 2 to 8 depending on the polymer (*18*). The stress is usually estimated by using an elastic model of contact stress (Hertzian contact) that predicts that the stress is proportional to the one-third power of the normal load P and the two-thirds power of the elastic modulus E. The relationships between fatigue wear rate, stress, modulus, and load are given in Equation 3.

$$\text{wear rate} = (\text{stress})^t = P^{t/3}\,E^{2t/3} \tag{3}$$

For a brittle polymer such as polymethyl methacrylate, $t = 8$.

The data in Figure 12 clearly show that the wear rate increased as the normal load increased. The wear rates and elastic moduli for the materials used in the wear studies are given in Table IV. As can be seen from the table, the wear rate decreased with the modulus. Thus, the variations in wear rate with the normal load and with the elastic modulus agree with the trends predicted by the fatigue-wear theory.

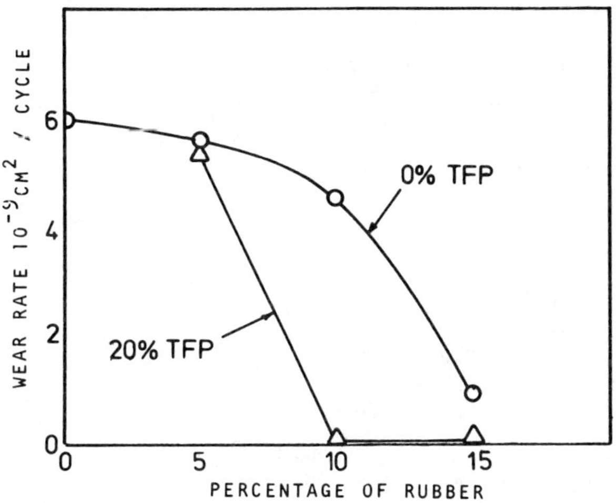

Figure 15. Interactive effect of percents rubber and TFP on wear rate at 10-N load.

An additional factor that could reduce the wear rate as the percent rubber increases is the reduction in surface energy caused by the rubber components. When sliding starts on the as-cast sample, the surface is already enriched with the rubber components that migrated to the surface during the curing process. After wear starts and the original surface is removed, the slider encounters the rubber domains and smears them over the epoxy surface, thus renewing the rubber-rich surface that existed on the as-cast surface. Such smearing

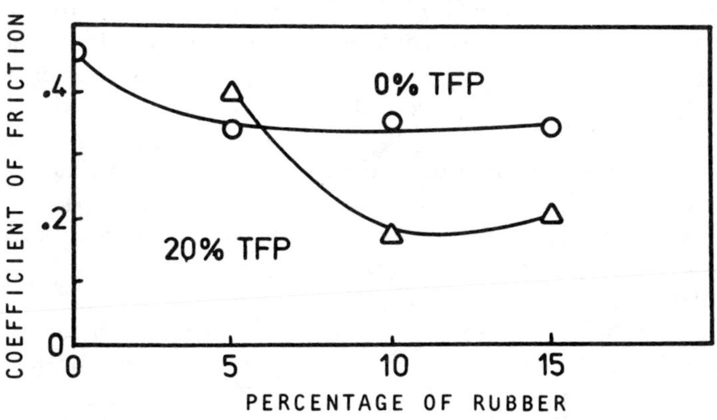

Figure 16. Interactive effect of percents rubber and TFP on coefficient of friction at 10-N load.

Table IV. The Characteristics of Polysiloxane Modifiers, Wear Rate, and Elastic Modulus

	Characteristics of the Polysiloxane Modifiers				
Sample No.	*Rubber (%)*	M_n	*TFP (%)*	*Wear Rate*[a] *(10^{-9} cm²/cycle)*	*Elastic Modulus (GPa)*
1	0	0	0	6.0	1.11
2	5	2190	0	5.5	1.01
3	10	2100	0	4.5	0.97[b]
4	15	2190	0	0.9	0.92
5	5	2330	20	5.3	1.14
6	10	2330	20	0	0.88
7	15	2330	20	0	0.88
8	10	2540	30	1.8	—
9	10	1500	70	7.1	1.00

[a] Average of wear rates at the normal loads 2, 5, 10 N.
[b] Sample for modulus measurement: 10–2190–0.

is suggested by SEM micrographs of the wear tracks. The lower surface energy of the rubber compared to that of the epoxy would then decrease the adhesive forces and thereby decrease the stress on the surface. This decreased stress results in a decrease in wear rate according to the fatigue-wear theory.

In work with 5C–, 10C–, and 15C–3880–17AN; 5C–3690–10AN; and 10A–1750–18AN at 5-N loads, wear tracks formed between 1.7 and 10.8 kc. These values are similar to the initiation times for most of the siloxane-modified epoxies. However, wear tracks were not initiated on the 10–2330–20F and 15–2330–20F formulations in 30 kc.

The reduction in friction coefficient with increasing percent rubber (Figure 16) is also consistent with a reduction in surface energy and tangential stress at the surface. One component of friction is the product of the shear strength of the interface and the area of contact at the interface. Thus, if the shear strength is reduced as a result of the presence of rubber on the surface, the friction should also be reduced.

Interactions between the percent rubber and percent TFP pertaining to the wear and friction are illustrated in Figures 15 and 16, respectively. No significant difference in wear or friction was observed for a resin of either 0 or 20% TFP content. However, at 10 and 15% rubber content, the 20% TFP oligomer reduces both friction and wear compared to the 0% oligomer. The wear results are consistent with the elastic modulus changes shown in Table IV. The friction data show that a minimum rubber content between 5 and 10% is necessary before the addition of TFP units significantly lowers the coefficient of friction.

For TFP percentages greater than 20, the wear rate increases and is a maximum at 70% TFP content. However, the data at 70% TFP must be analyzed with caution because the molecular weight of this modifier is much lower than that of the other oligomers. From these limited results, the elastic modulus at 70% TFP is seen to be greater than that at 20% TFP; thus, the wear results are still consistent with the changes in elastic modulus. The 70% TFP sample also has a significantly different morphology than the other friction and wear samples; the domains are smaller and more uniformly dispersed. (*See* Figures 2 and 3.)

Conclusions

Dimethylsiloxane, dimethylsiloxane–TFPsiloxane copolymers, and dimethylsiloxane–diphenylsiloxane copolymers with AEP end groups were prepared and used as modifiers for epoxy networks based on Epon 828 resin and PACM-20 hardener. T_g values of the TFP–dimethyl oligomers increased with TFP content. DSC studies of the TFPsiloxane-modified epoxies indicated that phase separation had occurred; however, high levels of TFP at relatively low molecular weights depressed T_g somewhat, thus partial miscibility was implied. SEM study shows that increasing TFP content in the siloxane modifiers lowered the size of the rubber domains in the modified resins. The smaller domains indicated again that increasing TFP content enhances the siloxane's compatibility with the resin. The effects of increasing TFP content are analogous to increasing AN content in ATBN- and CTBN-modified resins.

For all modifiers, flexural and Young's moduli decrease with increasing rubber content. The decrease is less severe as either TFP or AN content in the modifier is increased. The flexural and Young's moduli of the ATBN- and CTBN-modified epoxies were generally lower than those of the TFPsiloxane-modified resins. DSC data suggest that this difference resulted from the greater miscibility of the ATBN and CTBN oligomers with the cured epoxy resin system.

K_{Ic} values of the modified resins were most strongly influenced by the weight percent of the modifier and the apparent compatibility of the modifier with the resin. As TFP content was increased in the TFP–dimethyl oligomers, K_{Ic} values climbed from below the K_{Ic} of the unmodified control resin to values above that of the control. The greatest improvement in K_{Ic} occurred with a 40% TFPsiloxane, but siloxanes of higher TFP content produced reasonable improvement. K_{Ic} values of an epoxy modified with a 20% diphenylsiloxane were—at all weight percents of rubber—superior to the control's value. Low K_{Ic} values in the siloxane-modified resins were primarily associated with large heterogeneous particles that tore under experimental conditions. Good K_{Ic} values were predominantly traced to small homogeneous particles that dilated during the fracture process.

Incorporation of ATBN and CTBN oligomers into the glassy networks raised K_{Ic} relative to the control. The CTBN-modified resins exhibited higher K_{Ic} values than the ATBN-modified systems at 17–18% AN content; but, at 10% AN content, the opposite was true. This reversal may be related to a corresponding change in morphology. SEM study showed that the ATBN-modified resins contain small resin–rubber particles at both 10 and 18% AN content. In contrast, the CTBN-modified epoxies exhibited small homogeneous rubber domains at 17% AN content, but at the lower AN level, developed large resin–rubber particles that coexisted with the small homogeneous particles.

The wear of the siloxane-modified epoxies was dominated by a fatigue mechanism in which an initiation period of sliding occurred before wear started. The wear rate correlated positively with changes in the normal load and elastic modulus as predicted by the fatigue-wear theory. For the normal loads tested, the wear rate was lowest at 15% weight content of a rubber containing 20% TFP. The friction coefficient was reduced by increasing the percent rubber and, at 10 and 15% rubber, was reduced even further by increasing the percent TFP from 0 to 20%. The friction drop was assumed to be caused by the lowering of surface energy as rubber was spread over the epoxy during sliding. This particular conclusion is consistent with our earlier publication in this area (*15*).

Acknowledgment

The authors would like to acknowledge the generous financial support provided by the Office of Naval Research under Contract N00014-78-C-0629-P000005.

Literature Cited

1. Potter, W. G. "Epoxide Resins"; Springer-Verlag: New York, 1970.
2. "Epoxy Resin Chemistry and Technology"; May, C. A.; Tanaka, Y., Eds.; Marcel Dekker: New York, 1973.
3. "Epoxy Resin Chemistry"; Bauer, R. S., Ed.; ACS SYMPOSIUM SERIES No. 114; American Chemical Society: Washington, D.C., 1979.
4. Rowe, E. H., Siebert, A. R.; Drake, R. S. *Mod. Plast.* 1970, *47*, 110.
5. Riew, C. K., Rowe, E. H.; Siebert, A. R. In "Toughness and Brittleness of Plastics"; Deanin, R. D.; Crugnola, A. M., Eds.; ADVANCES IN CHEMISTRY SERIES No. 154; American Chemical Society: Washington, D.C., 1976; p. 326.
6. Bucknall, C. B.; Yoshii, T. *Br. Polym. J.* 1978, *10*, 53.
7. Riew, C. K.; *Rubber Chem. Technol.* 1981, *54*, 374.
8. Daly, J.; Pethrick, R. A.; Fuller, P.; Cunliffe, A. V.; Datla, P. K. *Polymer* 1981, *22*, 32.
9. Sayre, J. A.; Assink, R. A.; Lagasse, R. R. *Polymer* 1981, *22*, 87.
10. Bascom, W. D.; Cottington, R. L. *J. Adhes.* 1976, *7*, 3333.
11. Sultan, J. M.; McGarry, F. *Polym. Eng. Sci.* 1973, *13*, 29.
12. Bascom, W. D., Ting, R. Y.; Moulton, R. J.; Riew, C. K.; Siebert, A. R. *J. Mater. Sci.* 1981, *16*, 2657.
13. Kunz, S. C.; Sayre, J. A.; Assink, R. A. *Polymer* 1982, *23*, 1897.
14. Warrick, E. L.; Pierce, O. L.; Polmanteer, K. E.; Saam, J. C. *Rubber Chem. Technol.* 1979, *52*, 437.

15. Riffle, J. S.; Yilgor, I.; Banthia, A. K.; Tran, C.; Wilkes, G. L.; McGrath, J. E. In "Epoxy Resin Chemistry II"; Bauer, R. S., Ed.; ACS SYMPOSIUM SERIES No. 221, American Chemical Society: Washington, D.C., 1982, p. 21.
16. Okamoto, Y. *Polym. Eng. Sci.* 1983, *23*, 222.
17. Creed, K. E., Jr.; Neutron Devices Dept., General Electric Co., St. Petersburg, Fla.; INIS Atomindex 1980, 11(8), Abst. No. 517048, 1982.
18. Lancaster, J. K. *Plast. Polym.* 1973, *41*(*12*), 297.
19. Tran, C. Ph.D. Thesis, Virginia Polytechnic Institute and State University, Blacksburg, Va., 1984.
20. Riffle, J. S.; Yilgor, I.; Banthia, A. K.; Wilkes, G. L.; McGrath, J. E. *Am. Chem. Soc., Div. Org. Coat. Appl. Polym. Sci.* 1982, *46*, 397.
21. Yilgor, I.; Yilgor, E.; Banthia, A. K.; Wilkes, G. L.; McGrath, J. E. *Polym. Bull.* 1981, *4*, 323.
22. Yorkgitis, E. M. Ph.D. Thesis, Virginia Polytechnic Institute and State University, Blacksburg, Va., 1984.
23. Eiss, N. S. Jr., Milloy, S. C. In "Wear of Materials, 1983"; Ludems, K. C., Ed.; ASME: New York, 1983.
24. Manzione, L. T.; Gillham, J. K.; McPherson, C. A. *J. Appl. Polym. Sci.* 1981, *26*, 889, 907.
25. Knott, J. F. "Fundamentals of Fracture Mechanics"; John Wiley & Sons: New York, 1973; Chap. 5.
26. Kunz-Douglass, S.; Beaumont, P. W. R.; Ashby, M. F. *J. Mater. Sci.* 1980, *15*, 1109.
27. Gent, A. N.; Tobias, R. H. In "Elastomers and Rubber Elasticity"; Mark, J. E.; Lal, J., Eds.; ACS SYMPOSIUM SERIES No. 193, American Chemical Society: Washington, D.C., p. 367.
28. Gent, A. N.; Pulford, C. T. R. In "Developments in Polymer Fracture - 1"; Andrews, E. H., Ed.; Applied Science: London, 1979; pp. 164–66.
29. Bascom, W. D.; Moulton, R. J.; Rowe, E. H.; Siebert, A. R. *Org. Coat. Plast. Chem.* 1978, *38*, 164.
30. Bascom, W. D.; Cottington, R. L.; Jones, R. L.; Peyser, P. *J. Appl. Polym. Sci.* 1975, *19*, 2545.

RECEIVED for review November 18, 1983. ACCEPTED June 10, 1984.

11

An Alternative Liquid Rubber for Epoxy Resin Toughening

Improving Poly(*n*-butyl acrylate) Rubber–Epoxy Compatibility by Use of Acrylonitrile and Acrylic Acid Copolymers and Terpolymers

S. L. KIRSHENBAUM[1], S. GAZIT[2], and J. P. BELL

Institute of Materials Science, University of Connecticut, Storrs, CT 06268

*The solubility (compatibility) of poly(*n*-butyl acrylate) (PnBA) rubber in a bisphenol A–methylene dianiline epoxy resin has been increased by copolymerization and terpolymerization of acrylonitrile (AN) and acrylic acid (AA) with *n*-butyl acrylate (nBA) monomer. The general effect of rubber–epoxy compatibility on the precipitated rubber particle size and number has been investigated. Rubbers containing substantial amounts of AN or AA precipitate later from solution (during curing), and the particles are smaller in size. Some rubber remains dissolved in the matrix, and the total volume fraction of precipitated rubber decreases. The most compatible rubber (PnBA–19% AA), which did not precipitate from solution before gelation, resulted in a single-phase system. Tensile impact measurements (test speed ~ 8000 in./min) indicate that impact strengths for the rubber-modified epoxies are affected by rubber–epoxy compatibility, but impact strength improvements are limited by the brittleness of the high glass transition temperature matrix.*

In any rubber-modified system, the compatibility between the rubber and the epoxy before and during cure is an important chemical criterion for toughening of the matrix (*1*). Initially, the rubber should be completely soluble in the epoxy resin. This solubility depends upon the initial molecular weight of the rubber, the chemical composition and amount of the functional groups, and the solubility

[1] Current address: AVCO Specialty Materials Division, 2 Industrial Avenue, Lowell, MA 01851.

[2] Current address: Rogers Corporation, Lurie Research and Development Center, Rogers, CT 06263.

0065-2393/84/0208-0163/$06.00/0

parameters of the rubber and epoxy (*2, 3*). Without good initial rubber dissolution, the rubber does not contribute to the improvement of impact strength; in the case of poly(*n*-butyl acrylate) (PnBA) modified, methylene dianiline (MDA) cured epoxy systems (*4*), the rubber has a detrimental effect.

During epoxy cure, the compatibility of the rubber and epoxy should be such that the initially dissolved rubber precipitates as a discrete, randomly dispersed rubbery phase with a definite size and shape (*3, 5*). The degree of phase separation (i.e., volume fraction, domain size, and number of particles) is determined by the competing effects of the incompatibility between the rubber and the curing epoxy, the rate of particle nucleation and domain growth, and the quenching of morphological development by gelation (*6*). These effects in turn are controlled interdependently by the conditions of cure, the type and concentration of curing agent and resin, the type and concentration of rubber, and the rubber molecular weight (*3, 6–9*).

The toughening of epoxy resins has been attempted with liquid rubbers. The major area of interest is low molecular weight carboxyl- or amine-terminated butadiene–acrylonitrile (CTBN and ATBN, respectively) rubber modifiers. For the most part, use of these rubber modifiers in low glass transition temperature (T_g) epoxy systems has been successful in improving impact strength. Recently, a low molecular weight, carboxyl-terminated PnBA rubber has been synthesized and successfully incorporated as a toughening agent in a 2,5-dimethyl-2,5-hexanediamine-cured diglycidyl ether of bisphenol A (DGEBA) type system (*10, 11*). Unfortunately, in other epoxy–curing agent systems, the PnBA rubber exhibited poor compatibility, which limits its role as a toughness modifier.

PnBA rubber, which consists of a saturated backbone structure, is more stable chemically than the butadiene-based CTBN modifier. If the PnBA–epoxy compatibility could be improved, applications could become possible that could not be obtained with CTBN. For this reason, we wanted to develop a PnBA rubber type whose solubility parameter could be easily altered to fit different epoxy–curing agent systems. Our goal was to obtain a rubber that at 120 °C was completely soluble in the epoxy resin. This rubber would stay in solution with the addition of the curing agent and would not precipitate until the epoxy began to cure.

For this study we used acrylonitrile (AN) and acrylic acid (AA) separately, as copolymers, and in combination as terpolymers with the PnBA rubber. AN was considered as possible copolymer material because of its high intermolecular bonding forces and high solubility parameter. When copolymerized with butyl acrylate in small quantities (0–25% by weight), AN considerably alters the solubility parameter of the rubber. Addition of AA to the PnBA rubber not only changes the rubber solubility parameter, but also increases the number of reactive groups along the rubber chains. This effect allows greater rubber–epoxy interfacial bonding. Combining AN and AA as a terpolymer with the PnBA rubber allows us even greater control of rubber solubility and interfacial bonding.

Experimental

Copolymerization Method and Characterization of Rubber. The procedure for the copolymerization and terpolymerization of the PnBA–AN, PnBA–AA, and PnBA–AN–AA rubbers is similar to the procedure for the PnBA rubber (*4*). AN (0–42% by weight) and AA (0–20% by weight) were polymerized together and separately with butyl acrylate by using 4,4′-azobis(4-cyanovaleric acid) as the initiator and dithiodiglycolic acid as the chain transfer agent. The reaction proceeded by bulk polymerization under continuous mixing in a reaction vessel purged with nitrogen gas. The polymerization temperature was maintained at 72 °C because AN has a relatively low boiling point (77 °C). A reflux column above the reactor served as a condenser for possible evaporation. After polymerization, the rubbers were washed with an acetone–water separation technique (*12*) to remove all unreacted materials. Low polydispersity molecular weight fractions were obtained by slowly varying the acetone-to-water ratio during the wash cycle.

Number ($\overline{M}_n$) and weight ($\overline{M}_w$) average molecular weights of the rubbers were measured with gel permeation chromatography (GPC). The GPC was calibrated with low polydispersity PnBA–AN–AA rubber fractions. Actual $\overline{M}_n$ values for PnBA fractions were measured by using a Knauer vapor pressure osmometer. Rubber functionality was measured by KOH titration (*4*).

Preparation of Scanning Electron Microscope and High-Speed Tensile Test Samples. Scanning electron microscope (SEM) and high-speed tensile test samples were made by using a two-step cure cycle. Ten weight parts Epon 828 with a reaction catalyst, 1% tetra(*n*-butyl)ammonium iodide, was mixed with one weight part rubber and placed in an oven for 2 h at 120 °C. This cure step ensures the formation of the epoxy–rubber intermediate (*4, 12*) and promotes blending of epoxy and rubber prior to the final cure.

In the second step of the curing cycle, a stoichiometric amount of melted (approximately 100 °C) MDA was added to the epoxy–rubber mixture. After hand stirring, the solution was poured into small (1/8 × 1/8 × 1/2 in.) silicone rubber molds or cast between two glass plates (*4*) sprayed with Miller–Stephenson MS-136 Hot Mold Release agent and placed in the oven at 120 °C for 1 h. The temperature was raised to 150 °C, and the samples were postcured for 2 h before slowly cooling to room temperature. The cured SEM samples were fractured at room temperature in a vise and coated with a layer of gold, approximately 200 Å thick. All samples were observed under the SEM, and the number of rubber particles, their size and distribution, and rubber area were determined from the photographs.

High-Speed Tensile Impact Tester. The tensile impact tester used in this study is a modified version of the Plastechon Universal Tester (Plas-Tech Equipment Company) (*see* Reference 4 for a detailed description of instrumentation). Ram speeds were from 8000 to 10,000 in./min. Load and displacement were measured by a 1000-lb capacity load cell (Tyco) and a noncontacting displacement measuring system (Kaman Multi-Vit model DK-2300-10cu), respectively. Signals from the load and displacement transducers were stored on magnetic discs via an Explorer III digital oscilloscope (Nicolet Instrument Corporation). With the aid of a table-top computer interface (Hewlett Packard HP 9826), ultimate stress, ultimate strain, modulus, energy to break, and strain rate were calculated for all samples. To ensure statistical integrity, 6–12 samples from each group were broken. Average values, along with standard deviations, are reported.

Results and Discussion

Compatibility Study—PnBA–AN Rubber Series. Table I shows characterization data for the PnBA–AN series of rubber samples used in this solubility study. The solubility parameters, calculated by using a group contribution method (*13*, *14*), for PnBA, PAN, Epon 828, and MDA are 1.23, 12.78, 9.78, and 10.05 $(\text{cal/cm}^3)^{\frac{1}{2}}$, respectively. With the addition of 20% AN by weight to the PnBA, the calculated rubber solubility parameter is raised from 9.23 to 9.94 $(\text{cal/cm}^3)^{\frac{1}{2}}$. This value theoretically means the rubber should be more soluble in the epoxy. Addition of the curing agent (MDA) raises the calculated solubility parameter of the epoxy mixture, but not sufficiently to precipitate the rubber from solution before cure. Under experimental conditions, the calculated predictions hold true for the PnBA–20% AN–Epon 828 mixture. At 120 °C, the rubber readily dissolves and, with the addition of curing agent, initially remains in solution. During the first 5 min of cure the rubber begins to precipitate, and the system becomes cloudy. In the PnBA–AN (0–42% AN by weight) copolymer series, the rubber precipitated before gelation.

The average precipitated particle size and size distribution as a function of percent AN in the rubber are shown in Figure 1. With increasing AN composition, the average particle size decreases. The PnBA rubber alone (molecular weight 6500) is not completely compatible with the epoxy before or after the addition of the curing agent. The resultant modified epoxy, as shown in the SEM micrograph in Figure 2, exhibits a broad distribution of particle sizes. The larger particles are the result of rubber that did not dissolve, and the numerous small particles are rubber that partially dissolved and precipitated during cure.

The PnBA–5% AN and PnBA–10% AN rubber-modified sam-

Table I. Characterization Data for PnBA–AN Rubber Series

AN in Rubber (%)	$\overline{M}_n$ *(GPC)*	$\overline{M}_w/\overline{M}_n$	*Functionality (acid groups per chain, or eq/mol)*
0	6,500	7.00	1.60
5	6,690	6.03	1.12
10	7,672	7.12	1.66
20	5,614	6.54	1.23
26	5,715	8.54	1.14
42	14,412	4.56	1.76

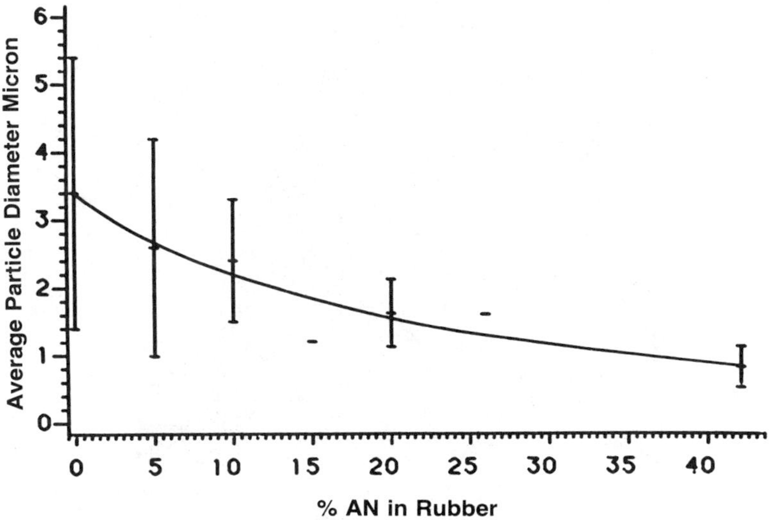

Figure 1. Rubber particle size vs. percent AN in the rubber (determined from SEM micrographs).

ples also exhibit a wide particle-size distribution. The 5 and 10% AN rubber samples initially dissolved in the epoxy but precipitated out of solution shortly after the addition of MDA. The 20–42% AN rubbers, all soluble in the Epon 828–MDA mixture, exhibit small particle size and narrow size distribution. An SEM micrograph of the 20% AN rubber-modified sample (Figure 3) is a good example of an initially compatible system.

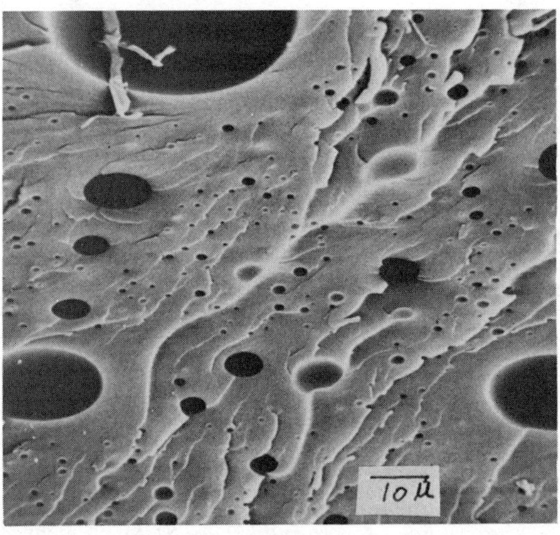

Figure 2. SEM micrograph of PnBA rubber-modified Epon 828 at stoichiometric equivalence of amine (MDA) to epoxy.

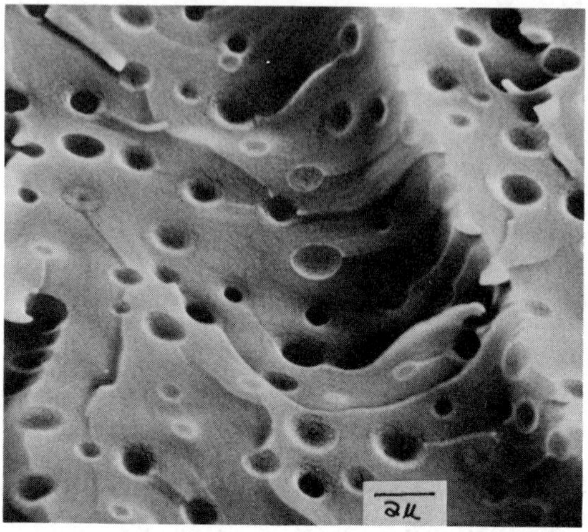

Figure 3. SEM micrograph of PnBA–20% AN rubber-modified Epon 828 at stoichiometric equivalence of amine (MDA) to epoxy.

In Figure 4, area fractions of the rubber phase are plotted on the basis of the total area of the fractured surface, related to increasing AN concentration in rubber. For all rubber-modified samples, the area fraction of the rubber exceeds the 10 weight parts per

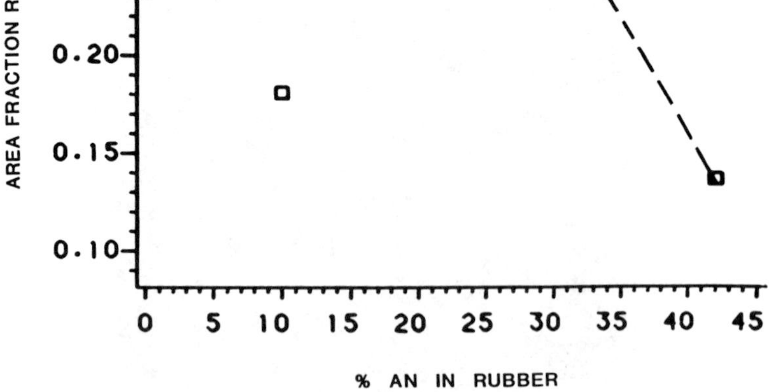

Figure 4. Area fraction of precipitated rubbery phase as a function of percent AN in the rubber (determined from SEM micrographs). Note: dotted line portion of curve is due to uncertainty because of the higher molecular weight for the PnBA–42% AN rubber.

Table II. SEM Data for PnBA–AN Rubber Series

AN in Rubber (%)	*Particles per cm² (×10⁶)*
0	3.0
10	3.1
20	14.8
26	15.1
42	15.1

hundred (pph) of rubber initially added to the epoxy. In agreement with previous reports (*3, 6, 15–18*), this observation may indicate the presence of epoxy inclusions that were trapped within the rubber particles. We must be careful, however, when making assumptions comparing the measured area fraction to the added weight or volume fraction. With the SEM, we cannot determine where in the rubber sphere the particle fractured, and thus cannot directly compare our measured rubber area to the actual volume fraction of precipitated rubber. The measured area fractions can only be compared relative to each other.

The number of particles per unit area of fracture surface is presented in Table II. Initially, as compatibility increases, the rubber particles become smaller in size and increase in number (0–20% AN in rubber). Copolymer rubbers, with AN concentrations above 20%, level off in the number of precipitated particles.

Compatibility Study—PnBA–AA Rubber Series. Characterization data for the PnBA–AA rubber series are shown in Table III. With the AA copolymer rubbers, particle size and distribution again decrease as the AA concentration increases (Figure 5). At a concentration of 19% AA, the rubber remains in solution throughout the cure and produces a single-phase system. The total number of precipitated particles (Table IV) related to AA concentration goes through a maximum at a concentration of 12% AA. The total area of rubber on the fracture surface, which is related to the size and

Table III. Characterization Data for PnBA–AA Rubber Series

AA in Rubber (%)	$\overline{M}_n$ *(GPC)*	$\overline{M}_w/\overline{M}_n$	*Functionality (acid groups per chain, or eq/mol)*
0	6,500	7.00	1.60
5	10,000	6.20	8.16
9	10,300	4.90	11.60
12	9,000	4.60	12.60
16	12,400	5.00	22.40
19	12,300	5.40	26.20

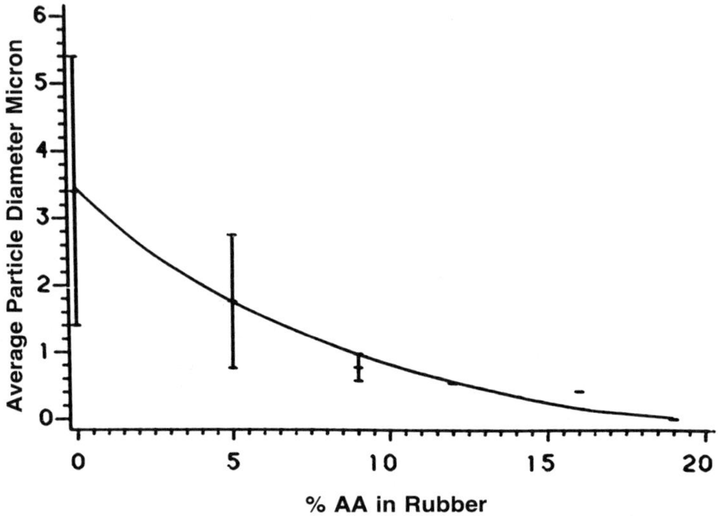

Figure 5. Rubber particle size vs. percent AA in the rubber (determined from SEM micrographs).

number of particles, decreases sharply with an increase in AA concentration (Figure 6).

In the PnBA–AA rubber composition studies, the amount of precipitated rubber is more sensitive to small AA concentration changes. An SEM micrograph of a fairly compatible PnBA–AA rubber-modified system (Figure 7) shows small and uniform size particles, which appear to be better bonded to the matrix. Compared to the PnBA–20% AN micrograph (Figure 3), the interface between rubber and epoxy is more difficult to distinguish. When compatibility is complete, the rubber remains in solution, and a single-phase system after cure is obtained (Figure 8).

Effect of Rubber Molecular Weight on Compatibility. Figure 9 shows a plot of particle size for four molecular weight fractions of PnBA–15% AN–2% AA terpolymers. As expected, the average par-

Table IV. SEM Data for PnBA–AA Rubber Series

AA in Rubber (%)	*Particles per cm² (×10⁶)*
0	3.00
5	5.98
9	17.49
12	42.80
16	8.83
19	0

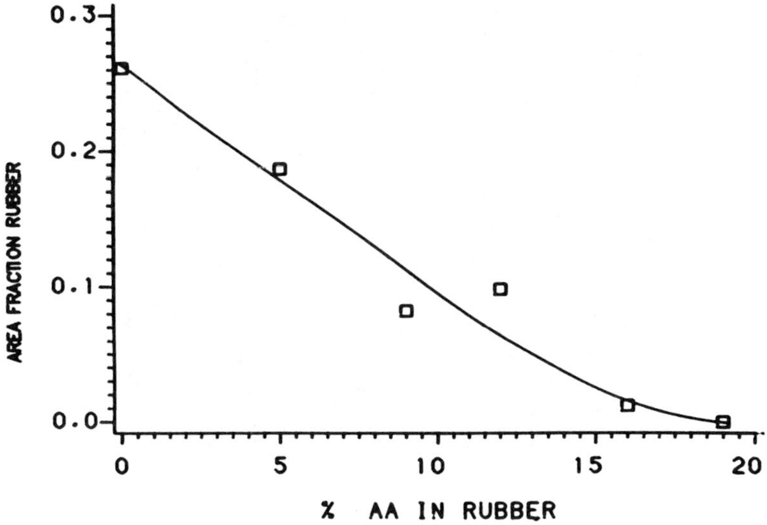

Figure 6. Area fraction of precipitated rubbery phase as a function of percent AA in the rubber (determined from SEM micrographs).

ticle size and total area of rubber (Figure 10) decrease with a decrease in molecular weight. The number of particles exhibits an inverse relationship with molecular weight; the number increases with a decrease in molecular weight (Table V). Interestingly, the effect of molecular weight on the particle size and area is greater at the lower molecular weights. Below average molecular weights of 4000, small decreases drastically affect the rubber precipitation characteristics.

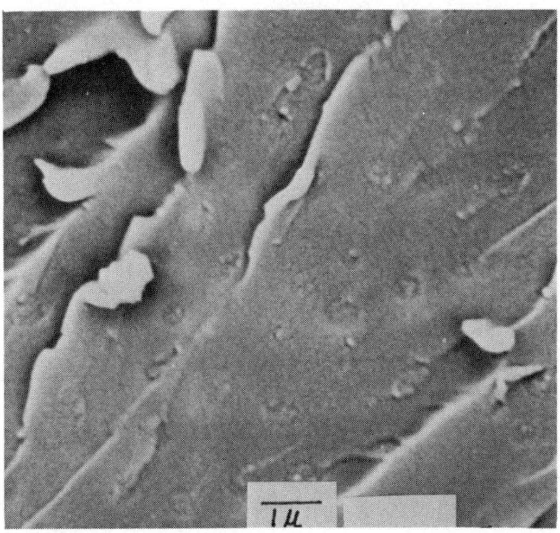

Figure 7. SEM micrograph of PnBA–12% AA rubber-modified Epon 828 at stoichiometric equivalence of amine (MDA) to epoxy.

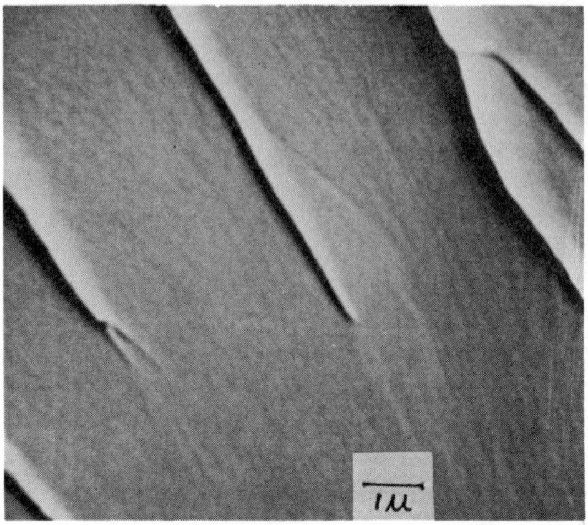

Figure 8. SEM micrograph of one-phase PnBA–19%AA rubber-modified Epon 828 system at stoichiometric equivalence of amine (MDA) to epoxy.

Effect of Rubber Compatibility on Tensile Impact Properties. By varying the composition of the rubber, we have shown that size and distribution, area, and number of the precipitated rubber particles are all related to the rubber–epoxy compatibility. Next, we examined the effect of compatibility on the impact strength of the rubber-modified epoxies. For this study we chose to break the samples by using a high-speed tensile impact tester. By breaking un-

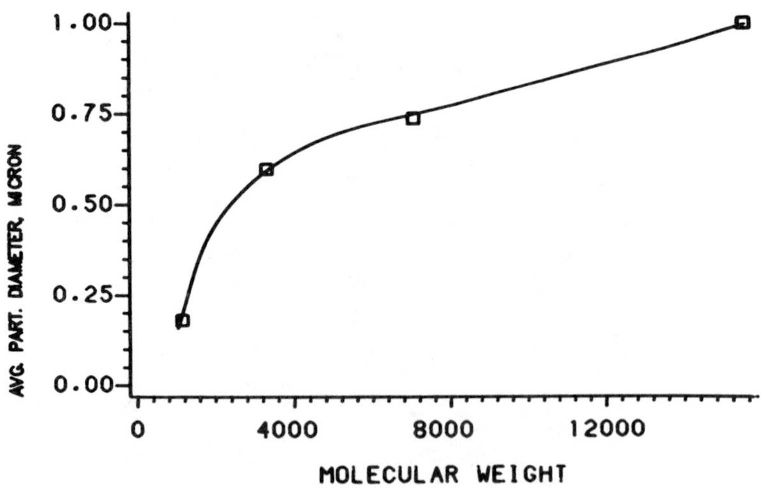

Figure 9. Rubber particle size as a function of rubber molecular weight for PnBA–15% AN–2% AA weight fractions (from SEM micrographs).

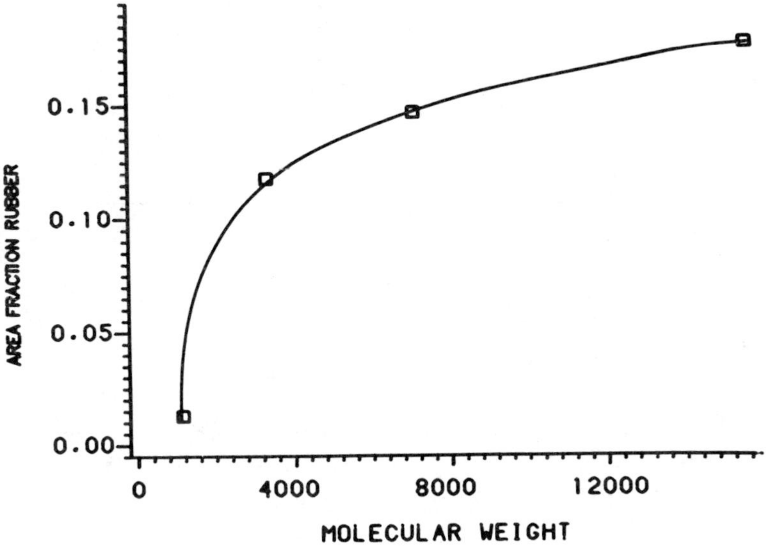

Figure 10. Area fraction of precipitated rubbery phase as a function of rubber molecular weight for PnBA–15% AN–2% AA weight fractions (from SEM micrographs).

notched ASTM dog-bone samples in tension at high speeds, we obtain a measure of the materials' combined strength to resist crack initiation and crack propagation. This test method gives a better indication of the total energy absorbed during impact failure than many other common measures.

The impact test results for the PnBA–AN series of rubber-modified epoxies are shown in Figure 11. The pure PnBA rubber-modified sample with an $\overline{M}_n$ of 6500 is not a very compatible system. It has a low impact strength, as illustrated by Figure 11, at 0% AN. Addition of AN to the rubber improves rubber–epoxy compatibility and impact strength. Although compatibility may still be good, at high AN concentrations (PnBA–42% AN) the rubber begins to lose its rubbery nature (the T_g of polyacrylonitrile is approximately 100

Table V. Characterization and SEM Data for PnBA–15% AN–2% AA Weight Fractions

AN–AA in Rubber (%)	$\overline{M}_n$ *(GPC)*	$\overline{M}_w/\overline{M}_n$	*Particles per cm² (×10⁶)*
15–2	15,416	3.28	22.6
15–2	7,041	2.10	34.1
15–2	3,298	2.02	38.5
15–2	1,140	1.53	50.1

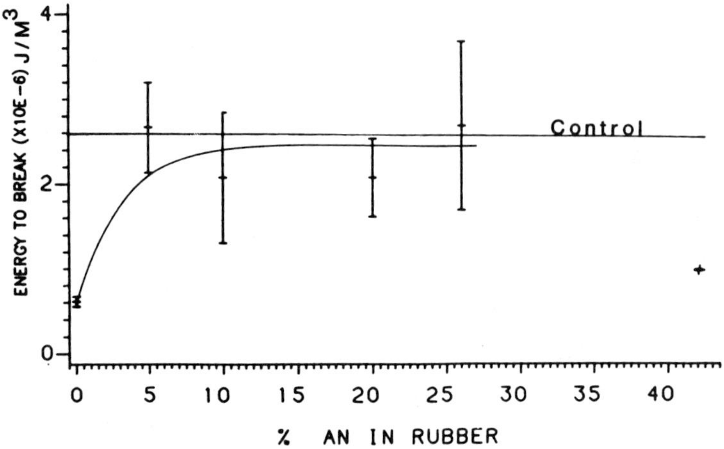

Figure 11. High-speed tensile impact test, energy to break vs. percent AN in the rubber; ram speed, 8000 in./min.

°C), and impact strength decreases. Interestingly, impact strengths of the rubber-modified samples can only approach or slightly exceed the impact strength of their equivalent control sample. This result indicates that poor rubber–epoxy compatibility will be detrimental to impact strength. Nevertheless, a compatible system does not ensure improved toughening. Possibly, the matrix viscoelasticity and ability to flow are the determining factors when trying to improve the impact strengths of brittle, high T_g epoxies. Results by Ochi and Bell (*12*) seem to further support this matrix viscoelasticity theory.

The high-speed tensile moduli for the PnBA–AN formulations (Figure 12) show no change in modulus with an increase in AN concentration in the rubber. Moduli for all rubber-modified compositions were slightly reduced compared to that of the control. The elongation-to-break curve (Figure 13) appears almost identical in shape to the impact-strength curve. Elongation of the rubber-modified samples is not too different from that of the controls, but ultimate stress values (Figure 14) for the rubber-modified samples are 10% less than the controls.

Impact strength, ultimate stress, elongation to break, and modulus results from our PnBA–AA series of rubber-modified epoxies (not shown) exhibit the same trend as that for the PnBA–AN–epoxy compositions. By varying the AA content of the rubber, we were able to vary the impact strength. However, the strength of the rubber-modified samples again only approached those of the controls at best.

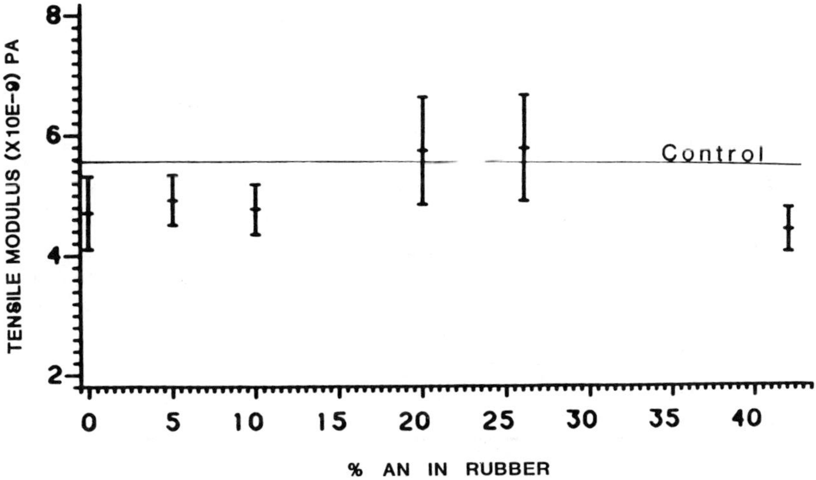

Figure 12. High-speed tensile impact test, modulus vs. percent AN in the rubber; ram speed, 8000 in./min.

Interestingly, the impact strength of samples with high-functionality, chemically well-bonded rubbers and single-phase systems (no precipitate) were virtually identical to the controls.

T_g values for the Epon 828–stoichiometric equivalent controls and a few of the rubber-modified systems are shown in Table VI. With increased rubber–epoxy compatibility, the T_g of the matrix decreases. This decrease is a result of an increased amount of rubber remaining in solution.

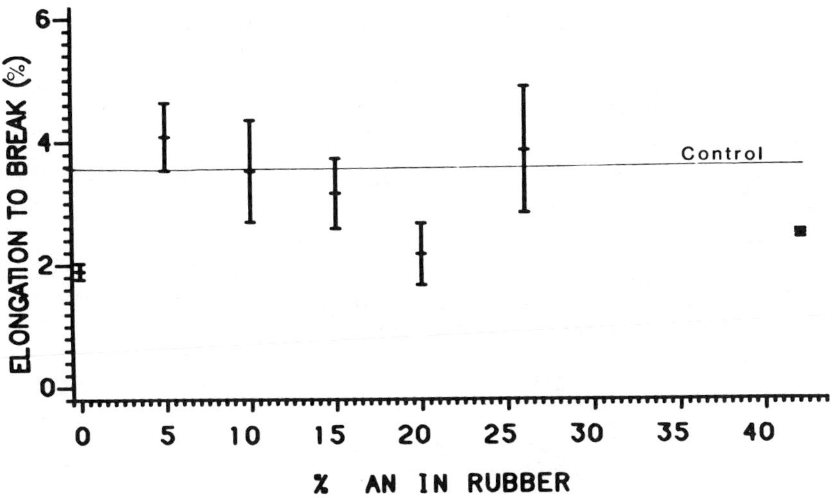

Figure 13. High-speed tensile impact test, elongation to break vs. percent AN in the rubber; ram speed, 8000 in./min.

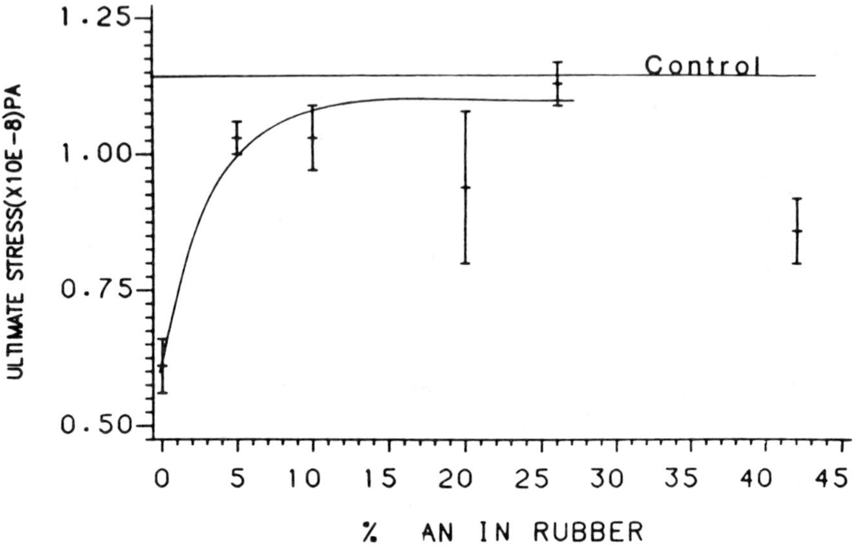

Figure 14. High-speed tensile impact test, ultimate stress vs. percent AN in the rubber; ram speed, 8000 in./min.

Conclusions

PnBA rubber with an $\overline{M}_n$ of 6500 is not completely soluble in Epon 828. Rubber–epoxy compositions containing PnBA have a wide particle-size distribution and exhibit poor impact strength. For toughening the rubber must dissolve completely in the resin before curing.

By copolymerizing AA or AN with *n*-butyl acrylate, the rubber solubility parameters can be altered. Epoxy–rubber compatibility, the precipitated rubber particle size, number of particles, and total precipitated rubber area can be controlled.

Using small concentrations of AN and AA with *n*-butyl acrylate as a terpolymer allows further control over the wide range of the rubber solubility parameters and functionality.

Varying rubber–epoxy compatibility will affect the ultimate impact strength of the compositions. However, with our high T_g matrix system, impact strength improvement was insignificant at best.

Table VI. DMA Glass Transition Temperatures for 10 pph Rubber-Modified Epon 828

PnBA–AN–AA in Rubber (%)	T_g (°C)
0–0–0	160
74–26–0	158
58–42–0	155
91–0–9	155
88–0–12	152

Literature Cited

1. Drake, R.; Siebert, A. *SAMPE Q.* July **1975.**
2. Sultan, J. N.; McGarry, F. J. *Polym. Eng. Sci.* **1973,** *13*, 29.
3. Bucknall, C. B.; Yoshii, T. *Br. Polym. J.* **1978,** *10*, 53.
4. Gazit, S. "Toughening of Epoxy Resin by Acrylic Elastomer", Ph.D. Thesis, Univ. of Connecticut, Sept. 1980.
5. Rowe, E. H.; Riew, C. K. *Plast. Eng.* March **1975.**
6. Manzione, L. T.; Gillham, J. K.; McPherson, C. A. *J. Appl. Polym. Sci.* **1981,** *26*, 889.
7. Siebert, A. R.; Riew, C. K. "The Chemistry of Rubber Toughened Epoxy Resin I"; presented at the 161st Natl. Meet. American Chemical Society, Washington, D.C., March 1971.
8. McGarry, F. J.; Willner, A. M. "Toughening of An Epoxy Resin by an Elastomeric Second Phase"; R68-8, MIT, March 1968.
9. Meeks, A. C. *Polymer* **1974,** *15*, 675.
10. Gazit, S.; Bell, J. P. In "Epoxy Resin Chemistry II"; Bauer, R. S., Ed.; ACS SYMPOSIUM SERIES No. 221; American Chemical Society: Washington, D.C., 1983; p. 69.
11. Ibid, p. 55.
12. Bell, J. P.; Ochi, M. *Polym. Mat. Sci. Eng. Prepr.* **1983,** *49*, 393.
13. Small, P. A. *J. Appl. Chem.* **1953,** *3*, 71.
14. Hoy, K. L. *J. Paint Technol.* **1970,** *42*, 76.
15. Manzione, L. T.; Gillham, J. K.; McPherson, C. A. *Org. Coat. Plast. Chem. Prepr.* **1979,** *41*, 371.
16. Rowe, E. H. Annu. Tech. Conf., Reinf. Plast./Compos. Div. SPI, Sec. 12-E, 26th 1971, 1.
17. Kunz-Douglass, S.; Beaumont, P. W. R.; Ashby, M. F. *J. Mater. Sci.* **1980,** *15*, 1109.
18. Manzione, L. T.; Gillham, J. K.; McPherson, C. A. *J. Appl. Poly. Sci.* **1981,** *26*, 907.

RECEIVED for review November 18, 1983. ACCEPTED March 14, 1984.

12

Morphology and Dynamic Mechanical Behavior of Rubber-Toughened Epoxy Resins

ALAN R. SIEBERT

Research and Development Center, The BFGoodrich Company, Brecksville, OH 44141

The use of either a carboxyl- or amine-terminated butadiene–acrylonitrile copolymer to modify an epoxy resin leads to a more crack-resistant (tougher) epoxy resin. This "toughness" is brought about by the formation of a second rubbery phase [with 5–15 phr (parts per hundred parts of resin) of the reactive liquid polymer] during cure and occurs with a variety of cure agents and epoxy resins. Generally, the carboxyl-terminated polymer is prereacted with the epoxy resin prior to cure. The phase separation is determined by a number of variables including initial compatibility, catalyst, and cure temperature. Both volume fraction and particle-size distribution of the dispersed phase are important in determining the optimum degree of toughness in a given system. Above 15 phr of liquid polymer a phase inversion occurs. A variety of techniques have been used to measure the presence of the second phase, including transmission and scanning electron microscopy, light microscopy, light scattering, cloud point measurement, X-ray scattering, and dynamic mechanical behavior.

INCORPORATION OF LOW LEVELS of a liquid carboxyl-terminated butadiene–acrylonitrile copolymer (CTBN) to a normally brittle epoxy resin significantly improves the crack resistance and impact strength without a reduction in other thermal and mechanical properties (*1*). This enhancement in crack resistance and impact strength (increased toughness) is brought about by the separation during cure of a predominately rubbery second phase. The size of this second phase is usually between 0.1 and 5 μm.

0065-2393/84/0208-0179/$06.00/0

These modified thermoset resins have found wide application in structural film adhesives for metal–metal bonding in aircraft, in paste adhesives for automotive and industrial application, in electronic encapsulation, in epoxy solvent and powder coatings, and in advanced aircraft and aerospace composites.

In most epoxy applications, the final properties are strongly dependent on the morphology generated during cure of these systems. The morphology is determined by a large number of variables, such as the compatibility of the rubber–resin system prior to cure, the cure agent, the time and temperature of cure, and prereaction of CTBN with the epoxy prior to cure. In some cases, optimum toughness is provided with a bimodal distribution of particle sizes. In the mid 1970s, liquid amine-terminated butadiene–acrylonitrile copolymers (ATBN) were produced. These copolymers provided another way to introduce rubber modification into a cured epoxy network (*2*). This chapter presents a systematic review of the work to date on these two-phase systems in neat resin systems, adhesives, and composites. This review is somewhat complicated by several factors. The amount of literature on these two-phase systems has increased in number and scope. For example, the work includes different types of cure agents (catalytic, primary amine, latent systems, and anhydrides) as well as a variety within each type. In admixed systems, the type of cure agent is important because of the reactivity and selectivity of the carboxyl–epoxy reaction. Many authors have resolved this problem by prereacting the CTBN and the epoxy resin in an alkylhydroxy esterification reaction. These acid adducts can now be cured with any cure agent because they only contain epoxy groups. A discussion of these prereactions and the catalysts used is given by Drake and Siebert (*3*). Different techniques are used to examine the morphology of these two-phase systems. Scanning (SEM) and transmission electron microscopy (TEM) have been used. In addition, some authors have measured morphology in the areas of fast-crack growth, and others have concentrated primarily in the area of crack initiation where considerable stress whitening occurs. An attempt will be made to highlight these differences.

Early Work

The pivotal work of McGarry and coworkers (*4–6*) demonstrated the necessity of developing a discrete, well-dispersed, rubbery second phase to provide enhanced crack resistance as measured by fracture-surface energy (FSE) measurements. They used a liquid diglycidyl ether of bisphenol A (DGEBA) epoxy resin, CTBN, and an amine catalyst. The authors then determined compositional and morphological effects related to FSE improvement of glassy, cross-linked resins. They also investigated the effect of CTBN molecular weight, relative solubility of rubber and resin, and rubber-phase particle size. Figure 1 is a TEM micrograph showing the presence of the second phase.

Rowe et al. (7) demonstrated a maximum in fracture energy in a liquid DGEBA epoxy resin–piperidine system when the bound acrylonitrile (AN) content of the CTBN was between 12 and 18%. They also showed a general decrease in average particle size from 3 μm at 12% bound AN to 0.2 μm at 25% bound AN.

Siebert and Riew (8) first described the chemistry of rubber-particle formation in an admixed model involving CTBN, a DGEBA liquid epoxy resin, and a piperidine catalyst. They proposed that the composition of the rubber particles in the dispersed phase critically depended upon the in situ formation of the epoxy–CTBN–epoxy adduct, which is then further chain-extended and cross-linked with additional epoxy resin. This progression provides a chemical bond between the dispersed rubber phase and the matrix resin and occurs with piperidine, a selective catalyst. Most other cure agents, however, favor either the epoxy–epoxy reaction or an epoxy–amine reaction, and the carboxyl–epoxy reaction is suppressed. Siebert and Riew showed that a nonreactive butadiene–acrylonitrile liquid rubber does show a second phase on cure even with a selective catalyst. However, the fracture energy did not improve for this system. This result demonstrated the need for chemical bonding between the dispersed phase and the matrix.

Riew and Smith (9) developed a new OsO_4 staining technique for optical or electron microscopy that aided in demonstrating the rubbery nature of the dispersed second phase.

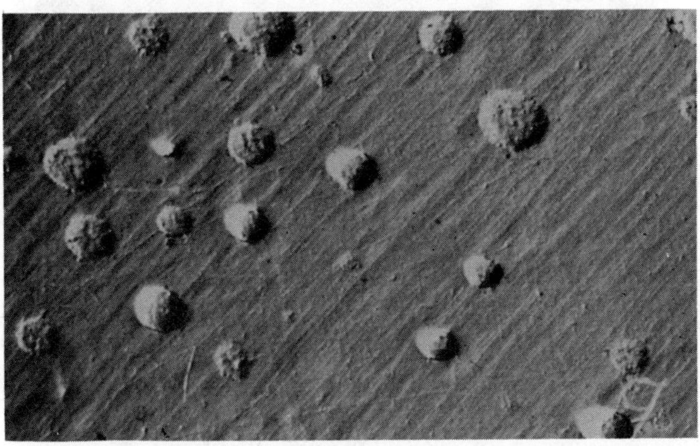

Figure 1. Electron micrograph showing liquid HYCAR rubber precipitated in the epoxy resin. (Magnification ×4800.)

Bimodal Particle Systems

Sultan and McGarry (*10*) demonstrated that different particle sizes of the dispersed phase in the epoxy matrix promote different deformation mechanisms. Electron micrographs of the fracture surfaces of systems with only large particles (0.5–5 μm) show microcavitation around the particle. According to the authors, this result is similar to the crazing observed in thermoplastic systems. However, no research has provided definitive evidence to date of pure crazing in a cross-linked epoxy resin. These authors also show that systems with only smaller particles (0.01–0.03 μm) exhibit shear banding.

Riew et al. (*11*) described an admixed model system based on CTBN–bisphenol A–DGEBA liquid epoxy resin–piperidine in which toughness synergism appears through the inclusion of a diphenol. This inclusion resulted in a bimodal distribution of rubber particles, and microscopy showed multiple failure sites (a term used by the authors to show that much more surface is created during failure of these systems). Figures 2 and 3 are micrographs showing the bimodal distribution (OsO_4-stained TEM) and multiple failure sites (TEM taken from the stress-whitened area), respectively. The physical properties of this system are given in Table I. They also define a set of chemical, morphological, and thermal-mechanical criteria for the toughened nitrile–epoxy systems (*see* Box).

Criterion C-3 was not directly demonstrated by the authors with FSE measurements. Subsequent work by Bascom and Cottington (*12*) and Hunston (*13*) shows the temperature and rate of test dependence of FSE for rubber-modified adhesives. Work by Bascom et al.

Figure 2. Negative of an electron micrograph of an OsO_4-stained microsection of a bisphenol A modified, CTBN-toughened epoxy resin. (Magnification × 29,700.) (Reproduced from Ref. 11. Copyright 1976, American Chemical Society.)

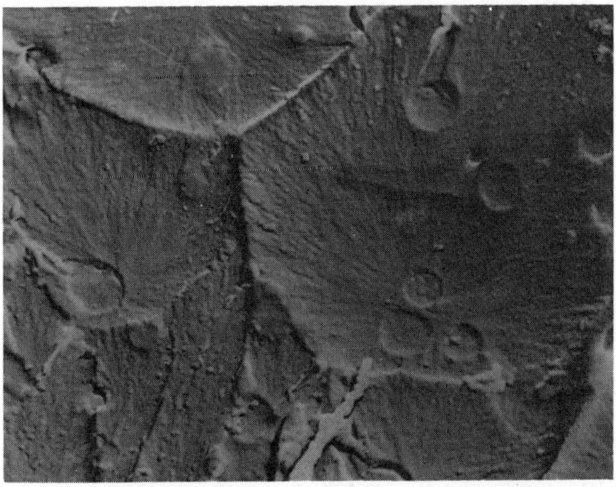

Figure 3. Electron micrograph of a bisphenol A modified, CTBN-toughened epoxy resin. (Magnification × 7000.) (Reproduced from Ref. 11. Copyright 1976, American Chemical Society.)

(*14*) shows the dependence of FSE on rate of test for rubber-modified bulk systems (*see* Figure 4).

Rowe and Riew (*15*) examined the fracture surface of a tensile specimen after stressing. The area examined showed considerable stress whitening and necking. Their micrograph (Figure 5) shows microvoid development around the rubbery particles. This microvoid area is primarily responsible for the stress whitening, and the microvoid regions disappear on heating above the heat distortion temperature (HDT) of the sample.

Table I. Thermal-Mechanical Properties of Bisphenol A Modified CTBN–Epoxy System

Property	*Unmodified Epoxy Resin System (Control)*	*Bisphenol A Modified CTBN–Epoxy System*
Tensile strength[a], MPa	65.5	64.1
Elongation at break[a], %	4.8	9.0
Modulus, GPa	2.8	2.7
Tensile strength[b], MPa	73.1	95.8
Elongation at break[b], %	7.3	11.3
Heat distortion temp, °C	83	83
Fracture energy, kJ/m^2	0.18	5.3–8.8
Gardner impact[c], J	6	23–34
Izod impact, J/m of notch	0.68	3.5

[a] 0.12 mm/s
[b] 6.35 m/s
[c] Nominal 0.635 cm thick samples
SOURCE: Reproduced from Ref. 11.

Toughening Criteria for Rubber-Toughened Thermosets

A. Chemical Criteria
  1. Two reactive end groups
  2. Chemical bonds between rubbery particles and the epoxy matrix
  3. Compatibility

B. Morphological Criteria
  1. Second particulate rubbery phase
  2. Definite size and shape
  3. Rubbery particles must be dispersed

C. Thermal-Mechanical Criteria
  1. Improved crack and impact resistance
  2. Retention of thermal-mechanical properties
  3. Insensitive to the rate of loading (or temperature)

SOURCE: Reproduced with permission from Ref. 3.

Yee and Pearson (*16*) made volumetric measurements on rubber-modified specimens similar in composition to those used by Rowe and Riew (*15*). These samples were subjected to constant crosshead rate uniaxial tensile tests. Optical microscopy revealed that the rubber particles had cavitated but not fractured and that shear bands connected the cavitated particles. They propose "that the toughening effect is caused by a sequence of mechanisms: sharp cracks are blunted by voiding; then the voids enhance shear localization between themselves. Thus, the rubber particles serve as void nucleants, but the toughness is derived largely from the shear plastic zone."

Bascom et al. (*14, 17*) demonstrated that the addition of a high molecular weight rubber (HYCAR 1472) along with CTBN does en-

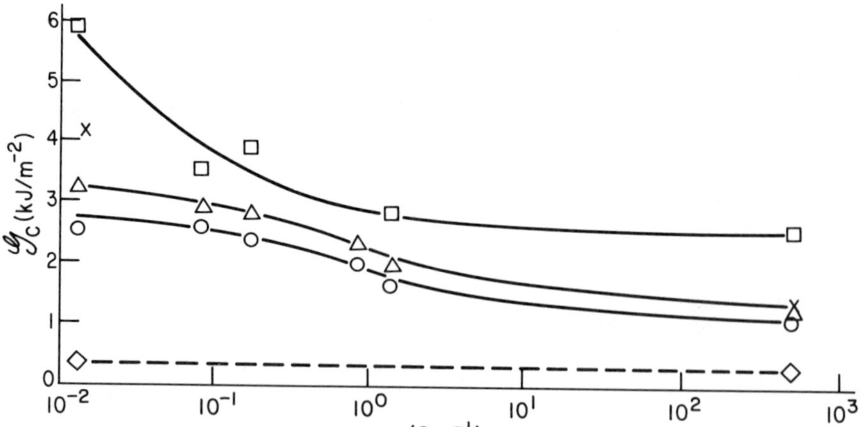

Figure 4. Fracture energy vs. strain rate for epoxy polymers. Key: ◇, Sample no. 205; ○, 206; ×, 207; △, 210; and □, 185. (Reproduced with permission from Ref. 14. Copyright 1981, Journal of Material Science.*)*

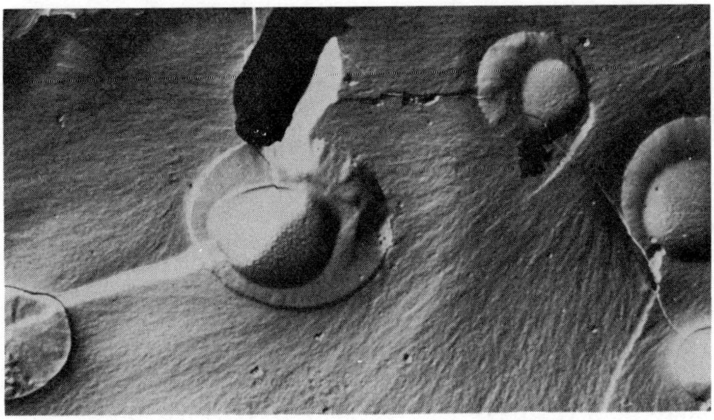

Figure 5. Micrograph of fracture surface. Microvoids are seen in association with rubber particles. (Magnification ×2580.) (Reproduced with permission from Ref. 12. Copyright 1976, Journal of Adhesion.*).*

hance the base fracture energy over that attained with the CTBN alone. Table II shows the compositions and fracture energies for these systems. These samples were prepared by the Hexcel Corporation; Hexcel F-185 is a commercial epoxy resin formulation. A latent catalyst (modified urea-accelerated dicyanodiamide) was used. In these systems both the CTBN and 1472 are prereacted with the epoxy resin to produce an acid-free adduct.

The enhanced toughness was observed over a wide range of strain rates (including impact testing), and the effect of adding the solid rubber was greatest at the lowest strain rate. This effect is demonstrated in Figure 4.

Phase Inversion

The work of Burhans and Soldatos (*18, 19*) showed enhanced toughness with CTBN in a cycloaliphatic epoxy resin cure with hexahydrophthalic anhydride. They also reported that, as the amount of CTBN was increased in the CTBN–ERL-4221–HHPA system, a phase inversion occurred at about 50 phr of CTBN. Their (TEM) micrographs showed that the CTBN–epoxy adduct became the continuous phase with domains of epoxy resin.

Table II. Fracture Energy ($\mathcal{G}_{Ic}$) of Epoxy Polymers

Sample No.	*Liquid Rubber (CTBN)*	*Solid Rubber (1472)*	$\mathcal{G}_{Ic}$ *(kJ/m²)*
205	none	none	0.27 ± 0.04
206	8.1	none	2.5 ± 0.3
207	8.1	1.0	4.1 ± 0.2
210	none	8.1	3.2 ± 0.2
185	8.1	5.4	5.1 ± 0.9

SOURCE: Reproduced with permission from Ref. 14.

Kalfoglou and Williams (*20*) in their dynamic mechanical analysis (DMA) (Rheovibron) of epoxy–rubber polyblends observed a phase reversal at a volume fraction of 0.5, in which an intermediate compound is formed.

Siebert (*21*) showed a similar phenomenon in CTBN–DGEBA liquid epoxy resin–amine catalyst systems. The extension of this phenomenon was a CTBN–DGEBA–amine castable system in which CTBN–DGEBA was the continuous phase, and the epoxy resin was the dispersed phase; a reinforced rubbery product was produced.

Dušek (*22*) also observed the phase separation in a CTBN–DGEBA–amine catalyst system. He did not, however, find the phase inversion in the CTBN–DGEBA–HHPA[1]–BDMA[2] system. This difference may be a result of the use of a DGEBA resin over a cycloaliphatic resin or because prior to use the CTBN was prereacted with the epoxy resin by using a chromite complex diisopropylsalicylic acid (Cr-DIPS).

Other Methods Used to Measure Presence of Second Phase

Kalfoglou and Williams (*20*) found that mechanical relaxations, for the most part, indicate degree of interaction–mixing state of these two-phase systems.

Bucknall and Yoshii (*23*) also measured DMA properties on rubber-modified epoxies with a Rheovibron. They used a variety of cure agents as well as liquid and solid epoxy resins with CTBN in admixed systems. They found that rubber-toughened systems exhibit a loss peak at the glass transition of the rubber phase; this peak coincides with the β transition of the epoxy resin. They further found that the magnitude of the loss peak is related to the volume fraction of dispersed rubber. They also showed that fracture resistance increased linearly with phase volume of separated rubber.

Manzione and Gillham (*24, 25*) used torsional braid analysis (TBA) and microscopy to measure phase separation in rubber-modified epoxies (admixed epoxies cured with piperidine, and prereacted epoxies cured with dicyanodiamide–monuron[3]). They determined factors affecting volume fraction, domain size, and the number of particles of phase-separated rubber. They showed that the processes involved in second-phase formation are phase separation, gelation, and vitrification.

These researchers also found that the rubber damping peak always occurs at or below the T_g of the unreacted CTBN. This phenomenon occurs even though the dispersed phase contains epoxy resin reacted with the CTBN. They suggest that differences in thermal shrinkage stresses may cause the domains to be constrained

[1] Hexahydrophthalic anhydride.
[2] Benzyldimethylamine.
[3] *N'*-(4-Chlorophenyl)-*N*,*N*-dimethylurea.

by the glassy epoxy matrix. The T_g of the rubber in the domain is depressed over that for the pure CTBN.

Numata and Kinjo (*26*) measured the viscoelastic behavior of CTBN-modified isocyanurate–oxazolidone resins. They observed an α_s-dispersion relaxation for the rubber domains at −60 °C, which is lower than the temperature (45 °C) obtained for unreacted CTBN. They also attribute this difference to thermal shrinkage stresses.

Daly et al. (*27*) used a variety of analytic methods on a series of prereacted rubber-modified epoxies. They found that the higher temperature transition appears to be associated with specific interactions involving the interface between the matrix resin and phase-separated rubber.

In their work on rubber-modified epoxies, Burhans and Soldatos (*18, 19*) used percent haze to measure the formation of the second phase. Boisserie and Marchessault (*28*) used solid-state light scattering from a rubber-modified DGEBA epoxy to measure the heterogeneity of these systems as a function of rubber content. They also noted a phase inversion above 20%.

Wang and Zupko (*29*) and Gillham and Chan (*30*) used light transmission (cloud point) measurements to determine the phase separation behavior of these two-phase systems. Gillham and Chan (*30*) found a significantly higher volume fraction of phase-separated rubber for an ATBN-modified compared to a CTBN-modified epoxy.

Volume Fracture and Composition of the Second Phase

The work of Sultan and McGarry (*10*) and Burhans and Soldatos (*18, 19*) showed the heterogeneity of the dispersed second phase. Bucknall and Yoshii (*23*) also showed this heterogeneity and also measured volume fraction of second phase as a function of hardener content and rubber content. They suggest that volume fraction may be the most important factor to property improvements.

Manzione et al. (*24, 25*) measured volume fraction of phase-separated systems with CTBNs of differing AN content. They correlated phase volume with the intensity of the rubber damping peak from TBA.

For an amine-cured rubber-modified epoxy, Sayre et al. (*29*) determined by using X-ray analysis of the dispersed rubber particle–matrix interfacial region an interface width less than 0.05 μm. They also found that the epoxy content of the dispersed phase is less than 17%. This concentration is less than that predicted if all the epoxy monomer units attached by prereaction to the rubber molecules were present in the dispersed phase. They also determined that the volume fraction of the dispersed phase was essentially that of the rubber plus the epoxy monomer units bonded directly to the rubber precipitates. Kunz et al. (*32*) attributed the diffuse appearance of the ATBN-modified epoxy interface to an irregularly shaped particle, whereas the CTBN-modified epoxy shows a more spherical-shaped particle. However, both systems produce the same toughness. In addition, Kunz et al. suggested "that the distribution of rubber particle sizes is more critical than the overall volume fraction to the toughness of the rubber-modified epoxies." Wang and Zupko (*29*) gave evidence for the completion of phase separation at the point of gelation, but showed that compositional changes continued to take place. Gillham and Chan (*30*) also stated that phase separation is complete before gelation.

Toughness of Adhesive and Composite Systems

Clarke (*33*) first demonstrated enhanced peel strength in a CTBN–epoxy prereacted system cured with a dicyanodiamide-hardened structural adhesive system. Micrographs showed a dispersed second phase with no reduction in T_g of the epoxy. Urethane-modifed epoxies showed no second phase, although they showed enhanced peel strength.

Bascom et al. (*34*), Hunston et al. (*35*), and Kinloch and Shaw (*36*) evaluated a model rubber-modified epoxy system in adhesive joints by using the methods of Mostovoy (*37*). They used an admixed system of CTBN, epoxy resin, and piperidine as catalyst. They measured adhesive FSE as a function of bond thickness, bond width, temperature rate of test, and mixed-mode fracture. In addition, they have evaluated bulk properties as a function of elastomer content. In all cases, they have found morphologies (as measured by SEM) in the adhesive joint comparable to those of the bulk systems and consequently found corresponding increases in FSE in the adhesives. The adhesive FSE shows a maximum when the bond thickness was approximately equal to the plastic-zone size. Bascom et al. (*34*) demonstrated a qualitative relationship between the fracture energy of a given system and the deformation zone size (Table III).

In work on rubber-modified epoxies, Kinloch and Hunston (*38, 39*) proposed a toughening mechanism in which a greater extent of energy-dissipating deformation occurs in the material in the vicinity

Table III. Bulk Fracture and Deformation Zone Size for CTBN-Modified Epoxy Resin

CTBN Concentration (wt%)	*Fracture Energy, $\mathcal{G}_{Ic}$* (J/m²)	*Deformation Zone Size, 2r$_c$ (cm × 10^{-3})*
0	121	0.8
4.5	1050	12.4
10	2720	21
15	3430	33
20	3590	70

SOURCE: Reproduced with permission from Ref. 34.

of the crack tip. The deformation processes are localized cavitation in the rubber or at the particle–matrix interface and plastic shear yielding (K_{Ic}) in the epoxy matrix. The predominate source of energy dissipation is the shear yielding. They suggest that the K_{Ic} is related to the square root of the crack-tip radius.

Hunston, Bascom, and Bitner (*40*) were the first to suggest important similarities among the failure behavior of neat polymers, adhesive bonds, and fiber composites.

For epoxy and polyester systems, McGarry and coworkers (*41*) determined that, except for a rubber-modified epoxy, the interlaminar fracture-surface work of fiberglass laminates was significantly greater than that of the unreinforced matrix. This observation was true even though the neat rubber-modified epoxy exhibited a ninefold increase in interlaminar toughness.

Scott and Phillips (*42*) evaluated carbon-fiber composites with rubber-toughened matrices. They used a model system of CTBN–epoxy resin–piperidine and found similar morphologies in the bulk and graphite composite but significantly reduced fracture energies in the composite compared to the bulk resin.

These studies (*41, 42*) in toughening composites may have been limited by the rather low range of toughening agents available at the time of their work. Work by Bascom et al. (*43*) showed a significant increase in interlaminar fracture energy for rubber-modified epoxy resins in both glass and graphite woven composites. In these studies, the neat rubber-modified epoxy showed about a 20-fold increase in FSE over the control epoxy. The matrix resin was Hexcel F-185, which contains both solid and liquid rubbers (*see* Table II) prereacted with the epoxy resin and cured with a modified urea-accelerated dicyanodiamide. The FSE data for these neat and composite materials are given in Table IV. About a fourfold increase in FSE was

Table IV. Effect of Elastomeric Modifiers on Resin and Composite Toughness

Material	*Fiber Volume (%)*	*Modulus (GPa)*	*Fracture Energy* ($\mathcal{G}_{Ic}$ (kJ/m^2))
		Tensile	Neat resin
205 Resin	0	2.8	0.27
F-185 Resin	0	2.2	5.1
		Flexural	Interlaminar
205/Glass cloth	59.9	31.6	1.0
F-185/Glass cloth	59.7	24.0	4.4
205/Graphite cloth	61.0	54.1	0.60
F-185/Graphite cloth	57.9	42.1	4.6
			Cross fiber
205/Glass cloth	59.9	31.6	11.1
F-185/Glass cloth	59.7	24.0	8.73

SOURCE: Reproduced with permission from Ref. 43.

observed for a glass composite, and about an eightfold increase was seen for a graphite composite.

They further found that the deformation processes identified in neat resin fracture also occur in composite interlaminar fracture. Hong et al. (*44*) reported on neat and unidirectional graphite composites with the same Hexcel F-185 resin matrix. They found differences in morphology between the neat resin and the composite by using small X-ray scattering. These differences indicated that phase separation may be changed by the presence of the unidirectional fibers.

Conclusions

The phase separation in rubber-modified epoxies has been studied by a number of researchers with a variety of cure agents and epoxy resins. Generally, the separation of the predominately rubbery phase is necessary to the generation of toughness properties in these systems (15 phr or less). Above about 20 phr of rubber, phase inversion takes place, and the dispersed phase becomes a cross-linked epoxy resin. The phase separation below 20 phr of rubber is determined by a number of variables including initial compatibility of the rubber and epoxy resin and proper choice of catalyst and cure temperature. Also, in most cases, the CTBN and epoxy resin must be prereacted to provide an acid-free system for subsequent cure. The type of cure agent and/or rubber–epoxy system used may determine which condition will produce the optimum toughness. The volume fraction of separated rubber, particle size, and particle-size distribution are of prime importance to optimizing toughness properties. The question of the relative importance of volume fraction and/or particle-size distribution of the precipitated phase to the toughness is still a matter of debate. Resolving this question is a problem because changing one parameter in these systems without affecting other parameters is difficult. Thus, definitive experiments are scarce. One can argue that to maximize the T_g of the matrix resin, all of the rubber must be precipitated in the second phase; thus, high volume fraction is of prime importance. More multidiscipline work is needed to resolve this complex problem in chemistry, physics, and kinetics. Finally, commercial products have been developed in which both particle-size distribution and volume fraction have been optimized, but we have no literature references to this work because it is all proprietary. The toughness properties in bulk can be carried through to adhesives and composites; however, the thickness of the resin phase becomes critical as the dimensions of the constraining layers approach those of the plastic-zone size.

Literature Cited

1. Drake, R. S.; McCarthy, W. J. *Rubber World* Oct. **1968**, 159.
2. Riew, C. K. *Rubber Chem. Technol.* **1981**, *54*, 374.
3. Drake, R. S.; Siebert, A. R. *SAMPE Q.* **1975**, *6*, No. 4, 7.

4. McGarry, F. J.; Willner, A. M. Research Report, School of Engineering, MIT, Cambridge, MA, 1966.
5. *Ibid.*, R68-8, 1968.
6. *Ibid.*, R69-35, 1969.
7. Rowe, E. H.; Siebert, A. R.; Drake, R. S. *Mod. Plast.* **1970,** *47*, 110.
8. Siebert, A. R.; Riew, C. K. *Org. Coat. Plast. Chem.* **1971,** *31*, No. 1, 552.
9. Riew, C. K.; Smith, R. W. *Polym. Sci., Part A-1* **1971,** *9*, 2737.
10. Sultan, J. N.; McGarry, F. J. *Polym. Eng. Sci.* **1973,** *13*, 29.
11. Riew, C. K.; Rowe, E. H.; Siebert, A. R. In "Toughness and Brittleness of Plastics"; Deanin, R. D.; Crugnola, A. M., Eds.; ACS ADVANCES IN CHEMISTRY SERIES No. 154; American Chemical Society: Washington, D.C., 1976; pp. 326–43.
12. Bascom, W. D.; Cottington, R. L. *J. Adhesion* **1976,** *7*, 333.
13. Hunston, D. L.; Rushford, J. L.; Bitner, J. L. *Org. Coat. Plast. Chem.* **1979,** *41*, 379.
14. Bascom, W. D.; Ting, R. Y.; Moulton, R. J.; Riew, C. K.; Siebert, A. R. *J. Mater. Sci.* **1981,** *16*, 2657.
15. Rowe, E. H.; Riew, C. K. *Plast. Eng.* March **1975,** *45*.
16. Yee, A. F.; Pearson, R. A. presented at the American Chemical Society Rubber Division meeting, Philadelphia, PA, Spring 1982.
17. Bascom, W. D.; Moulton, R. J.; Rowe, E. H.; Siebert, A. R. *Org. Coat. Plast. Chem.* **1978,** *39*, 164.
18. Burhans, A. S.; Soldatos, A. C. presented at the 25th Annu. Techn. Conf. RP/Compos. Div. SPI, 3-C, 1970.
19. Burhans, A. S.; Soldatos, A. C. In "Multicomponent Polymer Systems"; Platzer, N. A. J., Ed.; ACS ADVANCES IN CHEMISTRY SERIES No. 99; American Chemical Society: Washington, D.C., 1971; pp. 531–46.
20. Kalfoglou, N. K.; Williams, H. L. *J. Appl. Polym. Sci.* **1973,** *17*, 1377.
21. Siebert, A. R. *J. Elastomers Plast.* **1976,** *8*, 177.
22. Dušek, K. et al. *Chem. Prum.* **1980,** *130*, 591.
23. Bucknall, C. B.; Yoshii, T., *Br. Polym. J.* **1978,** *10*, 3.
24. Manzione, L. T.; Gillham, J. L.; McPherson, C. A. *J. Appl. Polym. Sci.* **1981,** *26*, 889.
25. *Ibid.*, 907.
26. Numata, S.; Kinjo, N. *Polym. J.* **1982,** *14*, 671.
27. Daly, J.; Pethrich, R. A.; Fuller, P.; Cunliffe, A. V.; Datta, P. K. *Polymer* **1981,** *22*, 32.
28. Boisserie, C.; Marchessault, R. H. *J. Polym. Sci.* **1977,** *15*, 1211.
29. Wang, T. T.; Zupko, H. M. *J. Appl. Polym. Sci.* **1981,** *26*, 2391.
30. Gillham, J. K.; Chan, L. C. *Org. Coat. Plast. Chem.* **1983,** *48*, 571.
31. Sayre, J. A.; Assink, R. A.; Lagasse, R. R. *Polymer* **1981,** *22*, 87.
32. Kunz, S. C.; Sayre, J. A.; Assink, R. A. *Polymer* **1983,** *23*, 1897.
33. Clarke, J. A. *Org. Coat. Plast.* **1971,** *31*, No. 2, 105.
34. Bascom, W. D.; Cottington, R. L.; Jones, R. L.; Peyser, P. *J. Appl. Polym. Sci.* **1975,** *19*, 2545.
35. Hunston, D. L.; Rushford, J. L.; Bitner, J. L.; and Oroshuik, J. Y. *J. Elastomers Plast.* **1980,** *12*, 133.
36. Kinloch, A. J.; Shaw, S. J. *J. Adhes.* **1981,** *12*, 59.
37. Mostovoy, S.; Ripling, E. J. *J. Appl. Polym. Sci.* **1971,** *15*, 641.
38. Kinloch, A. J.; Shaw, S. J.; Tod, A. D.; Hunston, D. L. *Polymer*, in press.
39. Kinloch, A. J.; Shaw, S. J.; Hunston, D. L. *Polymer*, in press.
40. Hunston, D. L.; Bascom, W. D.; Bitner, J. L. presented at the Proc. Adhes. Soc. Meet., Mobile, AL, February, 1982.
41. McKenna, G. B.; Mandell, L. E.; McGarry, F. J. presented at the 29th Annu. Techn. Conf. RP/Compos. Div. SPI, 13-C, 1974.
42. Scott, J. M.; Phillips, D. C. *J. Mater. Sci.* **1975,** *10*, 551.
43. Bascom, W. D.; Bitner, J. L.; Moulton, R. J.; Siebert, A. R. *Composites* **1980,** 9.
44. Hong, S.; Chung, S. Y.; Fedors, R. F. *Org. Coat. Plast. Chem.* **1983,** *48*, 581.

RECEIVED for review November 18, 1983. ACCEPTED March 21, 1984.

MORPHOLOGY AND FAILURE MECHANISMS

13

A Model for Phase Separation During a Thermoset Polymerization

R. J. J. WILLIAMS, J. BORRAJO, H. E. ADABBO, and A. J. ROJAS
Institute of Materials Science and Technology (INTEMA), University of Mar del Plata—National Research Council, J.B. Justo 4302, (7600) Mar del Plata, Argentina

A model was developed to predict the fraction and composition of the dispersed phase segregated during a thermoset polymerization (e.g., a rubber-modified epoxy system). This model can also be used to predict the mean radius and volumetric concentration of the dispersed particles. The location of equilibrium and spinodal curves is described by a Flory–Huggins equation that includes the thermoset conversion. Equations for nucleation, coalescence, and growing rates were derived and analyzed. The morphology development is assumed to be arrested by gelation of the thermoset matrix. The dispersed-phase fraction and the concentration of dispersed particles decrease with an increase in the curing temperature; however, the mean radius goes through a maximum. Accelerating the polymerization by adding a catalyst has the same effect as a temperature increase. Conditions leading to spinodal demixing are discussed.

A DISPERSED RUBBERY PHASE can improve the toughness and impact properties of cured thermosets. A typical example is the use of low levels of rubber copolymers of butadiene–acrylonitrile in epoxy resins. Initially the system is homogeneous; but, at a certain thermoset conversion, rubber-rich domains begin to be segregated from the matrix. The morphology development continues until the thermoset matrix reaches a conversion close to the gel point (*1*), or vitrifies (if the glass transition takes place before gelation).

The mechanical properties of the resulting specimen are strongly dependent on the morphology developed during the phase-separation process. Both formulation and curing conditions affect the resulting morphology (*1–3*). Therefore, prediction of convenient formulations and/or curing conditions to obtain desired morphologies is important.

0065-2393/84/0208-0195/$06.00/0

A model for phase separation during a thermoset polymerization is analyzed in this chapter. Although the model may be adapted to describe different systems, particular attention is given to rubber-modified epoxies. Moreover, for the sake of illustration purposes most of the model parameters are taken from a formulation used by Manzione et al. (2). This system was a low molecular weight bisphenol A type epoxy resin that was modified with a carboxyl-terminated polybutadiene–acrylonitrile (CTBN) with piperidine as the catalyst. For this model, we assume that the rubber is an inert component dissolved in a monodispersed epoxy resin that homopolymerizes in the presence of piperidine. Rubber-containing diepoxide resulting from the capping of carboxyl end groups of the CTBN with the epoxy resin is assumed to have no influence on kinetic and statistical parameters of the epoxy homopolymerization.

Thermoset Polymerization

The curing rate is assumed to follow a second-order kinetics,

$$dp/dt = A\,(1 - p)^2 \exp(-E/RT) \tag{1}$$

where p is the thermoset conversion. The specific rate constant (A) is $5 \times 10^7\ s^{-1}$ and $E/R = 10^4$ K; these values are consistent with reported data on time to gelation (2).

By assuming that the usual simplifying hypotheses are valid (equal reactivity, no substitution effects, and no intramolecular reactions in finite species), we can characterize the homopolymerization of a monomer with four reactive sites per mole (functionality = 4) by the following statistical parameters (*4–6*):

$$p_{gel} = 1/3 \tag{2}$$

$$M_{n_p}/M_{n_o} = 1/(1 - 2p) \tag{3}$$

$$M_{w_p}/M_{w_o} = (1 + p)/(1 - 3p) \tag{4}$$

M_n and M_w are the number and weight average molecular weights, respectively.

We have also assumed that, in this range of curing temperatures, the resin gels before vitrifying; the morphology development will thus be arrested at p_{gel}.

The thermoset viscosity, η, may be written as a function of M_w, because both parameters become infinite at the gel conversion,

$$\eta/\eta_o = (M_{w_p}/M_{w_o})^n \tag{5}$$

The following particular functionality will be used for illustration purposes:

$$\eta(\text{kg/m} \cdot \text{s}) = 2.24 \times 10^{-6}\,[(1 + p)/(1 - 3p)]^2 \exp(E\eta/RT) \tag{6}$$

where $E\eta/R = 4 \times 10^3$ K. This equation gives viscosity values close to reported experimental data at $p = 0$ (2).

Thermodynamics

The initial system is regarded as a solution of rubber (component 2) in an epoxy solvent (component 1) with a free energy per unit volume (ΔG_V^M) described by the Flory–Huggins equation.

$$\Delta G_V^M = (RT/\overline{V}_1)\,[(1 - \phi_2)\ln(1 - \phi_2) + (\phi_2/z)\ln\phi_2 + x\,\phi_2\,(1 - \phi_2)] \tag{7}$$

where $\overline{V}_1$ is the molar volume of the epoxy resin, ϕ_i is the volume fraction of component i, z is the ratio of molar volumes of both components ($\overline{V}_2/\overline{V}_1$), and x is the Flory–Huggins interaction parameter per mole of solvent (component 1).

The parameters $\overline{V}_1$, z, and x vary during the thermoset polymerization. The molar volume of epoxy resin changes according to

$$\overline{V}_1 = \overline{V}_{1_o}\,(M_{n_p}/M_{n_o}) = \overline{V}_{1_o}/(1 - 2p) \tag{8}$$

Then, the ratio of molar volumes becomes

$$z = \overline{V}_2/\overline{V}_1 = z_o\,(1 - 2p) \tag{9}$$

where $z_o = \overline{V}_2/\overline{V}_{1_o}$.

The Flory–Huggins interaction parameter per mole of solvent may be written as

$$x = x_o\,\overline{V}_1/\overline{V}_{1_o} = x_o/(1 - 2p) \tag{10}$$

Therefore, by combining Equations 7–10, we obtain

$$\Delta G_V^M = (RT/\overline{V}_{1_o})\{[(1 - 2p)(1 - \phi_2)\ln(1 - \phi_2)] + [(\phi_2/z_o)\ln\phi_2] + [x_o\,\phi_2(1 - \phi_2)]\} \tag{11}$$

When the epoxy matrix polymerizes, the free energy begins to increase because the absolute value of the configurational entropy decreases (the first two terms of the right-hand side of Equation 11). A reaction extent (p) is reached at which thermodynamics predicts that phase separation will lead to a more stable system. To illustrate this situation, the following parameters will be taken: $z_o = 10$, which fits several rubber–epoxy formulations (2, 7), and $x_o = 0.35 + 90/T$. The latter parameter illustrates the usual temperature dependence and gives reasonable values on an order-of-magnitude basis (8). The first term represents an entropic contribution, independent of temperature; the second term is an enthalpic contribution, inversely proportional to temperature. (The temperature is in Kelvins.)

Figure 1 illustrates Equation 11 for $T = 300$ K and various conversions up to the gel point. When p has advanced sufficiently, the shape of the free energy curves allows the choice of a pair of points at different ϕ_2 values with a common tangent leading to the same ordinate intersection. These conjugated points correspond to equilibrium compositions because the chemical potentials of each component (read at the ordinate) are the same in both phases. The locus of these conjugated points is the equilibrium or binodal curve (Figure 1). Also, in the same p region the free-energy curves show a pair of inflexion points. Their locus is the spinodal curve (Figure 1).

The stable region (homogeneous system) is located under the equilibrium curve. The area between equilibrium and spinodal

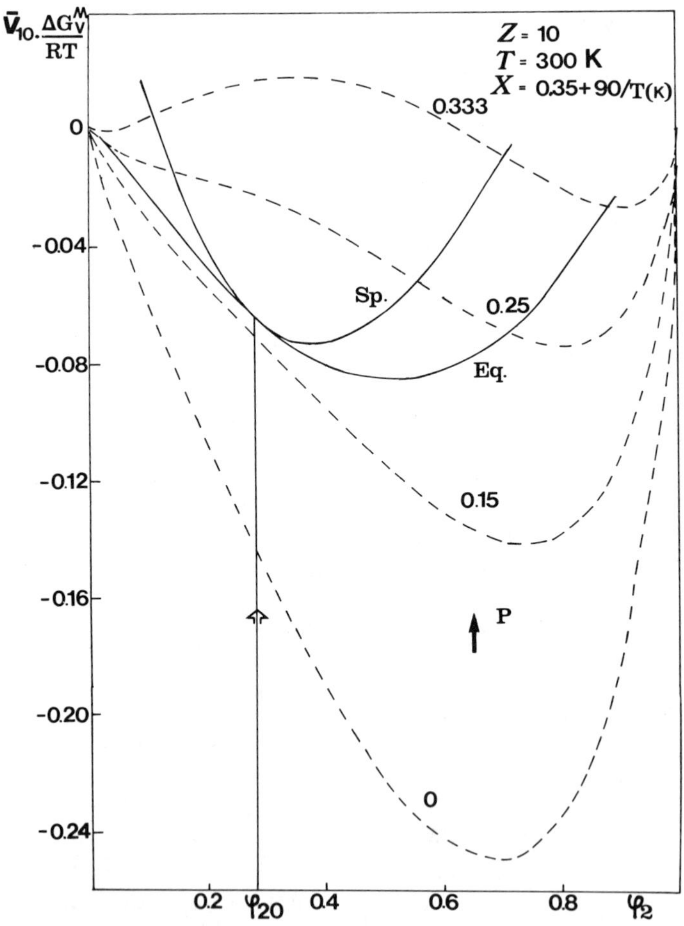

Figure 1. Dimensionless free energy of solution per unit volume as a fraction of rubber fraction volume, for different thermoset conversions. Key: Eq., equilibrium or binodal; and Sp., spinodal.

curves corresponds to the metastable region. In this area, phase separation may take place and depends on the rate at which the system evolves through this region (polymerization kinetics) and the intrinsic phase-separation rate. If the system meets the spinodal curve, an unstable condition results and spontaneous demixing will take place.

At 300 K, the initial rubber volume fraction (ϕ_{2_0}) that produces the critical point (common to both binodal and spinodal curves) is 0.28. This point is reached at $p = 0.16$. Let us consider a system with a ϕ_{2_0} less than 0.28. If polymerization is carried out until the equilibrium curve is reached (vertical trajectory in Figure 1), phase separation may begin. Then, the dispersed phase will have a composition close to that at the conjugated point (*see* the following section). Thus, dispersed domains rich in the rubber component, but always including some epoxy, will appear in the system. The presence of some epoxy in the dispersed phase explains experimental observation that the volume fraction of the dispersed domains at the end of separation may be greater than ϕ_{2_0} (*1, 7*).

If ϕ_{2_0} is greater than 0.28, phase separation will produce a dispersed phase rich in epoxy (i.e., an inversion in the nature of the system, which now becomes a reinforced elastomer). The predicted inversion is in agreement with experimental observations (*9*). Commercial formulations use rubber volume fractions in the range of $\phi_{2_0} = 0.05–0.15$, and this range always produces toughened thermosets.

Also, at 300 K, a system with $\phi_{2_0} = 0.1$ will be able to reach the metastable region but will never reach the spinodal before gelation (at $p = 1/3$) (*see* Figure 1). The influence of curing temperature on this situation is shown in Figure 2. For curing temperatures of practical significance, spinodal demixing is not possible. Moreover,

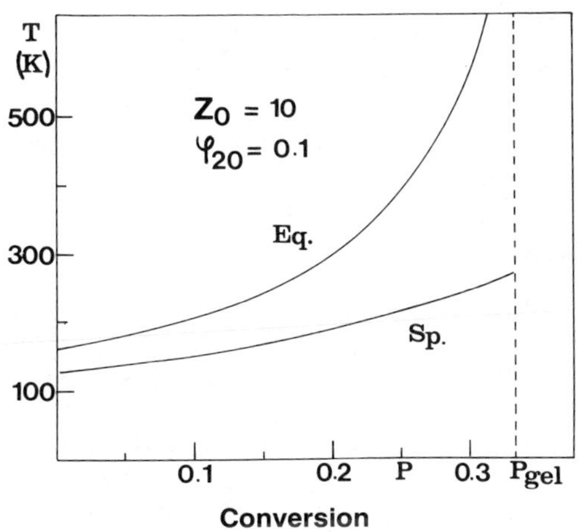

Figure 2. Temperature vs. conversion phase diagram showing location of equilibrium and spinodal curves. Key: Eq., equilibrium; and Sp., spinodal.

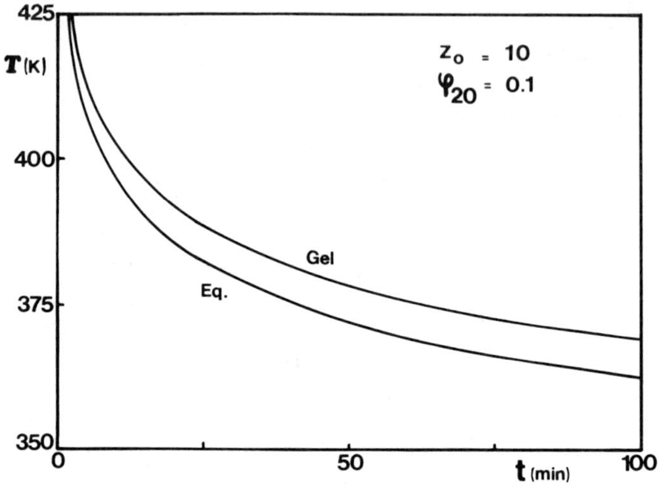

Figure 3. Temperature vs. time transformation diagram showing location of equilibrium and spinodal curves.

as the temperature increases, the conversion range for phase separation becomes narrower. This narrowing may also be seen in the temperature versus time transformation diagram (*10*) shown in Figure 3. This diagram was developed by using the relationship

$$t = p/[(1 - p) A \exp(-E/RT)] \tag{12}$$

arising from the integration of Equation 1 at constant temperature.

The influence of the initial rubber concentration on the location of equilibrium and spinodal curves, at constant temperature, is shown in Figure 4. At 373 K, spinodal demixing at ϕ_{2_o} values equal to or greater than 0.124 is thermodynamically possible. The actual achievement of this condition will still require a high polymerization rate compared to the phase-separation rate. Thus, the only possibility of producing spinodal demixing is to process a very reactive formulation with a high initial rubber fraction.

Phase Separation

When the system enters the metastable region, phase separation begins. Figure 5 shows a free energy versus composition curve for a system evolving through the metastable range [$\overline{V}_{1_o}$ was assumed to equal 340 cm^3/mol, a reasonable value for a low molecular weight bisphenol A type epoxy resin (*2, 7*)]. ϕ_2^α and ϕ_2^β are the equilibrium compositions (i.e., two conjugated points of the binodal curve), and ϕ_{2_c} is the actual rubber concentration in the continuous phase. Initially $\phi_{2_c} = \phi_{2_o}$; then, at a certain p, the system reaches equilibrium and $\phi_{2_c} = \phi_{2_o} = \phi_2^\alpha$. At this point, the value of ϕ_{2_c} varies from ϕ_2^α

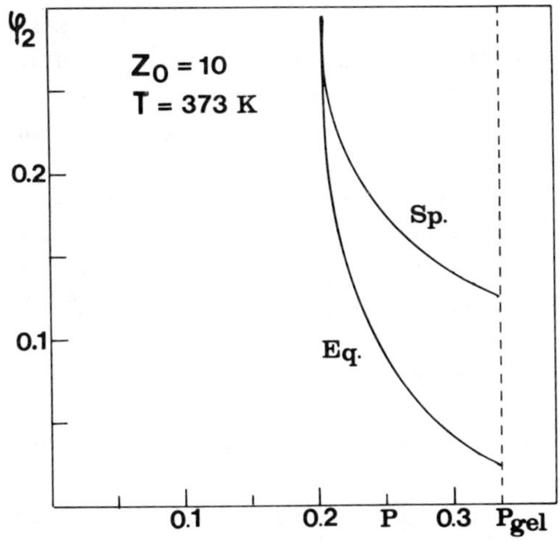

Figure 4. Location of equilibrium (Eq.) and spinodal (Sp.) curves in a rubber fraction vs. conversion diagram.

(i.e., the metastable region, or supersaturation, begins). The extent of variation depends upon the relationship between the intrinsic rate of phase separation and the polymerization rate. The significant departure of ϕ_{2_c} from ϕ_2^α (Figure 5) arises from a relatively high polymerization rate. The opposite situation would produce a ϕ_{2_c} value very close to the equilibrium value during the entire phase-separation process.

By definition, the intersections of the tangent at ϕ_{2_c} and both

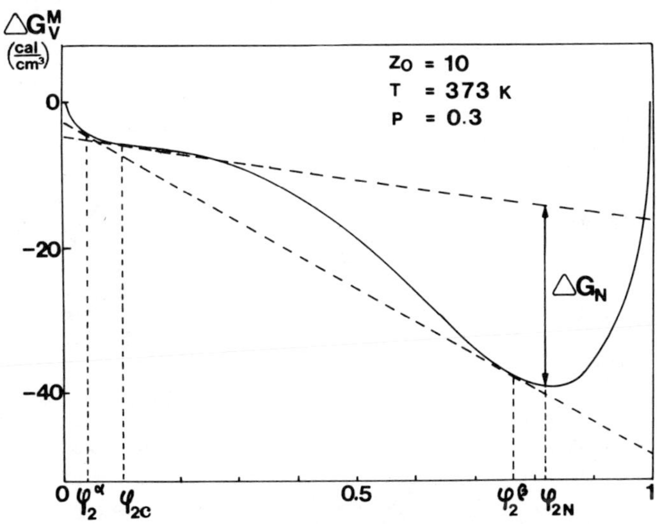

Figure 5. Free energy of solution per unit volume vs. rubber fraction volume. ΔG_N is the free energy change associated with phase separation.

ordinates represent the partial free energy per unit volume for the pure components in the continuous phase. The tangent line also gives the virtual free energy per unit volume for any solution with an arbitrary composition, ϕ_2, segregated from the solution with a composition ϕ_{2_c}. The continuous curve, ΔG_V^M vs. ϕ_2, gives the actual free energy value per unit volume for any solution. Therefore, the free energy change involved in the separation of a dispersed phase of any arbitrary composition, ΔG_N, is given by the distance between the tangent line at ϕ_{2_c} and the free energy function ΔG_V^M. For example, if the dispersed phase has the composition ϕ_{2N}, the free energy per unit volume will decrease by an amount ΔG_N, as is depicted in Figure 5.

All the values possible for ΔG_N are shown in Figure 6. For $\phi_2 = \phi_{2N}$, the maximum possible decrease in free energy is obtained. Thus, ϕ_{2N} will be the actual composition of the phase that will be segregated from a continuous phase with composition ϕ_{2_c}, at the conditions in Figures 5 and 6.

A negative value of ΔG_N may be regarded as the driving force for phase separation in free energy coordinates. This driving force develops when the system enters the metastable region. At the boundary (i.e., when $\phi_{2_c} = \phi_2^{\alpha}$), the same calculation results in $\Delta G_N \geqslant 0$. That is, there is no driving force for the segregation of a new phase, and the system becomes supersaturated.

To segregate a volume fraction with composition ϕ_{2N} from the continuous phase with composition ϕ_{2_c}, we may assume that some sort of composition fluctuations are necessary. As ϕ_{2_c} changes to ϕ_{2N} (Figure 6), ΔG_N increases slightly to a maximum value (E_F) and then decreases continuously. E_F may be regarded as an energy barrier for

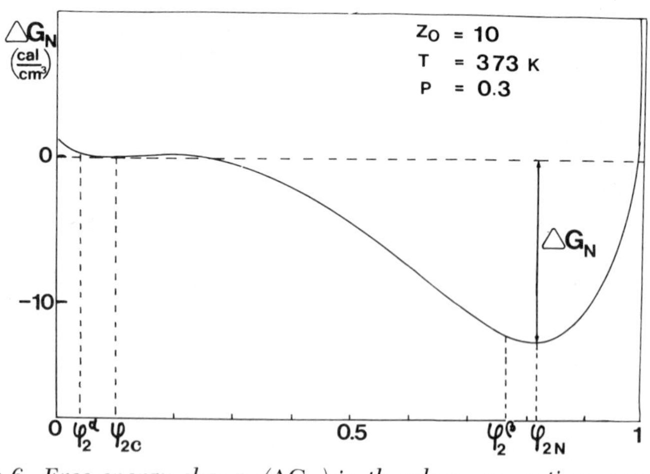

Figure 6. Free energy change (ΔG_N) in the phase-separation process vs. rubber concentration.

phase separation. Calculations show that E_F reaches a maximum value when $\phi_{2_c} = \phi_2^\alpha$ and disappears if the system reaches the spinodal curve; thus, spontaneous demixing occurs.

Nucleation

In order to develop a constitutive equation for the rate of composition fluctuations per unit volume (dF/dt), we assumed that dF/dt is proportional to the diffusion coefficient of rubber in the epoxy, D, and has an Arrhenius dependence on the activation energy, E_F

$$dF/dt = F_o\, D \exp(-E_F/RT) \tag{13}$$

where F_o is a proportionality constant.

The free energy change for the formation of spherical domains with composition ϕ_{2_N} is given by

$$\Delta G = (4/3)\pi r^3 \Delta G_N + 4\pi r^2 \sigma \tag{14}$$

where r is the radius of dispersed domains and σ is the surface tension. Any possible free energy change associated with elastic effects is neglected.

A plot of Equation 14 is shown in Figure 7. ΔG reaches a maximum (ΔG_c) for the critical radius, r_c. From Equation 14, the following values result:

$$r_c = 2\sigma/|\Delta G_N| \tag{15}$$

$$\Delta G_c = 16\pi\sigma^3/(3|\Delta G_N|^2) \tag{16}$$

Dispersed domains with a radius $r = r_c$ are called nuclei because their growth decreases the free energy, as shown in Figure 7. Then,

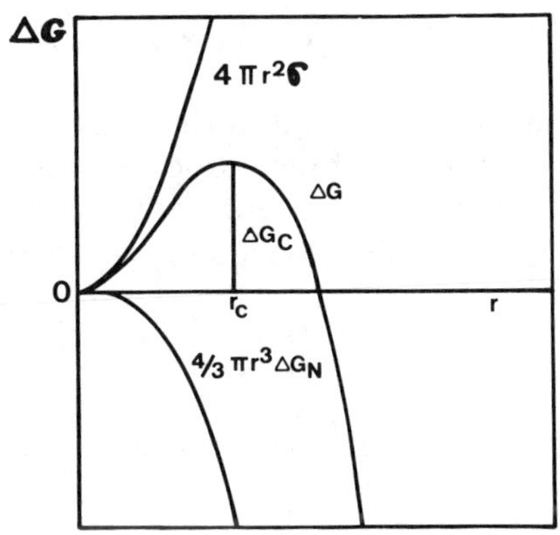

Figure 7. Different contributions to the free energy change associated with nucleation.

only those fluctuations reaching the free energy barrier, ΔG_c, will become nuclei. Therefore, the nucleation rate may be written as

$$dN/dt = (dF/dt)\exp(-\Delta G_c/RT) \quad (17)$$

To give an explicit expression for the nucleation rate, we must adopt a constitutive equation for the diffusion coefficient. A Stokes–Einstein functionality is assumed to be true, in which $D = Q \cdot T/\eta$. In this equation, η is given by Equation 6, and Q is a constant that is adjusted to predict a correct order of magnitude. For the epoxy–rubber system, we assumed that $D = 10^{-11}$ m^2/s at $T = 363$ K and $p = 0$. Using Equation 6, we obtain

$$D = 5 \times 10^{-14}\, T/\eta \quad (18)$$

Equation 17 may be rewritten by using Equations 1, 6, 13 and 18, and by using the conversion p as the independent variable

$$dN/dp = (F'_o\, T/A)\,[(1 - 3p)/(1 - p^2)]^2 \exp\{[E - (E_\eta + E_F + \Delta G_c)]/RT\} \quad (19)$$

where F'_o is an adjustable parameter. At $p = p_{gel}$, Equation 19 predicts that $dN/dp = 0$ (i.e., nuclei production is arrested at gelation). This result occurs because, at p_{gel}, $\eta \rightarrow \infty$ and $D \rightarrow 0$.

Concentration of Dispersed-Phase Particles

The concentration of dispersed-phase particles per unit volume, P, increases with nucleation but decreases through coalescence. This last process is proportional to the square concentration of particles and inversely proportional to viscosity (*11*). Therefore, the rate of variation of the concentration of particles per unit volume may be written as

$$dP/dt = dN/dt - (4/3)(kT/\eta)\, P^2 \quad (20)$$

Although the numerical factor used in the coalescence term may not be exact (*11*), a correct order of magnitude is predicted.

Because time and reaction extent are related through polymerization kinetics, Equation 20 may be rewritten by using p as the independent variable

$$dP/dp = dN/dp - (4/3)(kT/A\eta)\, P^2[\exp(E/RT)/(1 - p)^2] \quad (21)$$

Because $\eta \rightarrow \infty$ at $p = p_{gel}$, the coalescence is arrested at gelation.

Particle Growth

When the system evolves through the metastable region, particle growth occurs because of the driving force $(\phi_{2_c} - \phi_2^\alpha)$, which tries to restore the system to equilibrium conditions. The growth rate (G.R.), defined as the increase in volume fraction per unit time, may be assumed to be proportional to the interfacial area per unit volume and the driving force

$$\text{G.R.} = k_\phi \, 4\pi \overline{R}^2 \, P \, (\phi_{2_c} - \phi_2^\alpha) \tag{22}$$

where k_ϕ is a mass transfer coefficient and $\overline{R}$ is the mean radius of dispersed-phase particles ($4\pi\overline{R}^2 P$ is the surface area per unit volume).

The mass transfer coefficient for a sphere in a stagnant medium, expressed in terms of a volumetric fraction driving force, is given by (*12*)

$$k_\phi = D/\overline{R} \tag{23}$$

Because $D = 0$ at p_{gel}, the growth rate is arrested at gelation.

Mean Particle Radius and Volume Fraction of Dispersed Phase

The mean particle radius is given by

$$\overline{R} = (3\, V_D/4\pi P)^{\frac{1}{3}} \tag{24}$$

where V_D is the volume fraction of the dispersed phase at any p. The change in this fraction with p is given by

$$dV_D/dp = (4\pi/3)(r_{c_p})^3\,(dN/dp) + k_\phi \, 4\pi\overline{R}^2 \, P \, (\phi_{2_c} - \phi_2^\alpha)(dt/dp) \tag{25}$$

The amount of dispersed phase increases with nucleation and particle growth. The composition of the new phase segregated at any p is ϕ_{2_N} for both mechanisms (*see* Figures 5 and 6).

Compositions of Both Phases

Although the instantaneous composition of the phase that is being segregated is ϕ_{2_N}, the average rubber concentration in the dispersed phase must take into account the history of the ϕ_{2_N} variation during the phase-separation process. Thus,

$$\overline{\phi}_{2_D} = \int_0^{V_D} \phi_{2_N} \, dV_D/V_D \tag{26}$$

The rubber concentration in the continuous phase may be obtained from

$$\phi_{2_o} = V_D \, \overline{\phi}_{2_D} + (1 - V_D)\, \phi_{2_c} \tag{27}$$

Therefore,

$$\phi_{2_c} = (\phi_{2_o} - V_D \, \overline{\phi}_{2_D})/(1 - V_D) \tag{28}$$

Results and Discussion

Differential Equations 19, 21, and 25 were solved with a fourth-order Runge–Kutta method. The only adjustable parameter of the model was selected as $F_o' = 10^{11}$ cm^{-3}K^{-1}s^{-1} to fit the order of magnitude of particle concentration with reported experimental results (2). The surface tension, σ, was varied from 4×10^{-4} to 2×10^{-2} N/m^2, but no relevant effect was noted.

Figures 8 and 9 illustrate the evolution of the rubber concentration in each phase for two curing temperatures. At low temperatures (T = 363 K), the instantaneous and equilibrium compositions are practically the same. This result implies that the rate of phase separation is much higher than the polymerization rate. On the other hand, the evolution of the average rubber concentration in the dispersed phase, $\overline{\Phi}_{2D}$, suggests that most of the phase segregation takes

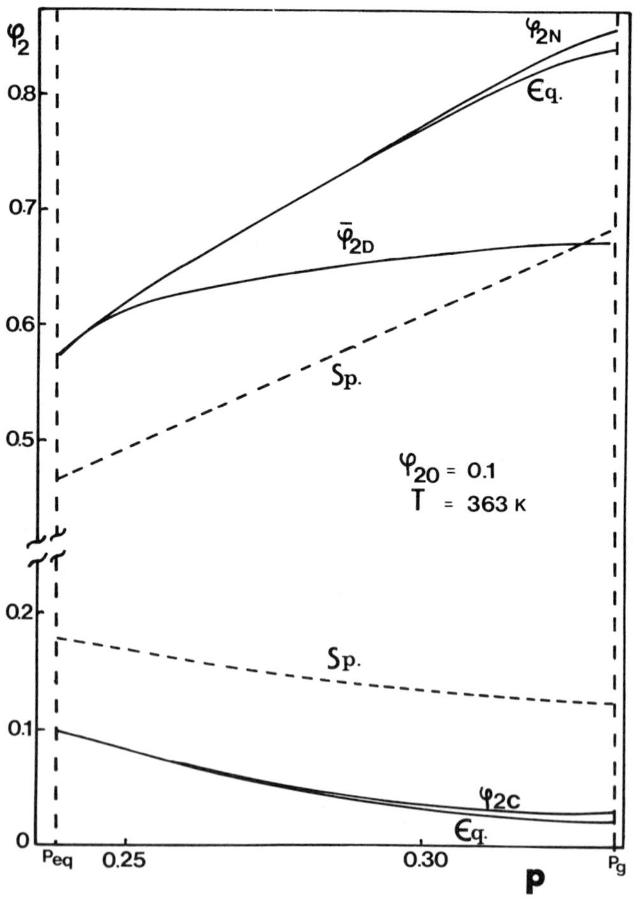

Figure 8. Evolution of the rubber fraction volume in both phases as a function of the thermoset conversion (low curing temperature).

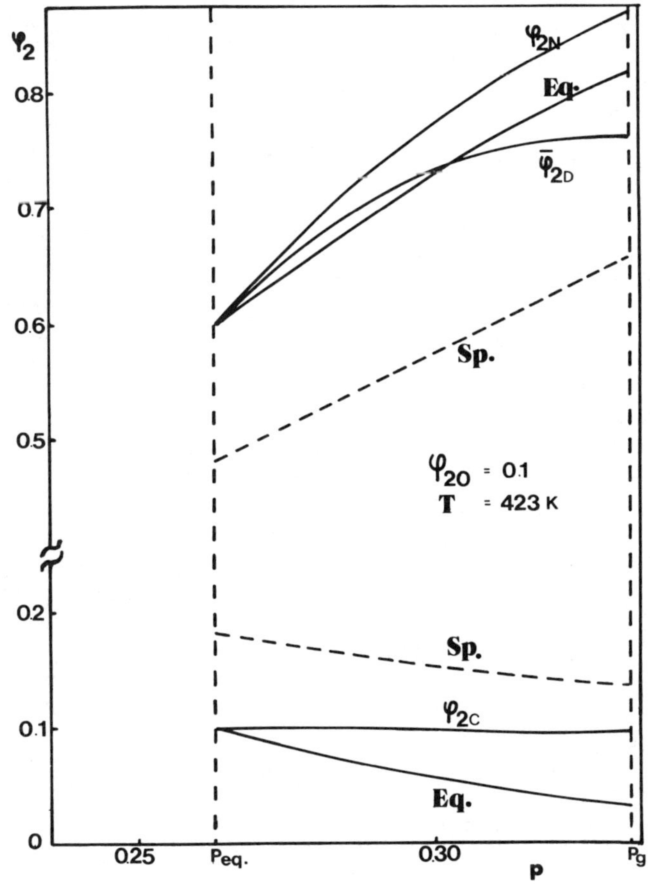

Figure 9. Evolution of the rubber fraction volume in both phases as a function of the thermoset conversion (high curing temperature).

place well before gelation, in agreement with experimental results (*1*). At high curing temperatures ($T = 423$ K), the instantaneous compositions are different from the equilibrium values, and the system has advanced into the metastable region. Now, phase separation does not restore the system to the equilibrium condition because of the high polymerization rate. However, the spinodal curve is not attained because of thermodynamic restrictions (*see* Figure 2). Significantly, ϕ_{2_c} remains very close to the initial value, ϕ_{2_0}. Therefore, the volume fraction of the dispersed phase must be negligibly small.

Results show that producing spinodal demixing in a rubber-modified epoxy is not easy. From Figures 2, 4, 8, and 9, the necessary conditions seem to be high initial rubber content (ϕ_{2_0} close to the critical value), low curing temperatures (thermodynamic requirement), and high reaction rates (kinetic condition).

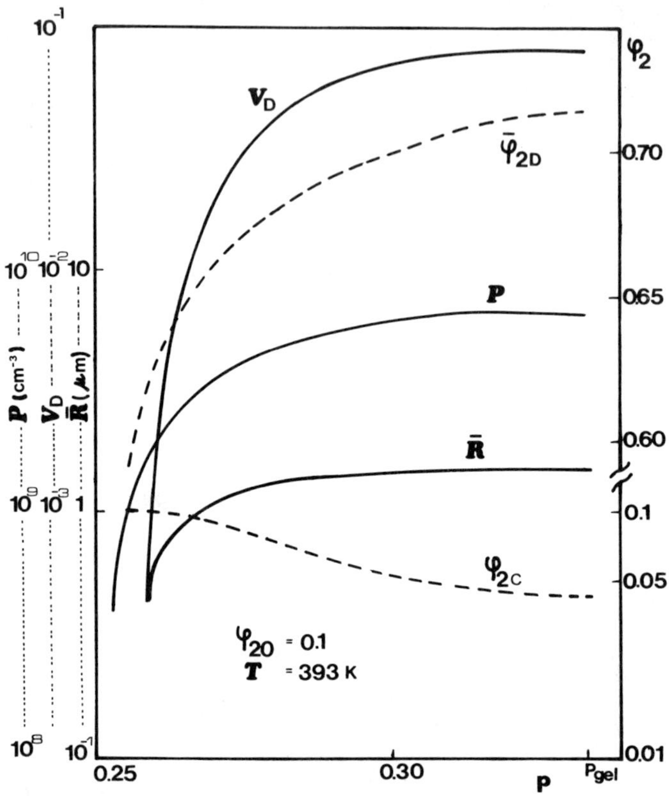

Figure 10. Evolution of different parameters that characterize the morphology as a function of the thermoset conversion (intermediate curing temperature).

Figure 10 shows the evolution of the different parameters that characterize the morphology for an intermediate curing temperature. The mean radius of particles at $p = p_{gel}$ is $\overline{R} = 1.5$ μm, a value attained well before gelation. This value lies in the range of reported experimental results (2, 7). The volume fraction of the dispersed phase reaches a final value, $V_D = 0.08$. This fraction includes slightly more than half of the rubber initially added to the formulation. One possibility of increasing this amount is to operate with a thermoset that gels at a high conversion (i.e., and epoxy–diamine system for which p_{gel} is close to 0.58).

The influence of the curing temperature on the resulting morphology at p_{gel} is shown in Figure 11. The concentration of dispersed particles decreases continuously with temperature because the polymerization rate increases more rapidly than the nucleation rate (in Equation 19, $E > E_\eta + E_F + \Delta G_c$). The coalescence rate did not seem to be relevant when compared to the nucleation rate, and E_F and ΔG_c were very small values.

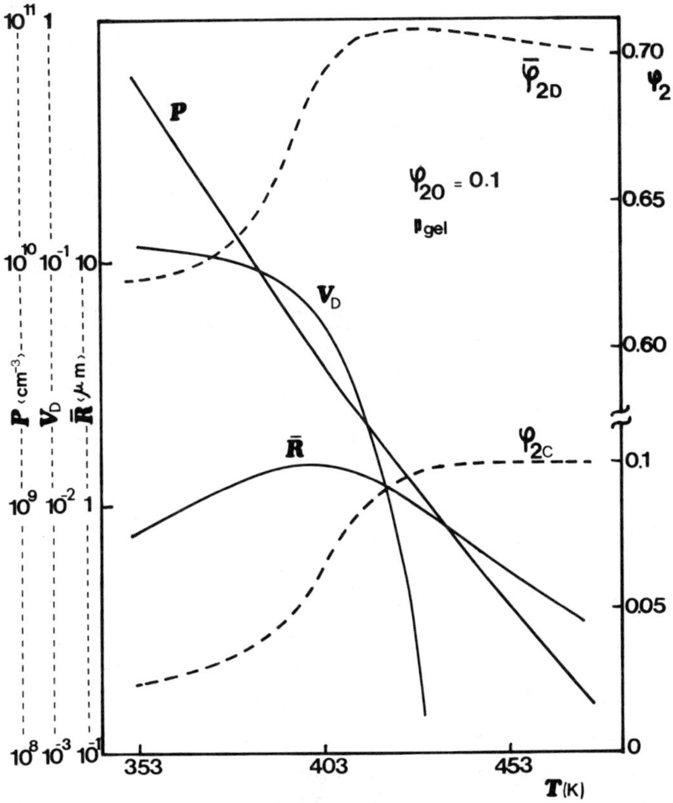

Figure 11. Final values of different parameters that characterize the morphology vs. curing temperature.

The volume fraction of dispersed phase, V_D, is almost constant at low temperatures, because it is limited only by the binodal curve and the gelation conversion (thermodynamic restriction). However, at high temperatures, V_D drops abruptly because of the high polymerization rate (kinetic restriction). As a consequence, the mean radius, $\overline{R}$, which depends on the V_D/P ratio, goes through a maximum as the temperature is increased. This result agrees with experimental observations (2). All these trends depend on the fact that $E > E_\eta$. A thermoset for which $E_\eta > E$ will show opposite effects when the curing temperature is increased.

An increase in the polymerization rate at constant temperature (i.e., by adding a catalyst) has the same effect as a temperature increase (Figure 12), because both parameters increase the polymerization rate in relation to the phase-separation rate.

Finally, increasing the initial rubber amount in the formulation leads to a considerable increase in the volume fraction of dispersed phase and concentration of dispersed particles (Figure 13). This ob-

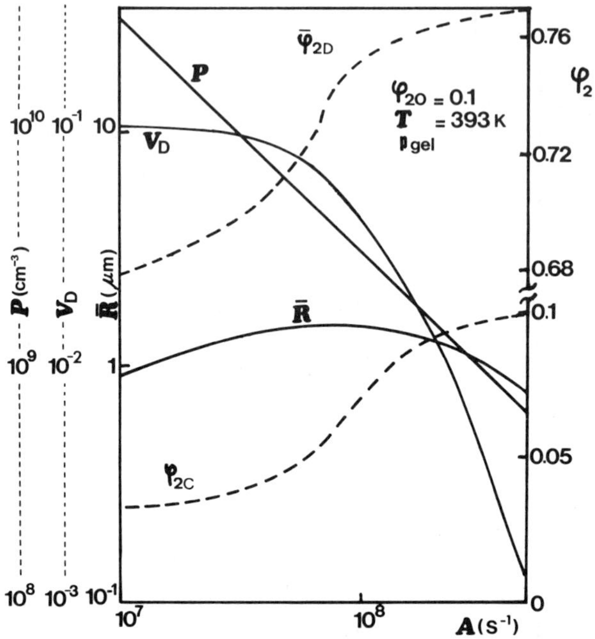

Figure 12. Final values of different parameters that characterize the morphology as a function of the polymerization rate.

servation is in agreement with experimental results (7). However, the average rubber concentration of the dispersed phase, $\bar{\phi}_{2D}$, drops significantly. Both effects occur because the binodal is reached at lower conversions when ϕ_{2_0} is increased (Figure 4). This significant change in the composition of dispersed domains may have a bearing on the toughening behavior.

Summary

We have modeled the process of phase separation during a thermoset polymerization by using simple thermodynamic, kinetic, and statistical arguments, as well as related constitutive equations. The model may be adapted to different thermoset formulations and operating conditions.

For rubber-modified epoxies cured under isothermal conditions, the model predicts the following trends:

- Phase separation probably proceeds through a classic nucleation-growth mechanism rather than through spinodal demixing. Spinodal demixing would require the use of a formulation with a high initial rubber amount, low curing temperatures, and high reaction rates.
- The influence of the curing temperature on phase separation is strongly dependent on the difference of chemical and

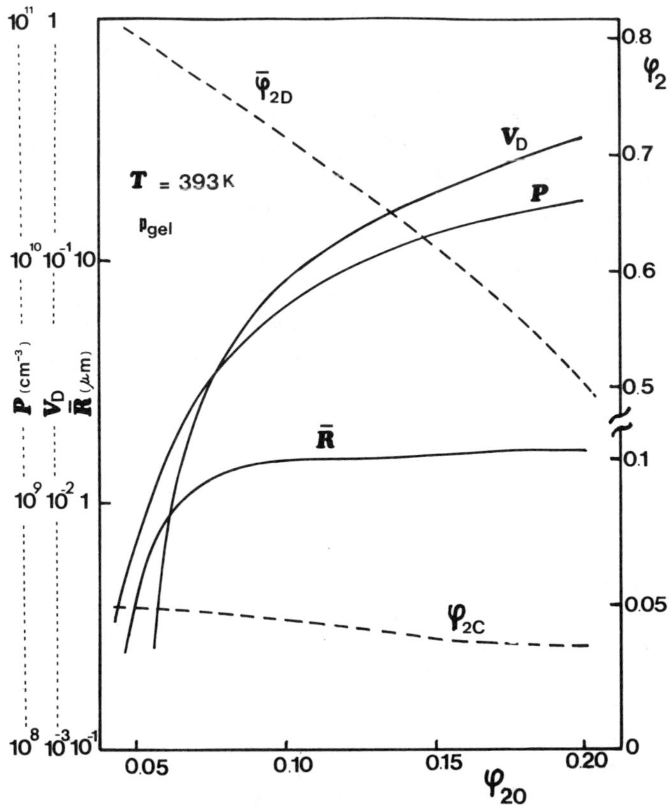

Figure 13. Final values of different parameters that characterize the morphology as a function of the initial rubber concentration.

viscous activation energies, $(E - E_\eta)$. If this difference is high, as is the usual situation, a temperature increase will cause a considerable decrease in the concentration of dispersed-phase particles, a maximum in the mean radius, and a significant drop in the volume fraction of the dispersed phase at high temperatures. These effects are exactly the same when the polymerization rate is increased at constant temperature. In both situations, the explanation relies on the increase of the polymerization rate with respect to the phase-separation rate.

- Increasing the initial rubber amount, ϕ_{2_0}, leads to a considerable increase in the volume fraction and concentration of dispersed-phase particles. However, the rubber concentration in the segregated domains drops significantly. If ϕ_{2_0} is greater than a critical value, an inversion in the nature of the segregated phase will result.

Most of these trends agree with experimental observations (*1, 2, 7, 9*).

List of Symbols

A	Specific rate constant for thermoset polymerization
D	Diffusion coefficient of rubber in the thermoset
E	Activation energy of polymerization
E_F	Activation energy associated with composition fluctuations
E_η	Activation energy of viscosity
F	Composition fluctuations per unit volume
F_o	Proportionality constant
F'_o	Adjustable parameter
G.R.	Particle growth rate
ΔG	Free energy change in the nucleation process
ΔG_c	Free energy barrier for the formation of a nucleus with critical radius
ΔG_N	Free energy change associated with phase separation
ΔG_V^M	Free energy per unit volume
k	Boltzmann constant
k_ϕ	Mass transfer coefficient
M_n	Number average molecular weight
M_w	Weight average molecular weight
N	Nuclei concentration per unit volume
p	Thermoset conversion (reaction extent)
P	Concentration of dispersed-phase particles per unit volume
r	Nucleus radius
r_c	Critical radius of nucleus
R	Gas constant
$\overline{R}$	Mean radius of dispersed-phase particles
$\overline{V}$	Molar volume
V_D	Volume fraction of dispersed phase
x	Flory–Huggins interaction parameter per mole of thermoset
z	Ratio of molar volumes of rubber and thermoset

Greek Letters

η	Viscosity of the continuous phase
σ	Surface tension
ϕ	Volume fraction

Subscripts

o	Initial value
1	Thermoset resin
2	Rubber
c	Continuous phase
D	Dispersed phase (average value)
N	Segregated phase (instantaneous value)

Superscripts

α, β	Equilibrium compositions

Literature Cited

1. Chan, L. C.; Gillham, J. K. *Proc. 185th Am. Chem. Soc. Natl. Meet., Org. Coat. Appl. Polym. Sci.* **1983,** *48*, 571.
2. Manzione, L. T.; Gillham, J. K.; McPherson, C. A. *J. Appl. Polym. Sci.* **1981,** *26*, 889.
3. Manzione, L. T.; Gillham, J. K.; McPherson, C. A. *J. Appl. Polym. Sci.* **1981,** *26*, 907.
4. Flory, P. J. *J. Am. Chem. Soc.* **1941,** *63*, 3083.
5. Flory, P. J. *J. Am. Chem. Soc.* **1941,** *63*, 3097.
6. Stockmayer, W. H. *J. Chem. Phys.* **1943,** *11*, 45.
7. Kunz, S. C.; Sayre, J. A.; Assink, R. A. *Polymer* **1982,** *23*, 1897.
8. Champetier, G.; Monnerie, L. "Introduction à la Chimie Macromoléculaire"; Masson: Paris, **1969;** p. 135.
9. Gillham, J. K.; Glandt, C. A.; McPherson, C. A., In "Chemistry and Properties of Crosslinked Polymers"; Labana, S. K., Ed.; Academic Press: New York, **1977;** p. 491.
10. Naé, H. N.; Gillham, J. K. *Proc. 185th Am. Chem. Soc. Natl. Meet., Org. Coat. Appl. Polym. Sci.* **1983,** *48*, 566.
11. Moelwyn-Hughes, E. A. *"Physical Chemistry"*, 2nd Ed.; Pergamon Press: London, **1965;** p. 1212.
12. Sherwood, T. K.; Pigford, R. L.; Wilke, Ch. R. "Mass Transfer"; McGraw-Hill: New York, **1975;** p. 215.

RECEIVED for review November 18, 1983. ACCEPTED March 21, 1984.

14

Effect of Rubber Cross-link Density and Tear Energy on the Toughness of Rubber-Modified Epoxies

J. A. SAYRE, S. C. KUNZ, and R. A. ASSINK
Sandia National Laboratories, Albuquerque, NM 87185

The relationship between the cross-link density of the rubber phase and the fracture toughness of rubber-modified epoxies has been investigated. Diethanolamine-cured and polyoxypropylamine-cured rubber-modified epoxies were exposed to gamma radiation and their fracture toughness was measured. (Gamma radiation cross-links rubbers but has a negligible effect on epoxies at the doses used.) The results are consistent with two factors governing changes in composite toughness: particle tear energy and the particulate phase relaxation time. Initially, as the rubber phase is cross-linked, the particle tear energy increases and the composite toughness increases. At high cross-link densities, the particles lose their ductility and are unable to deform in the time frame of the propagating crack. The rubber particles therefore fail in a brittle manner. Correspondingly, the composite toughness decreases at high doses. Thus, qualitative agreement is seen with the mechanism that proposes a direct relationship between particle tear energy and composite toughness.

Several mechanisms have been proposed to explain the increased toughness that results from a dispersion of rubber particles in an epoxy matrix. These explanations include crazing (*1*) and shear banding of the matrix (*2*), and tearing (*3*) and triaxial dilatation of the rubber particles (*4*). Investigation of the relationship between cure state of the rubber and toughness of the composite is appropriate because these mechanisms must depend to some degree on the physical state of the rubber phase (i.e., state of strain and modulus and/or tear energy of the rubber). A previous study focused on the role of the particulate rubber phase by observing fracture through a microscope. This study revealed that strained rubber particles bridge the opening crack faces and fail by tearing (*3*). Quantitative formu-

0065-2393/84/0208-0215/$06.00/0

lation of this mechanism predicted that the toughness of rubber-modified epoxies depends on the tear energy of the rubber particles.

This chapter describes the results of experiments designed to probe further the toughness–particle tear energy model. In particular, we examine the relationship between state of cure of the rubber phase (cross-link density) and toughness of the composite. The cross-link density of the rubber phase was varied without affecting its volume fraction, the composition of either matrix or particulate phase, or the particle-size distribution. This effect was accomplished by exposing cured rubber-modified epoxies to gamma radiation. Samples of compounds that separately model both the matrix and the rubber phases of the composite were also prepared and exposed to radiation. Increases in fracture toughness of the rubber-modified epoxies could then be compared to increases in tear energies of model rubber samples and to toughness values of unmodified epoxy as a function of radiation. We also made independent measurements of the changes in tear energy of the rubber phase by measuring the tear strains of rubber particles in situ during fracture of the composite. These values were converted to tear energies by using the Kunz et al. model (3). This approach provided an added means of correlating changes in rubber particle tear energy with changes in composite toughness.

Experimental

Materials Formulation. Three rubber-modified epoxies were examined. Table I lists the proportions in parts by weight (pbw) for all the materials studied. Epon 828, a diglycidyl ether of bisphenol A with an average molecular weight of 370 g/mol (Shell Chemical Co.), was the epoxy used in all experiments. Two liquid rubbers were used: HYCAR 1300X8 and HYCAR 1300X16 (The BF Good-

Table I. Formulations of Rubber-Modified Epoxies

System	*Epoxy (Epon 828)*	*CTBN Rubber (HYCAR 1300X8)*	*ATBN Rubber (HYCAR 1300X16)*	*Diethanol-amine (DEA)*	*Polyoxy-propylamine (T-403)*
CTBN-modified–DEA-cured	90	10	—	12	—
Unmodified–DEA-cured	100	—	—	12	—
CTBN-modified–T-403-cured	90	10	—	—	36
Unmodified–T-403-cured	100	—	—	—	36
ATBN-modified–DEA-cured	90	—	10	12	—
CTBN rubber–DEA-cured	17.7	82.3	—	0.3	—
CTBN rubber–T-403-cured	17.7	82.3	—	—	0.9

NOTE: Data are given as parts by weight (pbw).

rich Company). HYCAR 1300X8 is a carboxyl-terminated butadiene–acrylonitrile (CTBN) random copolymer containing ~17% acrylonitrile, with a manufacturer-listed number average molecular weight of 3300 g/mol. HYCAR 1300X16 is an amine-terminated butadiene–acrylonitrile rubber (ATBN) formed by reacting aminoethylpiperazine with 1300X8 CTBN. Diethanolamine (DEA) and a polyoxypropylamine with an average molecular weight of 400 g/mol (T-403, Jefferson Chemical Company) were the curing agents used. All materials underwent a 16-h cure at 71 °C.

The rubber in the CTBN-containing materials was endcapped with epoxy by reacting the rubber carboxylic acid end groups with epoxy for 3 h at 150 °C. The reaction was carried out in a large (~100:1) excess of epoxy for the rubber-modified epoxies. For the model rubber compounds, the reaction was carried out with a ratio of 2 mol of epoxy to 1 mol of rubber. The model rubbers were dissolved in methyl ethyl ketone to facilitate mixing of the curing agents, and the solvent was then stripped in a vacuum before curing. For the ATBN-modified epoxies, the rubber was added directly to the epoxy with the curing agent, and the system was cured in the same way as the CTBN composites.

The composition of the unmodified epoxies and the model rubber compounds were chosen to simulate the matrix and particulate phase, respectively, of the rubber-modified epoxies. Previous studies (5, 6) have indicated that for the rubber-modified epoxies used in this study, the unmodified epoxies are good models for the matrix. With ^{13}C-NMR data, these studies have also shown that a 2:1 molar ratio of epoxy to rubber is the correct model for the mobile portion of the rubber phase. To simulate the particulate phase, DEA and T-403 were added to the rubber in the same stoichiometry as in the rubber-modified epoxies. The stoichiometry was based on unreacted epoxide groups and is 20% of theoretical for DEA and 100% for T-403. This combination of materials allowed us to study the rubber-modified epoxies and their constituent phases independently.

We exposed all the samples (composites, unmodified epoxies, and model rubber compounds) to gamma radiation from a cobalt-60 source in a nitrogen atmosphere. Three dose rates were used: 230 krads/h, 660 krads/h, and 1 Mrad/h. We ruled out any effect of dose rate after exposing rubber-modified and unmodified samples to gamma radiation at dose rates between 65 and 660 krads/h to a total dose of ~60 Mrads and observing no significant differences in fracture toughness.

Physical Testing. Fracture toughness, K_{Ic}, was measured from three-point bend tests with single-edge notched samples at a crosshead speed of 0.25 cm/min. K_{Ic} values were converted to toughness, $\mathcal{G}_{Ic}$, by using the Irwin relation, $K_{Ic}^2 = E\mathcal{G}_{Ic}$, where E is Young's modulus of the composite. Young's modulus was obtained from three-point bend tests with a span-to-depth ratio of 16 and a rate of 0.25 cm/min. The shear moduli of the model rubber compounds were measured on a forced oscillation torsion instrument at 1 Hz.

In situ measurements of particle tear strains were made on rubber-modified epoxies at selected radiation doses. Samples were fractured by manually screwing a wedge into a crack with a special rig; crack propagation was viewed simultaneously in a transmission optical microscope. The length of rubber stretched between the crack faces, Δl, (Figure 1) was measured typically at 400× magnification. The extension ratio at failure, λ, is defined as the total length of stretched rubber particle divided by the particle diameter, d.

$$\lambda = \frac{\Delta l + d}{d} \tag{1}$$

Values of Δl were measured at the first indication of tearing for the range of particle sizes observed in each sample.

Differential scanning calorimetry (DSC) was used to measure the glass tran-

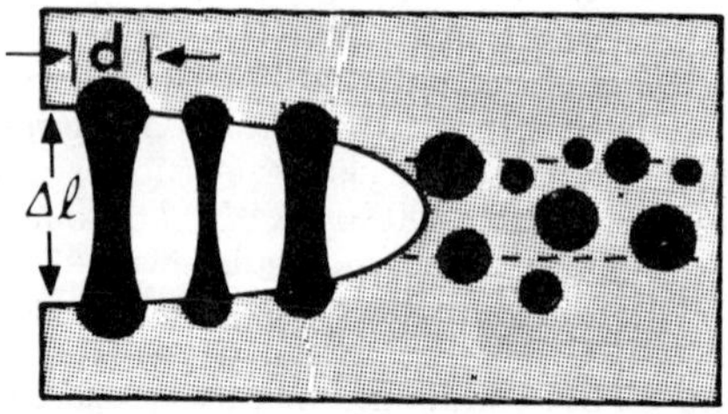

Figure 1. Model of crack propagation in a rubber-modified epoxy.

sition temperatures (T_g) of the rubber samples and of the particulate phase in the rubber-modified epoxies. The measurements were made at a scan rate of 20 °C/min.

Spin–spin NMR relaxation times were used to monitor the effect of radiation on the molecular dynamics of the rubber phase. The rubber-modified epoxies exhibited a two-component (5) free induction decay. The fast and slow components correspond to the epoxy matrix and mobile portion of the rubber phase, respectively. The spin–spin relaxation times were determined from the entire free induction decays of the model rubber compounds and from the slowly decaying portion of the free induction decays of the rubber-modified epoxies. For a given material, the spin–spin relaxation time is known to increase monotonically with the degree of molecular motion (7).

Results

Morphology. Osmium tetroxide stained transmission electron photomicrographs of the three rubber-modified epoxies studied are shown in Figure 2. Selection of these three materials allows the effect of three distinct ranges of particle sizes to be examined. The epoxy modified with ATBN rubber and cured with DEA has a narrow distribution of particle sizes with an average diameter of less than 20 nm. A previous study has shown that the epoxy modified with CTBN and cured with T-403 has a bimodal distribution of particle sizes; particles range in diameter from 20 nm to 1 μm (6). The same CTBN-modified epoxy cured with DEA has a particle-size distribution that is bimodal in nature; particles range in diameter from 20 nm to 20 μm (5).

Fracture Toughness and Tear Energy. The effect of radiation on the fracture toughness of the three rubber-modified materials and their unmodified counterparts is shown in Figures 3–5. Examination of these figures shows that, in the absence of oxygen, gamma radiation has little effect on the fracture toughness of the unmodified epoxies. For both DEA-cured rubber-modified epoxies, the effect of radiation on fracture toughness is pronounced despite the fact that these materials differ radically in particle size. The fracture toughness of the T-403-cured rubber-modified epoxy shows very little change with radiation until the material has been exposed to high doses (≥ 70 Mrads).

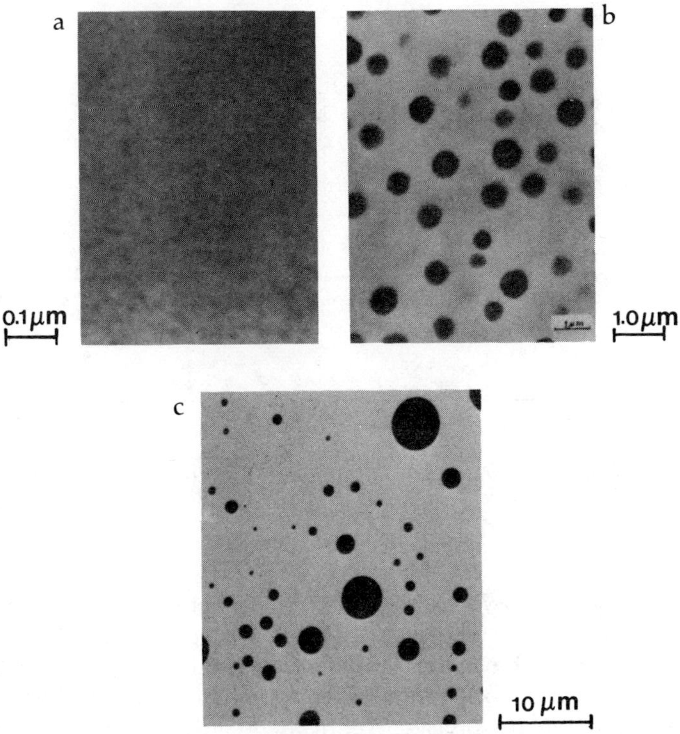

Figure 2. Transmission electron photomicrographs of 828–ATBN–DEA (a), 828–CTBN–T-403 (b), and 828–CTBN–DEA (c).

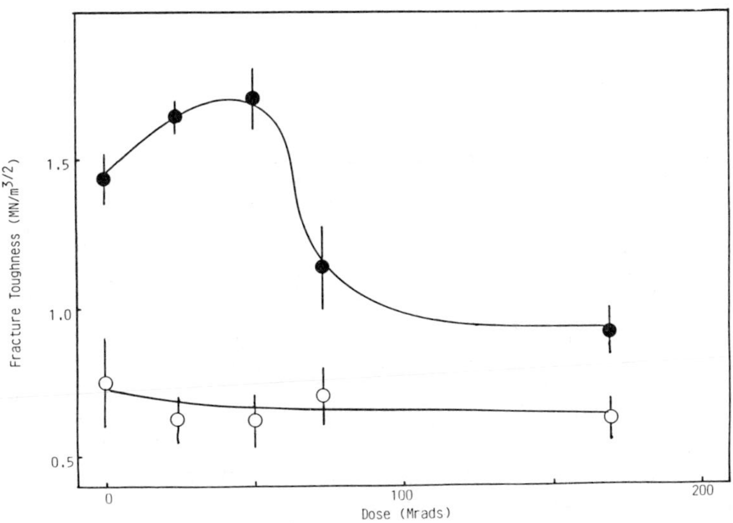

Figure 3. Fracture toughness vs. radiation dose for ATBN-modified and unmodified DEA-cured epoxy. Key: ●, 828–ATBN–DEA; and ○, 828–DEA.

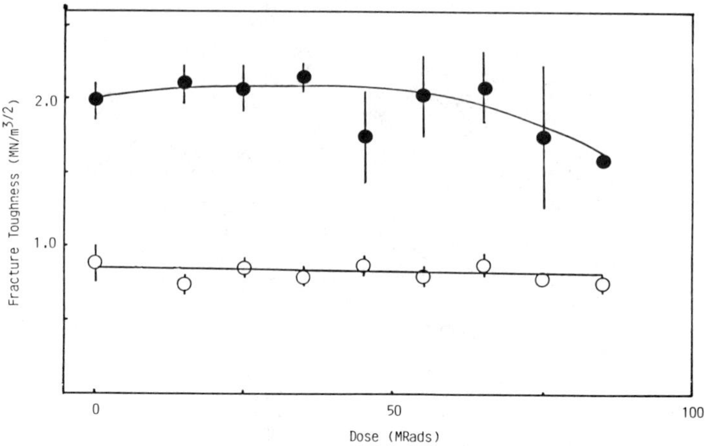

Figure 4. Fracture toughness vs. radiation dose for CTBN-modified and unmodified T-403-cured epoxy. Key: ●, 828–CTBN–T-403; and ○, 828–T-403.

In Figures 6 and 7, we compare the effect of radiation on the T_g of the rubber phase with the T_g of the corresponding rubber model compound of the DEA-cured and T-403-cured rubbers. In both systems, the T_g of the model rubber compound is typically higher than that of the particulate phase, except at the highest radiation levels. This effect is not attributed to differences in particle size; according to Bares (8), for particles greater than 0.1 μm (which account for the majority of rubber in these materials), particle size has a negligible effect on T_g.

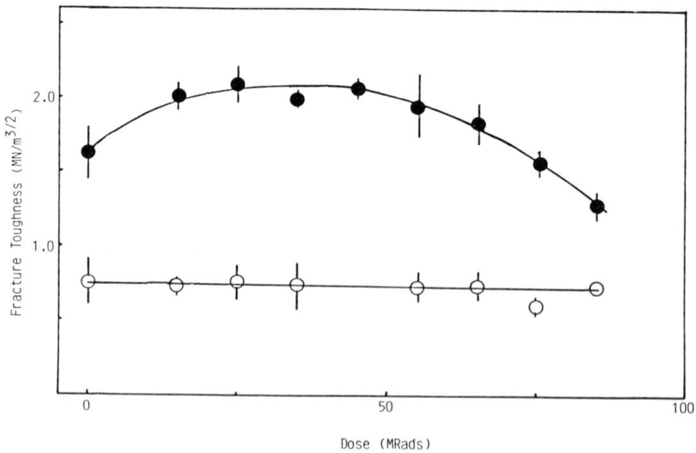

Figure 5. Fracture toughness vs. radiation dose for CTBN-modified and unmodified DEA-cured epoxy. Key: ●, 828–CTBN–DEA; and ○, 828–DEA.

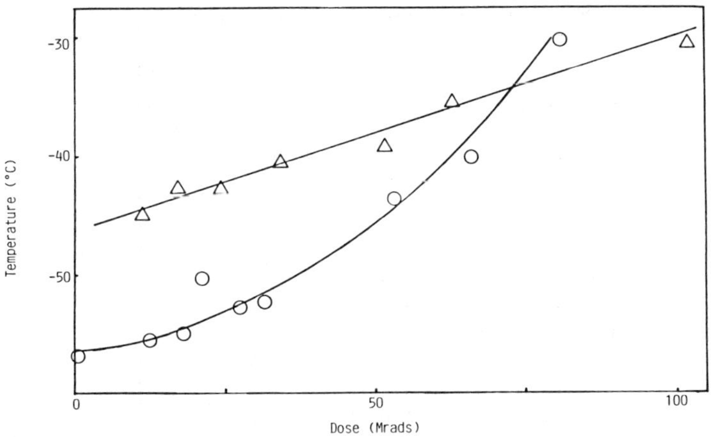

Figure 6. T_g *as a function of radiation dose for the dispersed-phase (○) and model rubber (△) compounds cured with DEA.*

The results presented in Figures 6 and 7 indicate that the model compounds do not exactly duplicate the particulate phase. The figures also show that the effect of radiation on both the particulate phase and the model rubber compounds is the same regardless of the curing agent used. This observation indicates that the curatives are not playing a role in the radiation cross-linking. Instead, the data are consistent with network formation via free-radical cross-linking through the butadiene unsaturation in the rubber backbone. According to the proton-NMR data in Figures 8 and 9, exposure to gamma radiation leads to a monotonically decreasing degree of molecular mobility for the particulate phases and the rubber model com-

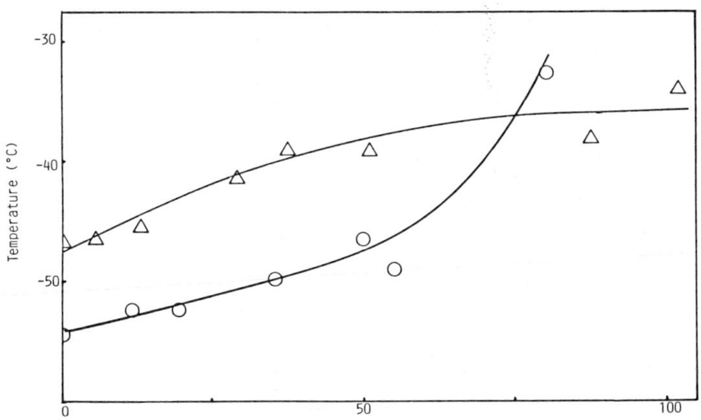

Figure 7. T_g *as a function of radiation dose for the dispersed-phase (○) and model rubber (△) compounds cured with T-403.*

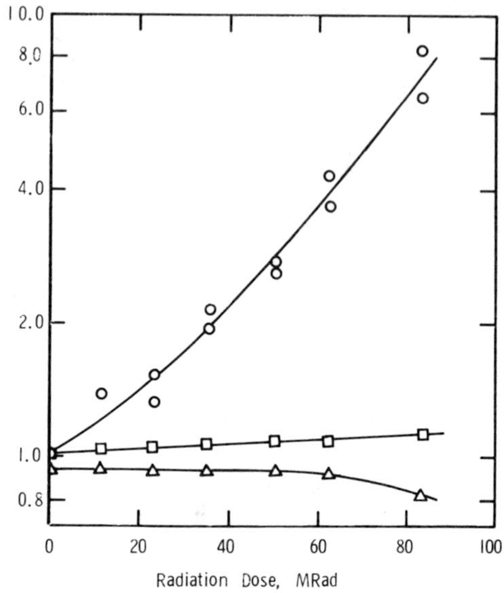

Figure 8. Normalized relaxation parameters of the 828–CTBN–DEA free induction decay, M(t), *expressed in the form*

$$M(t) = M_H exp(-t^2 M_2/2) + M_S exp(-t/T_2)$$

where M_H *and* M_S *are the fractions of the decay corresponding to the hard (epoxy) and soft (rubber) phases, respectively, and* $M_2/2$ *and* $1/T_2$ *are associated relaxation parameters. Key:* ○, $1/T_2$; □, $M_2/2$ *normalized to 1.0 for the unradiated sample; and* △, M_H.

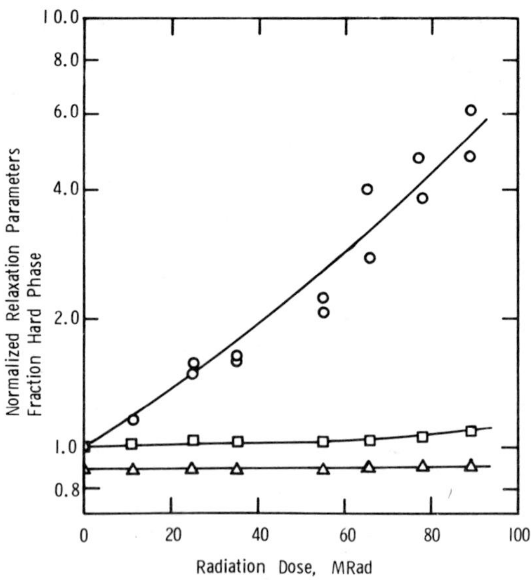

Figure 9. Normalized relaxation parameters of the 828–CTBN–T-403 free induction decay, M(t). (See *Figure 8 for explanation of the terms and symbols.*)

pounds. This decrease indicates that cross-linking of the rubber is the dominant reaction taking place and that chain scission is not a major reaction.

The tear energy versus dose measurements for the model rubber compounds are shown in Figures 10 and 11. The DEA-cured CTBN model rubber compound was too liquid to test following cure. From this apparent low cross-link density, the tear energy was assumed to be small at low doses. In contrast, the T-403-cured CTBN model rubber was a highly extensible rubber following cure and showed good tear strength at zero dose. The difference between these two rubbers stems from the mechanisms by which the two curatives cross-link epoxies. T-403 has three primary amine groups and cross-links epoxies through amine–epoxy reactions; it is used at 100% of stoichiometry. DEA is a secondary amine with one hydrogen available for reaction with epoxy groups; thus, it cannot cross-link via simple amine–epoxy reactions. Instead, DEA is believed to cause cross-linking by forming a tertiary amine salt and catalyzing homopolymerization; it is used at 20% of stoichiometry. This reaction is evidently made less favorable by the presence of large butadiene–acrylonitrile rubber molecules. Thus, the epoxy-endcapped rubbers appear to form a more completely cross-linked rubber with T-403 than with DEA. This observation is supported by the relative appearances of both model rubbers immediately following cure.

Toughening Model. Calculations of particle tear energies were made by using strains of rubber particles measured at the onset of tearing observed through a microscope. The model that quantifies the contribution to toughness of the stretching and tearing of rubber particles, described in detail elsewhere (3), can be summarized as follows. A crack in the composite propagates through the brittle epoxy matrix and leaves rubber particles bridging the crack as it opens (Figure 1). These particles are stretched to large strains before failing by a tearing mode. The toughening model is based on these

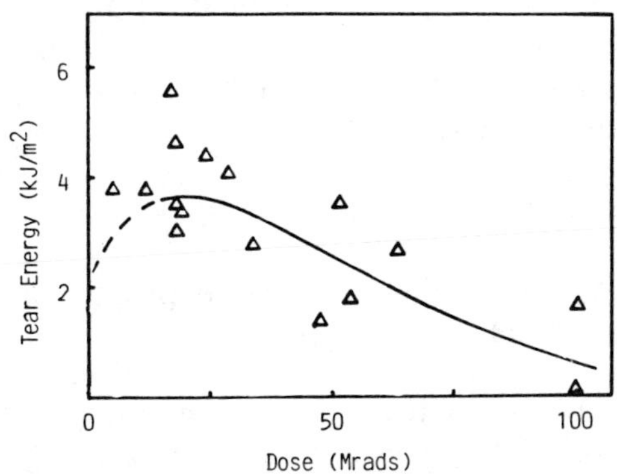

Figure 10. Effect of radiation on the tear energy of DEA-cured model rubber compound (828–CTBN–DEA).

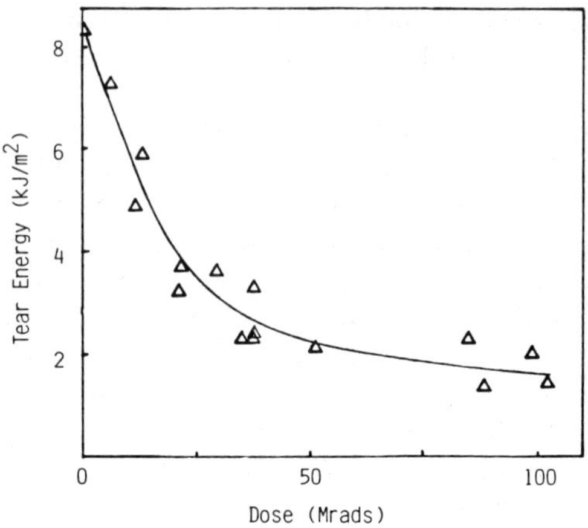

Figure 11. Effect of radiation on the tear energy of T-403-cured model rubber compound (828–CTBN–T-403).

observations and on the concept that elastic energy stored in the particle during straining is dissipated irreversibly (as heat) during the tear failure. Quantitative formulation of this failure model leads to an expression for the critical extension ratio, λ, for failure of the rubber particles:

$$\lambda^2 - \frac{4}{\lambda} + 3 = \frac{4\Gamma}{Gr} \tag{2}$$

Equation 2 predicts that the tearing strain of a particle depends directly on the rubber tear energy, Γ, and varies inversely with the particle radius, r, and the effective rubber shear modulus, G. An increase in the composite toughness, $\Delta\mathcal{G}_{Ic}$, can be identified with the energy stored in a volume fraction of particles, V_p, which is irreversibly dissipated upon failure.

$$\Delta\mathcal{G}_{Ic} = G\left(\lambda^2 + \frac{2}{\lambda} - 3\right)V_p r \tag{3}$$

Combining Equation 2 with Equation 3 gives the toughening contribution of the rubber particles.

$$\Delta\mathcal{G}_{Ic} = \left(1 - \frac{6}{\lambda^2 + \lambda + 4}\right)4\Gamma\, V_p \tag{4}$$

The model predicts a toughness increase proportional to the tear energy and volume fraction of the particles.

Figure 12 shows a typical photomicrograph illustrating the mechanism of rubber particles spanning an open crack during fracture in rubber-modified epoxies. If Equation 2 is rewritten in logarithmic form (*see* Equation 5), plotting $\ln(\lambda^2 - 4/\lambda + 3)$ versus $\ln(r)$ should lead to a straight line with a slope of -1.

Figure 12. Optical micrograph of crack in rubber-modified epoxy showing strained rubber particles bridging open crack.

$$\ln\left(\lambda^2 - \frac{4}{\lambda} + 3\right) = -\ln r + \ln\frac{4\Gamma}{G} \tag{5}$$

Values of λ were obtained (Equation 1) from measurements of DEA-cured and T-403-cured CTBN-modified epoxies at several radiation doses. Figure 13 shows a typical log–log plot of corresponding values of $\lambda^2 - 4/\lambda + 3$ against r. A line with a slope of -1 provides a reasonable fit of the data. Figure 14 summarizes the results for DEA-

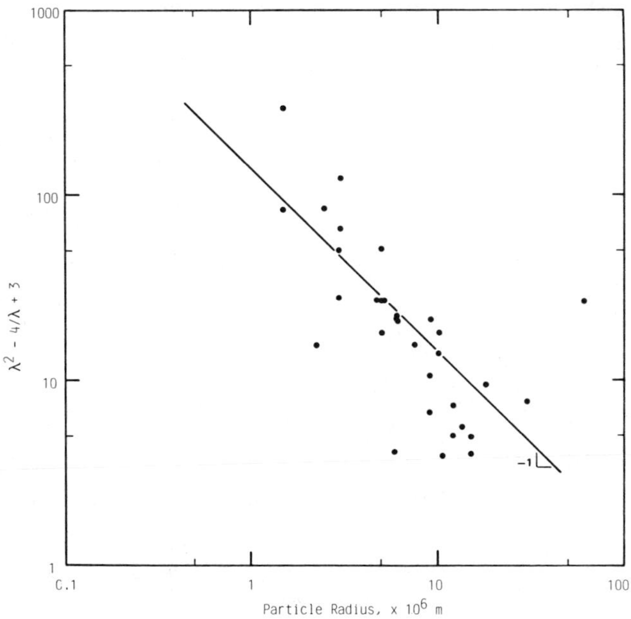

Figure 13. Typical plot showing inverse linear relationship between particle strain function and particle size (Equation 5) for a rubber-modified epoxy (828–DEA–CTBN) (0 Mrads).

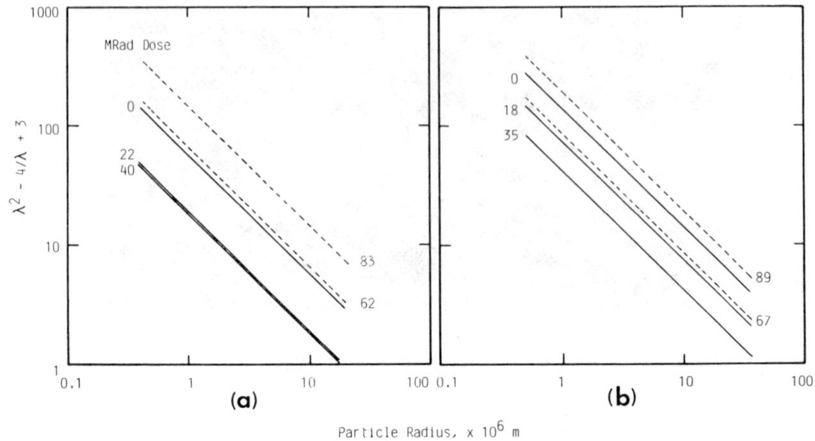

Figure 14. Plot showing inverse linear relationship between particle strain function and particle size (Equation 5) for T-403-cured (a) and DEA-cured (b) rubber-modified epoxies at different radiation doses.

cured and T-403-cured CTBN composites at different radiation levels.

Equation 5 is a straight line with the vertical intercept equal to ln $4\Gamma/G$. Values of tear energy, Γ, can therefore be extracted from Figures 13 and 14 as a function of dose if the shear modulus, G, of the dispersed phase can be estimated at each dose. The shear moduli of the model rubber compounds were taken as best estimates of the particulate phase moduli. Figure 15 shows the dependence of shear

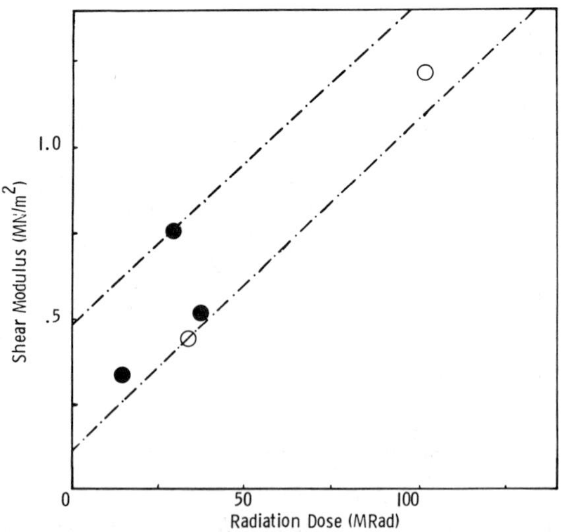

Figure 15. Shear modulus of DEA (●) and T-403 (○) model rubber compounds vs. radiation dose.

modulus on radiation for both model rubber compounds. An envelope was used to describe the data, which showed high scatter. The slopes for the upper and lower bounds of the envelope were derived from a linear-regression fit through the entire data set. Examination of the samples used for tear energy measurements led to the conclusion that the DEA-cured model rubber had a noticeably lower modulus. For this reason, the lower-bound value was taken as the DEA shear modulus and the upper-bound value as the T-403 shear modulus. Values of $4\Gamma/G$ obtained from Figure 14 were then combined with the G values to give tear energies of the particles, Γ. (Values for $4\Gamma/G$, G, Γ are given in Table II).

Equations 4 and 5, with $r = 5$ μm and a corresponding $\lambda \approx 4$, give $\Delta\mathcal{G}_{Ic} = 3\Delta\Gamma V_p$. Figure 16 gives these predicted values of toughness increase for both of the CTBN-modified epoxies examined. For comparison, Figures 17 and 18 show the toughness values for the CTBN-modified epoxies as calculated from the measured K_{Ic} values (Figures 4 and 5) by using the Irwin relation ($K_{Ic}^2 = E\mathcal{G}_{Ic}$). The E values are shown in Figure 19 as a function of radiation level.

Discussion

Exposure of rubber-modified epoxies to gamma radiation has been shown in this work to affect only the particulate phase. The toughness increase of the composites, attributed to cross-linking of the rubber particles, is illustrated in Figures 3–5. Our experiments show an increase in the cross-link density of the rubber phase that is directly proportional to the amount of irradiation to which the sample is exposed. Changes in the cross-link density and thus the physical prop-

Table II. Tear Energies of Rubber-Modified Epoxies

System	*Radiation Dose (Mrads)*	$\frac{4\Gamma}{G}$ ($\times 10^4$ m)	G (MN/m^2)	Γ (J/m^2)
828–DEA	0	1.40	0.13	4.6
	18.5	0.74	0.30	5.6
	29.7	0.43	0.41	4.4
	39.8	0.43	0.50	5.4
	66.8	0.90	0.75	16.9
	82.9	0.70	0.90	15.8
	88.6	2.00	0.94	47.0
828–T-403	0	0.77	0.48	9.2
	22.1	0.20	0.68	3.4
	31.7	0.47	0.77	9.0
	60.0	0.67	1.03	17.3
	62.1	0.87	1.05	22.8
	82.9	1.50	1.25	46.9

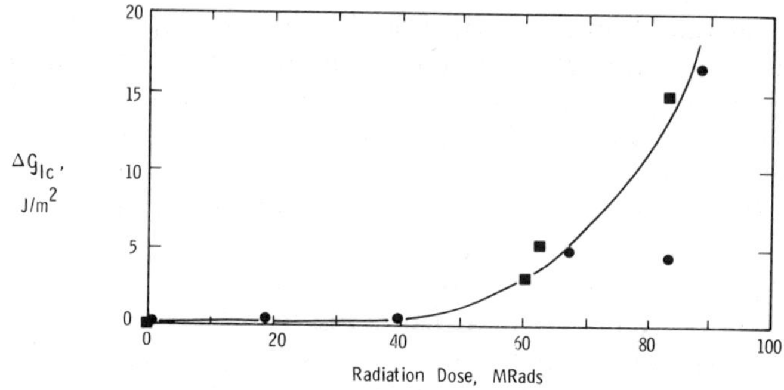

Figure 16. Predicted toughness increase (Equation 4) as a result of radiation cross-linking of the rubber particles in DEA-cured and T-403-cured rubber modified epoxies. Key: ●, 828–DEA–CTBN; and ■, 828–T-403–CTBN.

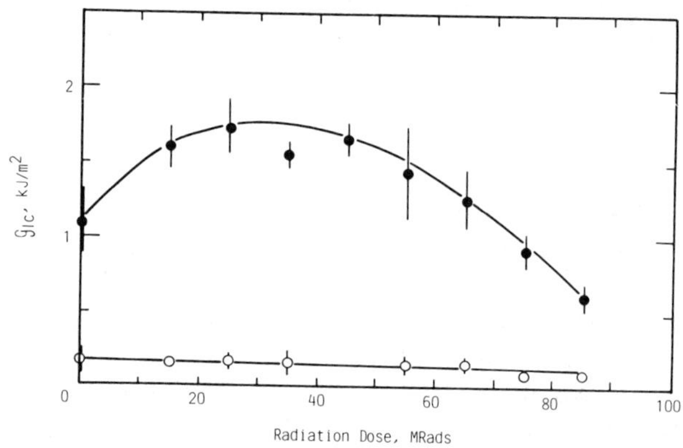

Figure 17. Toughness of DEA-cured rubber-modified epoxy calculated from K_{Ic} values in Figure 5 and from E values in Figure 19 by using Irwin relation $K_{Ic}^2 = E\mathcal{G}_{Ic}$. Key: ●, 828–DEA–CTBN; and ○, 828–DEA.

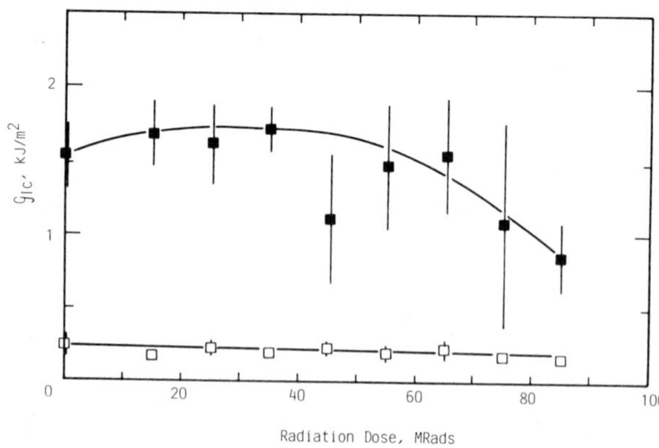

Figure 18. Toughness of T-403-cured rubber-modified epoxy calculated from K_{Ic} values in Figure 4 and from E values in Figure 19 by using Irwin relation. Key: ■, 828–T-403–CTBN; and □, 828–T-403.

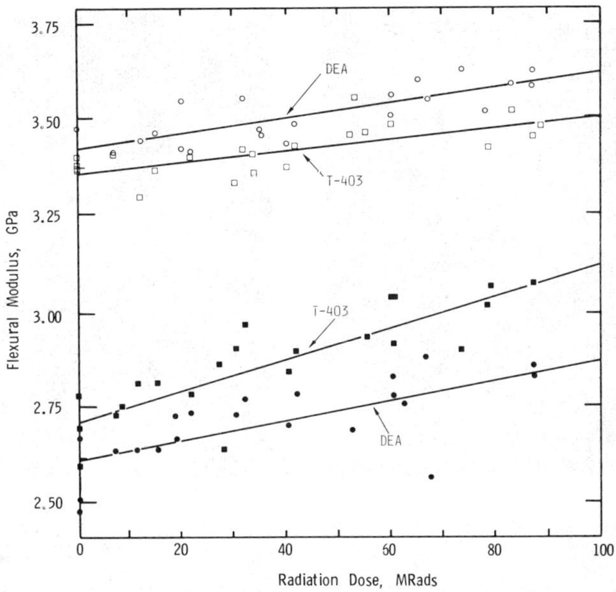

Figure 19. Flexural Young's modulus for DEA-cured and T-403-cured rubber-modified epoxies as a function of radiation dose. Key: ○, 828–DEA; □, 828–T-403; ■, 828–CTBN–T-403; and ●, 828–CTBN–DEA.

erties and stress state of the rubber particles have the potential for altering the toughness by several mechanisms. In this discussion we divide these mechanisms into two categories: toughening as a direct result of the rubber particles and toughening as a result of stress applied to the matrix by the particles.

Toughening by the Particulate Phase. In probing the applicability of the particle tear energy toughening model to rubber-toughened epoxies, two questions must be addressed. First, are the physical details of the proposed mechanism borne out as predicted. Second, how do the quantitative predictions of toughening based on the mechanism compare with the measured increases. Photomicrographs (Figure 12) that show rubber particles spanning a propagating crack in rubber-modified epoxies and tearing at critical strains confirm the postulated physical mechanism. Also, the increases in particle tear energies calculated from the measured particle strains as a function of radiation are consistent with the measured increases in both T_g and E of the dispersed rubber phase.

According to the tear energy model, the composite toughness should show a monotonic increase with radiation for a given volume fraction of particles. Figure 20 illustrates why this predicted trend is not followed exactly. Scanning electron microscopy (SEM) photographs of composite fracture surfaces reveal that the failure mode of the particles changes as the radiation level is increased. Unirradiated

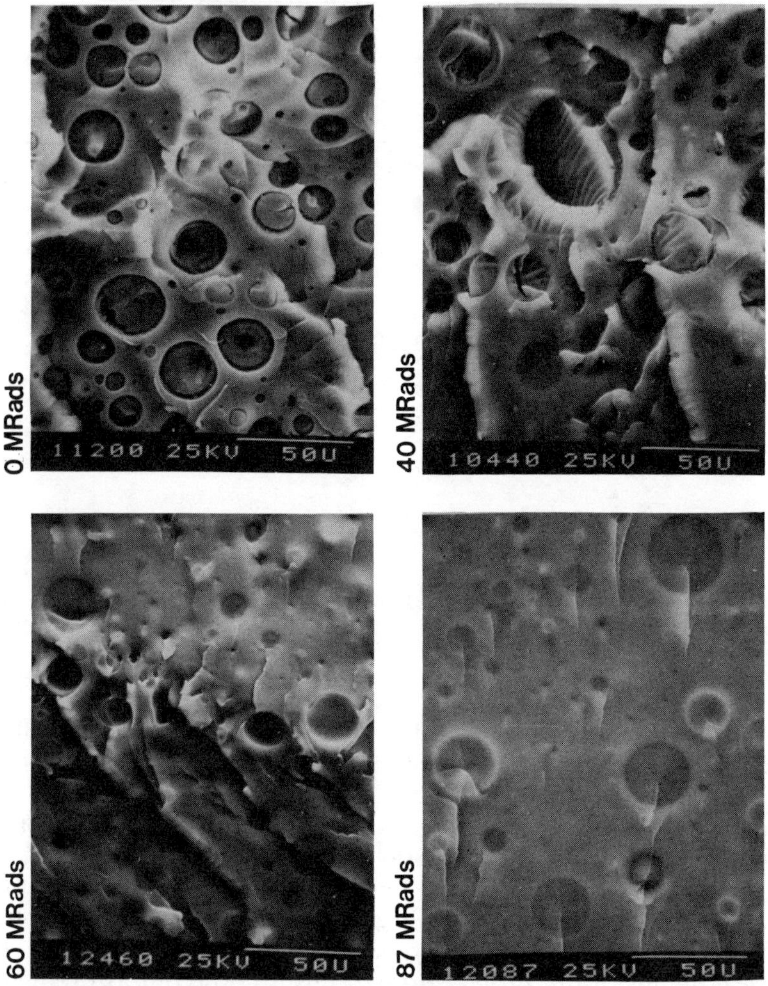

Figure 20. SEM photographs of fracture surfaces of DEA-cured rubber-modified epoxy showing transition from ductile to brittle fracture of rubber particles with increase in cross-link density or radiation exposure of rubber. (828–DEA–CTBN at 500× magnification).

rubber particles undergo large extensions and are torn partially away from their matrix sockets before failing in a characteristically ductile tearing mode. Particles exposed to 40 Mrads show definite indications of lower ductility; fewer particles show interfacial separation from the matrix and the flat failure surfaces of some are characteristic of low-strain, brittle fractures. Such loss in ductility is clearly evidenced by the exclusively flat particle fractures at 60 Mrads and, particularly, at 87 Mrads. At these dosages, the markings of a propagating crack visibly extend undisrupted from the matrix into the rubber particles.

Thus, the transition from ductile to brittle failure for the rubber in the composites occurs in the 40- to 70-Mrad range. This observation is consistent with the downturn in the toughness values observed in Figures 17 and 18.

The apparent inconsistency of the toughening model prediction of a monotonic increase in toughness with radiation (cross-link density) versus the observed increase and decline at high radiation levels can be resolved by considering the two critical parameters of the model: particle strain, λ, and particle tear energy, Γ (Equation 4). The tear energy of the rubber particles increases with cross-link density. This increase was shown experimentally (Table II) and is expected from the characteristic tearing behavior of rubber under quasi-static conditions (*10*). An increase in tear energy reflects an increase in the internal viscosity of the rubber; this increase produces a decrease in molecular mobility and, hence, an increase in the relaxation time of the rubber. Thus, although the tear energy (and, hence, toughening potential) of the particles increases with cross-link density, the ability of the particles to extend and absorb energy is limited by their response time relative to the propagating crack velocity. As the cross-link density increases with radiation dose, the extensibility (λ) of the particles decreases significantly and the rubber behaves in an increasingly brittle manner. This limited strain capacity of the particles severely limits their toughening potential. In addition, increased flaw sensitivity of the particles because of their brittle nature at high radiation levels will also decrease the extensibility of the rubber. As is apparent from the SEM photographs in Figure 20, many of the particles contain voids or other flaws that will severely limit stretching of the particles as they become brittle. This limitation effectively reduces the particle volume fraction that is active in toughening the composite.

The result of the counteracting effects of increasing tear energy and decreasing particle extensibility with cross-link density is illustrated schematically in Figure 21. At low cross-link densities, the particles exhibit large extensibility and high tear energy; therefore, the toughness rises. At higher radiation levels, the rubber is no longer capable of large strains. As a result, the overall toughening contribution of the particles decreases, despite high tear energy of the rubber.

Comparison of Figures 16 and 17 reveals that the predicted increase in toughness from particle tearing is lower than the observed toughness increase in rubber-toughened epoxies by more than an order of magnitude. Differences in experimental conditions can account for much of this discrepancy. The tearing rate of rubber particles in a composite that fractures catastrophically at a velocity between 1 and 100 m/s is in the range of 0.01 to 1 m/s (9). In contrast, when a crack is opened under quasi-static conditions, the rubber

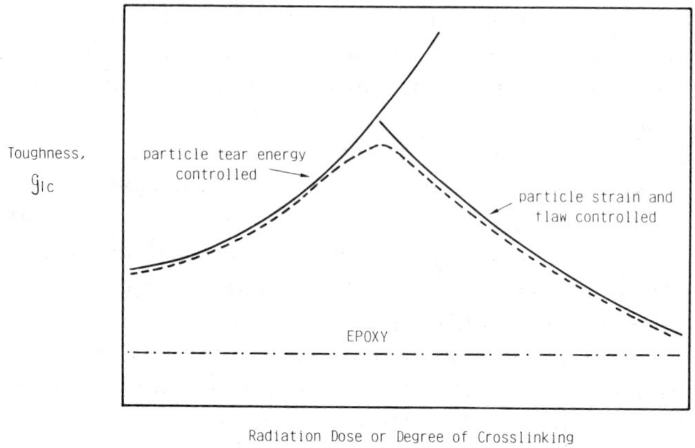

Figure 21. Schematic of toughness dependence of rubber-modified epoxy on cross-link density of the rubber phase. Toughness is predicted to increase with cross-link density until the particles are no longer extensible and become flaw sensitive; after this point toughness decreases.

particles are stretched at an approximate rate of 10^{-6} m/s. Thus, the tearing rates of the two experimental conditions differ by a factor of between 10^4 and 10^6. A three to four order of magnitude decrease in tearing rate reduces the tear energy of styrene–butadiene rubber by a factor of 10 (9). Accordingly, the measured effective tear energy of the quasi-statically strained particles is probably at least an order of magnitude lower than that of the rubber particles during fracture. Therefore, the curves of the predicted contribution to toughness by particle tearing shown in Figure 16 probably underestimate the actual contribution by at least a factor of 10.

Comparison of the composite toughness dependence on radiation (Figures 17 and 18) with the model rubber tear energy dependence on radiation (Figures 10 and 11) shows a difference in the shapes of the curves. The model rubber compounds exhibit several shortcomings as replicas of the particulate phase. First, ^{13}C-NMR data were used in selecting the molar composition of the model rubber compounds. Solid state ^{13}C-NMR spectroscopy, without magic-angle spinning, probes only the mobile species present. Thus, any immobile epoxy segments in the particulate phase are not detected and, therefore, were not incorporated in the model rubber compounds. Because different morphologies are then likely, the reinforcing (tear energy increasing) effect from hard segments is absent in the model rubbers. Second, a 2:1 molar ratio of epoxy to rubber was used in synthesizing the model rubbers. Because both the epoxy and rubber are difunctional, a series of oligomers with a broad molecular weight distribution will result. In contrast, rubbers in the rubber-modified epoxies are endcapped with epoxy by using a large excess of epoxy (~100:1). This endcapping leads to the formation of almost all epoxy–rubber–epoxy rubbers with few higher oligomers. Third, as with the particle-strain measurements, large differences in strain rate make comparison of the model rubber tear energy with composite toughness difficult. The model rubbers were torn at 25 cm/min compared to particles torn at crack velocities between 1 and 100 m/s for the composite.

Further evidence for the rubber tear energy contribution to toughening can be seen by examining the effects of the epoxy curative on the degree of cure of the rubber phase. Figures 3–5 show the effect of radiation on the fracture toughness of the three rubber-modified epoxies studied. Examination of these figures shows that below 30 Mrads, both DEA-cured systems undergo an increase in fracture toughness with increasing dose whereas the T-403-cured samples remained essentially unchanged. The apparent reason for this difference is an increase in the tear energy of the DEA-cured rubber (Figure 10). Examination of the glass transition of the rubber phase as a function of dose (Figures 6 and 7) shows that the DEA-cured systems do not achieve the T_g of the unirradiated T-403-cured material until 30 Mrads. In the model rubber studies, the unirradiated DEA-cured rubber was a liquid with presumed low tear energy. The T-403-cured model rubber at zero dose was a highly elastic, high tear energy rubber. If these differences also exist in the composites, then the changes in composite toughness do indeed parallel the changes in cure state of the respective particulate phases.

Toughening by Matrix Deformation. Several mechanisms have been predicated on the idea that the toughening resulting from the presence of a particulate rubber phase in epoxies is due to plastic deformation of the matrix. This deformation is believed to be enhanced by stress applied on the matrix by the particulate phase. In general during cure, the rubber shrinks and thereby increases the triaxial tension on the rubber. The increased triaxial tension results in increased radial tension and hoop compression in the matrix. For radiation cross-linking, the degree of cure of the rubber phase increases monotonically with exposure to radiation. The stress on the matrix should therefore increase monotonically with dose, regardless of the initial state of cure of the particulate phase. The observed decrease in toughness at high radiation doses indicates, instead, that the particles play a critical limiting role in the toughening mechanism. The results of this study clearly dispel theories that ignore the importance of state of cure and mechanical properties of the particles in toughening.

Changes in toughness of the composite appear to be directly related to the ability of the rubber phase to deform and tear. The composite with lightly cross-linked (cured) rubber increases in toughness as the cure progresses until the rubber loses its extensibility. This result is true regardless of the particle-size distribution. The composite with the initially highly cross-linked (cured) rubber remains relatively insensitive to further cure until the rubber loses its extensibility.

Acknowledgments

K. T. Gillen, Sandia National Laboratories Albuquerque (SNLA), suggested the use of gamma radiation to cross-link the dispersed phase. A. C. Davis, (SNLA) prepared the samples and performed the fracture toughness and tear energy measurements. D. A. Schneider, (SNLA) assisted in the NMR relaxation experiments. Research was supported by the U.S. Department of Energy under contract No. DE–AC04–76–DP00789.

List of Symbols

d	Particle diameter
Δl	Length of rubber stretched between crack faces
r	Particle radius
E	Young's modulus of composite
G	Effective rubber shear modulus
$\mathcal{G}_{Ic}$	Toughness
K_{Ic}	Fracture toughness
V_p	Volume fraction of particles
λ	Extension ratio at failure
Γ	Rubber tear energy

Literature Cited

1. Bucknall, C. B.; Yoshsii, T. presented at the *Third Int. Conf. on Deform. Yield and Fract. of Polym.*, PRI, Paper 13, Churchill College, Cambridge, April, 1976.
2. Sultan, J. N.; McGarry, F. J. *Polym. Eng. Sci.* **1973,** *27.*
3. Kunz-Douglass, S.; Beaumont, P. W. R.; Ashby, M. F. *J. Mater. Sci.* **1980,** *15*, 1109.
4. Bascom, W. D.; Cottington, R. L.; Jones, R. L.; Peyser, P. J. *Appl. Polym. Sci.* **1975,** *19*, 2545.
5. Sayre, J. A.; Assink, R. A.; Lagasse, R. R. *Polymer* **1981,** *22*, 87.
6. Kunz, S. C.; Sayre, J. A.; Assink, R. A. *Polymer* **1982,** *23*, 1897.
7. Slonim, I. Yu; Lyubimov, A. N. "The NMR of Polymers"; Plenum: New York, 1970.
8. Bares, J. *Macromolecules*, **1975,** *8*, 244.
9. Kunz, S. C. "The Toughening of Epoxy-Rubber Particulate Composites", Ph.D. Thesis, Univ. of Cambridge, England 1978.
10. Mullins, L. *Trans. Inst. Rubber Ind.*, **1959,** *35*, 213.

RECEIVED for review December 23, 1983. ACCEPTED April 18, 1984.

15

Rubber-Modified Epoxies

Cure, Transitions, and Morphology

L. C. CHAN[1,3], J. K. GILLHAM[1], A. J. KINLOCH[2,4], and S. J. SHAW[2]

[1] Polymer Materials Program, Department of Chemical Engineering, Princeton University, Princeton, NJ 08544

[2] Ministry of Defence, Royal Armament Research and Development Establishment (Waltham Abbey), Essex EN9 1BP, United Kingdom

A methodology for investigating cure and its relationship to the development of morphology and transitions for rubber-modified thermosetting systems has been developed. An aromatic tetrafunctional diamine-cured diglycidyl ether of bisphenol A epoxy resin [maximum glass transition temperature ($_{E}T_{g\infty}$) = *167 °C] was modified separately with two reactive liquid rubbers—a prereacted carboxyl-terminated rubber and an amino-terminated rubber. In order to develop fully cured but distinct cure-dependent morphologies, the systems were cured isothermally at different temperatures until reactions ceased and then were postcured by heating above* $_{E}T_{g\infty}$. *The maximum glass transition temperature of the matrix and the volume fraction of dispersed phase for the amino-terminated rubber-modified system were particularly sensitive to cure conditions because of complications in the cure chemistry.*

UNMODIFIED CURED EPOXY RESINS, like other thermosets with high glass transition temperatures (T_g), are brittle materials. The crack resistance can be improved by the addition of reactive liquid rubber to uncured neat epoxy systems (*1–3*). In situ phase separation occurs during cure; the cured rubber-modified epoxy resins consist of finely dispersed rubber-rich domains (~0.1–5 μm) bonded to the epoxy matrix. Improvement of fracture energy is dependent on the particle size, volume fraction, and size distribution of the dispersed phase, and on the chemical structures of the matrix and the dispersed phase (*4–14*). Shear deformation (*2, 7*), void formation (*6, 7*), crazing (*2–*

[3] Current address: Bell Laboratories, Whippany, NJ 07981

[4] Current address: Department of Mechanical Engineering, Imperial College, London University, London, SW7 2BX United Kingdom

0065-2393/84/0208-0235/$07.25/0

4), and rubber tear (*9, 10*) have been proposed as toughening mechanisms.

Cure of an initially homogeneous solution of epoxy resin–curing agent–rubber generally involves the sequential processes of phase separation, gelation, and vitrification. During cure, phase separation of the rubber occurs as a result of the increase in molecular weight, which lowers the compatibility of the rubber with the epoxy system. Separation is eventually quenched by the high viscosity accompanying gelation or vitrification (*5*). The development of morphology is dependent on the rates of nucleation and growth of the dispersed phase, compatibility of the rubber, and the rate of cure reactions (*5, 15*). Because these factors are temperature dependent, cure conditions may have a profound effect on the morphology and, consequently, on the mechanical behavior of the cured resin. Despite the potential implications, the effect of cure conditions on the properties of rubber-modified epoxies has scarcely been studied. If a range of morphologies can be obtained from a single rubber-modified epoxy formulation through control of cure conditions, relationships between mechanical properties and morphology may be better understood.

The main objective of this chapter is to discuss the cure process and its relationship to the development of morphology and transitions for two rubber-modified epoxy systems using an unmodified neat system as a control sample. The effect of cure conditions on the mechanical properties for the cured specimens is discussed in Reference 16. In an attempt to develop and arrest fully cured but distinct cure-dependent morphologies, the systems were cured isothermally at different temperatures (T_{cure}) to well beyond gelation and vitrification. They then were postcured by heating above the maximum glass transition temperture (${}_{E}T_{g\infty}$) of the system to complete the reactions of the matrix. An aromatic tetrafunctional diamine-cured diglycidyl ether of bisphenol A (DGEBA) type epoxy resin was selected as the neat system because of its high ${}_{E}T_{g\infty}$ (167 °C). The two rubbers were a prereacted carboxyl-terminated rubber and an amino-terminated rubber; both were made from the same copolymer of butadiene and acrylonitrile (AN) containing 17% AN. The chemistry of cure would be expected to be the same for the neat and prereacted carboxyl-terminated modified systems, whereas competing cure reactions should be introduced with the amino-terminated rubber. A preliminary report has been published (*17*).

A methodology for understanding and comparing cure behavior and properties of the cured state for rubber-modified thermosets has been developed. The first part involves monitoring cure by torsional braid analysis (TBA) to give gelation and vitrification times at different cure temperatures, assessing the kinetics of phase separation by complementary turbidity measurements, and summarizing the data in a time–temperature–transformation (TTT) isothermal cure diagram.

The second part summarizes transition temperatures after extended isothermal cure (and after postcure) in plots of transition temperatures versus the isothermal cure temperature. The third part involves assessing the relative amounts of dissolved and phase-separated rubber from the depression of the glass transition temperature of the fully cured system (${}_ET_{g\infty}$). The amount of the phase-separated rubber is then compared with the amount obtained from morphological examination with transmission electron microscopy (TEM). An apparent inconsistency in the amount of dissolved versus phase-separated rubber for the amino-terminated modified system led to considerations of the influence of time–temperature cure paths on the chemistry of cure of the system.

Experimental

Materials. The base resin and curing agent for the two rubber-modified epoxy systems were a diglycidyl ether of bisphenol A (DGEBA) resin (DER 331, Dow Chemical Company, epoxy equivalent weight = 190 g) and propyl 1,3-bis(4-aminobenzoate) (trimethylene glycol di-*p*-aminobenzoate, TMAB, Polacure 740M, Polaroid Corporation, NH equivalent weight = 78.5 g), respectively. DER 331–TMAB was selected because of the high T_g of the fully cured neat system (${}_ET_{g\infty}$ = 167 °C) and the nonvolatility (mp = 125 °C; bp > 250 °C), solubility, and health characteristics of TMAB.

The first system, denoted DTK-293, was modified with a commercial, prereacted, carboxyl-terminated butadiene–acrylonitrile (CTBN) copolymer containing 17% AN (K-293, Spencer Kellog Company, epoxy equivalent weight = 340 g). The K-293 rubber had been made by reacting a carboxyl-terminated rubber (CTBNx8, The BFGoodrich Chemical Company, AN content = 17%, COOH equivalent weight = 1850 g) with an excess of DGEBA resin. One mole of K-293 contains approximately 0.44 mol of CTBNx8 rubber and 0.56 mol of DGEBA resin. The second system, denoted DTAx16, was modified with a commercial amino-terminated butadiene–acrylonitrile (ATBN) copolymer containing 17% AN (ATBNx16, The BFGoodrich Chemical Company, NH equivalent weight = 850 g). The ATBNx16 rubber had been made by reacting the CTBNx8 rubber with *N*-(2-aminoethyl)piperazine (AEP) (*13*). The ATBN rubber used contained a residual amount of AEP (3% by weight) from its synthesis. Because K-293 and ATBNx16 were made from the same CTBN rubber, the molecular weights and the T_g values of the rubbers are approximately the same ($T_g \cong -50$ °C). The formulations for the neat system and the two rubber-modified epoxy systems containing 15 parts per hundred parts resin (phr) of rubber are given in Table I.

Specimen Preparation. A mixture of liquid epoxy resin and liquid rubber (for modified systems) was heated to 120 °C in an open beaker, and the solid curing agent was added and dissolved with the aid of mechanical stirring for 5 min. The solution was degassed at 100 °C for 25 min in a preheated vacuum oven at a pressure of about 1 torr. A small part of this degassed mixture was dissolved in methyl ethyl ketone in a volume ratio of 1:4 for the TBA experiments. The main batch of the degassed mixture was then poured into a preheated (at T_{cure}) mold [precoated with a release agent (QZ 13, Ciba-Geigy Chemical Company) (*18*)] and cured in an air oven to prepare a casting (220 × 220 × 6 mm) for morphological and mechanical studies (*16*). The open end of the

Table I. Chemical Formulations of the Neat and Rubber-Modified DTK-293 and DTAx16 Systems

Component	*Neat*	*DTK–293*	*DTAx16*
DER 331	100.0	100.0	100.0
TMAB	41.0	51.0	39.9
K–293[a]	—	42.0	—
ATBNx16	—	—	15.0

NOTE: Neat system (no rubber)—1 epoxy end group/1 amine hydrogen. Rubber-modified systems—15 phr of rubber/100 phr of unreacted epoxy and 1 free epoxy end group/1 amine hydrogen [it was assumed that all epoxy end groups in DTK–293 and in DER 331 react with TMAB and that all NH in the ATBN rubber (including the NH in the residual AEP) and the TMAB react with the epoxy].

[a] 1 mol of K–293 contains 0.44 mol of rubber and 0.56 mol of epoxy.

mold was sealed with a plug made of silicone rubber (RTV, General Electric Company) in an attempt to minimize exposure to air during cure.

After prolonged isothermal cure, TBA specimens were postcured by heating to and from 240 °C at a rate of 1.5 °C/min. In contrast, the casting was postcured by heating the mold from the cure temperature to 170 °C at 1.5 °C/min and holding at 170 °C for 5 h. Values of ${}_{E}T_{g\infty}$ obtained from film specimens (*16*) that had been machined from the casting were the same as those obtained from the TBA experiments. Therefore, the casting had been fully cured by the postcure step. The morphology of the cured specimens was not expected to be altered by the postcure process because the cure reactions had proceeded well beyond gelation, and a slow heating rate had been used in postcuring.

Torsional Braid Analysis. The transformation of a liquid epoxy formulation to solid thermoset polymer during the process of cure was monitored by using a composite specimen in an automated torsional pendulum instrument [Torsional Braid Analysis (TBA), Plastics Analysis Instruments, Inc.] (*19, 20*). The specimen was an inert, multifilamented glass braid impregnated with the reactive liquid. The pendulum was intermittently set into motion to generate a series of freely damped waves with a natural frequency of 0.05–5 Hz. The change of material behavior of the specimen was monitored as a function of time and/or temperature by measuring two dynamic mechanical properties, relative rigidity and logarithmic decrement, obtained from the frequency and decay constants that characterize each wave. The relative rigidity [$(1/P^2)$, where P is the period in seconds] is directly proportional to the in-phase portion of the shear modulus (G'). The logarithmic decrement [$\Delta = \ln(A_i/A_{i+1})$, where A_i is the amplitude of the ith oscillation of a freely damped wave] is directly proportional to the ratio of the out-of-phase portion of the shear modulus (G'') to G' ($\Delta \cong \pi G''/G' = \pi \tan\phi$, where ϕ is the phase angle between the stress and strain). At a selected temperature (T_{cure}), the specimen was cured until reactions were essentially quenched as indicated by the leveling off of the rigidity curve (although changes were still occurring in the mechanical damping). Transformations associated with changes from one state to another during cure were identified by the time of maxima (and the associated frequencies) in the logarithmic decrement plot. After prolonged isothermal cure, the dynamic mechanical spectrum of a cured specimen was obtained by cooling from T_{cure} to −170 °C and heating to 240 °C at a rate of 1.5 °C/min. A subsequent scan from 240 to −170 °C was considered to provide the spectrum of the fully cured specimen. Transition temperatures were identified by the temperature of maxima (and the associated frequencies) in the logarithmic decrement plot during a temperature scan. All the TBA experiments were performed in helium.

Turbidity Study. In situ phase separation was monitored by changes in light transmission through a sample consisting of a degassed epoxy formulation in a capillary glass tube (i.d. 1 mm) sandwiched between two heated copper blocks. The light beam was a He–Ne laser (λ = 632.8 nm, 0.95 mW; beam diameter ~1 mm). The signal-to-noise ratio was amplified by using a lock-in detection device (*21*) that consisted of a chopper (chopper frequency = 200 Hz) and a lock-in amplifier (Princeton Applied Research Company, Model 191 and 121, respectively). The chopper was placed between the laser and the sample. The intensity of light transmitted through the sample, which was monitored by a photodiode (Silicon Detector Company, Model SD-444-11-11-171) connected to the lock-in amplifier, was recorded on an x–y strip-chart recorder. The light transmission apparatus was covered by black cloth to minimize stray light from the background.

The initial mixtures of the two rubber-modified systems were optically clear within the cure temperature range (100–200 °C); therefore, each of the two rubbers was considered to be compatible with the neat epoxy system at the beginning of cure. During cure, an increase in molecular weight of the system lowered the compatibility of the rubber, and the mixture turned from optically clear to cloudy. The cloud point, which was defined as the first indication of decrease of light intensity transmitted through the sample, signaled the onset of phase separation. The end of phase separation was indicated by the leveling off of the light transmission curve. The light transmitted through the sample was constant throughout cure for the neat system.

Morphological Study. Morphologies of cured specimens were examined with scanning electron microscopy (SEM) and transmission electron microscopy (TEM). The SEMs were obtained from the fracture surfaces of cured specimens that had been coated with a thin layer (~600 Å) of gold by using a high vacuum sputterer. The TEMs were obtained from specimens that had been stained with OsO_4 and microtomed at room temperature. The OsO_4 reacted with the unsaturated double bonds of the rubber molecules, and the rubbery domains appeared as black sections in the TEMs. Two analytical methods were used to convert the distribution of the sections in the TEMs to the corresponding distribution of particles; the volume fraction and mean diameter of the particles were calculated from the particle distribution. The first method (Schwartz–Saltykov) (22) is based on the distribution of section diameters in the TEM micrographs. The second method (Spektor) (22) is based on the distribution of chord intercept lengths along random straight lines drawn in the TEMs. In general, good overall agreement is to be expected from using the two methods; however, the details of the results may differ because of the inherent differences in the analyses (22). On average, 20–30 TEMs were examined for each cure condition.

Results and Discussion

Neat Epoxy System (DER 331–TMAB). Transformations from liquid-to-rubber and rubber-to-glass were measured by gelation and vitrification times, respectively. Representative isothermal TBA cure spectra from low to high temperatures of cure (100–200 °C) of DER 331–TMAB are shown in Figure 1.

At low cure temperatures, three events are apparent in the logarithmic decrement (demonstrated by the 100 and 120 °C isothermal

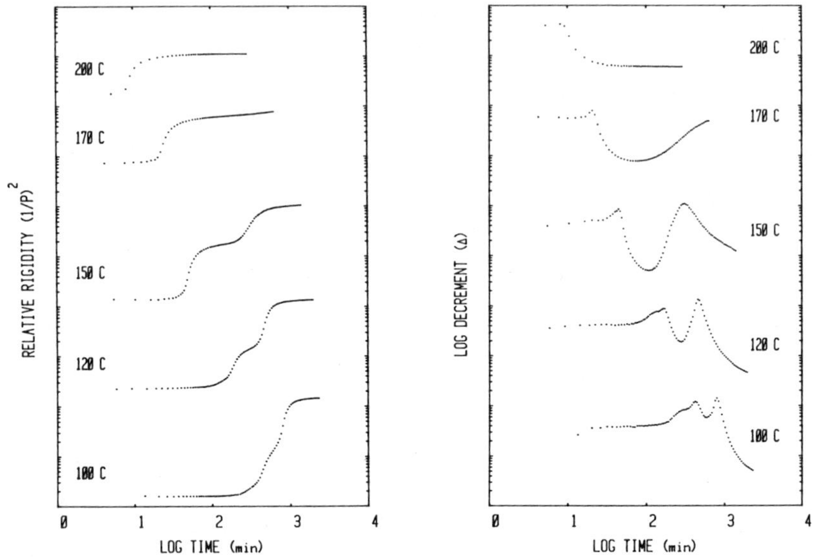

Figure 1. Representative TBA isothermal-cure spectra for the neat system (DER 331–TMAB).

cure spectra). The first peak has been designated as gelation (liquid-to-rubber transformation) and generally corresponds to the onset of insolubility (which accompanies the incipient formation of the cross-linked network) as measured in gel-fraction experiments (*23*). The shoulder before the gelation peak (which is less apparent at higher temperatures) has been attributed to an isoviscous event that is due to the interaction of the liquid and the braid in the TBA experiment (*19*); therefore, it will not be considered further. The second peak in the logarithmic decrement, which is located in the middle of the rubber-to-glass transition region (indicated by the relative rigidity plot), has been designated as vitrification (rubber-to-glass transformation) (*19*). Physically, the vitrification peak in the TBA isothermal cure spectrum corresponds to the state at which the ${}_{E}T_g$ of the epoxy resin has risen to T_{cure} (*14, 23*). Beyond gelation and approaching vitrification, the increases in the cross-link density and average molecular weight cause a reduction of molecular mobility and chemical reactivity; therefore, the reactions of the epoxy system become diffusion controlled. Eventually, the reactions are quenched when the resin has become a glass. This phenomenon is demonstrated by the continuous increase and the subsequent leveling off of the relative rigidity plot (modulus) beyond the vitrification peak. The ${}_{E}T_g$ has then risen to a temperature of $T_{cure} + \delta T$. The difference (δT) between ${}_{E}T_g$ of the isothermally cured resin and T_{cure} arises from the distinction between the quenching of chemical reactions and the particular method of measurement of both ${}_ET_g$ and vitrification (*24*). δT is a function of T_{cure} and of time at T_{cure}. If T_{cure} is higher than the ${}_ET_{g\infty}$ of the DER 331–TMAB epoxy system (167 °C), vitrification cannot occur and only the gelation peak is observed, as seen in the 170 and 200 °C isothermal cure spectra; therefore, reactions of the epoxy resin will not be quenched. Completion of the reactions can only be attained at temperatures above (${}_ET_{g\infty} - \delta T$); in contrast, reactions will be quenched if T_{cure} is below (${}_ET_{g\infty} - \delta T$).

Gelation and vitrification times of the DER 331–TMAB epoxy system at different cure temperatures are presented in Figure 2 in the form of a TTT isothermal-cure diagram (*19, 23*). The TTT diagrams presented do not account for the changes that occurred during the preparation of the specimens (i.e., 120 °C for 5 min and 100 °C for 25 min). The TTT diagram shows that the gelation and vitrification curves cross at 75 °C. This value was obtained from the intersection of the log time versus $1/T$ plots for gelation and vitrification. The temperature at which the time to gelation and the time to vitrification are the same has been denoted as ${}_{gel}T_g$ (*19*). If T_{cure} is below (${}_{gel}T_g - \delta T$) but above (${}_ET_{g_0} - \delta T$) (${}_ET_{g_0} = {}_ET_g$ of the initial resin), the resin will not gel on cure but will vitrify to form an ungelled thermoplastic glass of low molecular weight. ${}_ET_{g_0}$ for the DER 331–TMAB epoxy system is 2 °C as determined from a TBA thermomechanical spectrum. If T_{cure} is between (${}_{gel}T_g - \delta T$) and ${}_ET_{g\infty}$, the resin will gel to form a gelled glass. However, if T_{cure} is above ${}_ET_{g\infty}$, the resin gels to form a gelled rubber but does not vitrify at the temperatures of cure

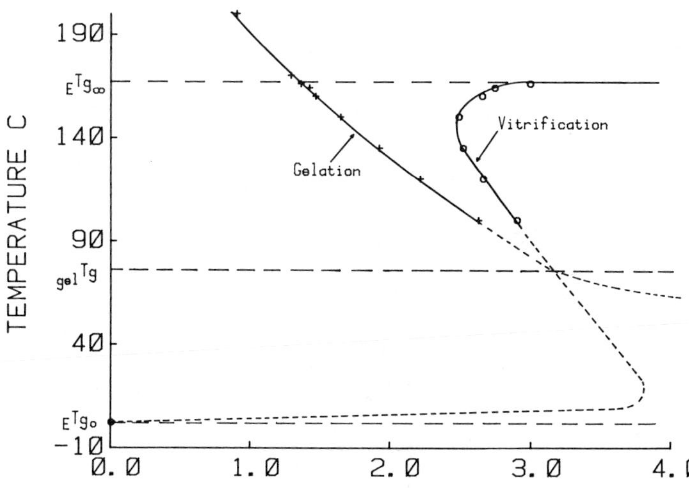

Figure 2. TTT isothermal-cure diagram for the neat system (DER 331–TMAB). Key: +, gelation and ○, vitrification.

(as discussed). The vitrification curve in TTT isothermal-cure diagrams is generally S-shaped, with ${}_{E}T_{g\infty}$ and ${}_{E}T_{go}$ as the upper and lower boundaries, respectively (*19*). The time to vitrify reaches a minimum at a T_{cure} just below ${}_{E}T_{g\infty}$ because of the competing effects of increasing reaction rate constant and increasing extent of conversion for vitrification with increasing temperature. In contrast, the time to vitrify reaches a maximum at a T_{cure} just above ${}_{E}T_{go}$ because of the competing effects of decreasing viscosity and increasing reactivity with increasing temperature. For the DER 331–TMAB epoxy system, only the upper part of the S-shaped vitrification curve in the TTT isothermal-cure diagram has been obtained experimentally because of the low reactivity of the neat system at low temperature. However, ${}_{E}T_{go}$ was measured directly in a temperature scan. In principle, the S-shaped vitrification curve and the gelation curve can be computed from the reaction kinetics, the conversions at gelation and vitrification, and the physical parameters of the system (${}_{E}T_{go}$, ${}_{E}T_{g\infty}$, and ${}_{gel}T_{g}$) that are dependent on the chemical structure of the reactive constituents of the system (*23*).

Representative TBA thermomechanical spectra, obtained after prolonged isothermal cure (100–200 °C), are shown in Figure 3. Thermomechanical spectra were obtained by cooling from T_{cure} to −170 °C and heating to 240 °C. If T_{cure} is below (${}_{E}T_{g\infty}$ − δT), the

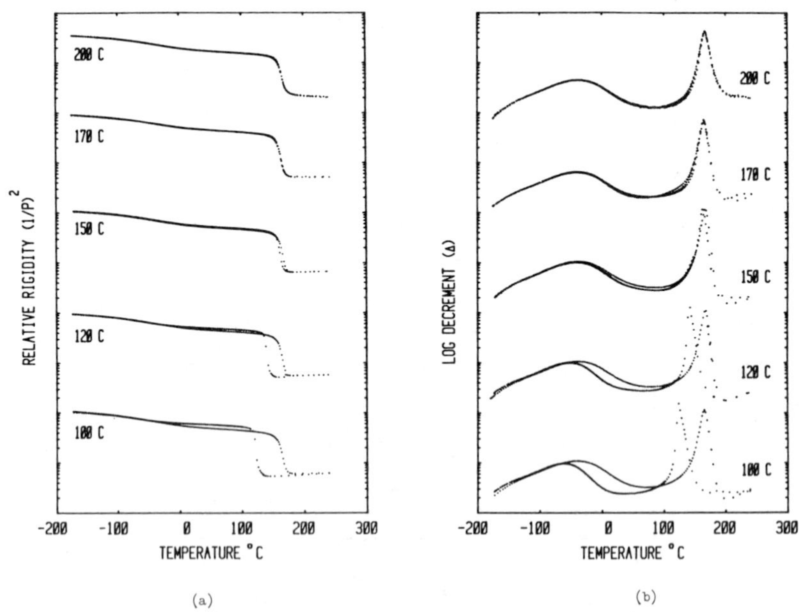

Figure 3. Representative TBA thermomechanical spectra after isothermal cure for the neat system (DER 331–TMAB).

partially cured resin can be postcured at higher temperatures; a subsequent scan from 240 to −170 °C will provide the spectrum of the fully cured resin. Three relaxations are observed: the glass transition ($_{E}T_g$) at higher temperatures; a subzero secondary transition ($_{E}T_{sec}$), which has been attributed to the motion of the $-CH_2-CH(OH)-CH_2-O-$ group in the epoxy (*25*); and a weak transition below −100 °C. The maximum values of $_{E}T_g$ and $_{E}T_{sec}$ obtained after heating to 240 °C for the DER 331–TMAB system were 167 °C ($_{E}T_{g\infty}$) and −39 °C ($_{E}T_{sec\infty}$), respectively. Both $_{E}T_{g\infty}$ and $_{E}T_{sec\infty}$ are independent of T_{cure}, whereas $_{E}T_g$ and $_{E}T_{sec}$ increase with T_{cure} (Figure 4). For $T_{cure} << {}_{E}T_{g\infty}$, the values of $_{E}T_g$ after cure at T_{cure} are about 20 °C higher than T_{cure}. The measured difference ($_{E}T_g - T_{cure}$) varies with epoxy systems (*26, 27*).

The relative rigidity plot for a fully cured specimen crosses over that for the partially cured specimen. This result is demonstrated in Figure 3a for the specimens cured at 100 and 120 °C. Therefore, the modulus at room temperature (which is below the crossover temperature) for a cured specimen decreases as the $_{E}T_g$ value and extent of conversion (cross-linking density) increase. For other high T_g epoxy systems, the decrease in room temperature (RT) modulus with increasing extent of cure is paralleled by corresponding decreases in RT density and increases in RT equilibrium levels of absorbed water

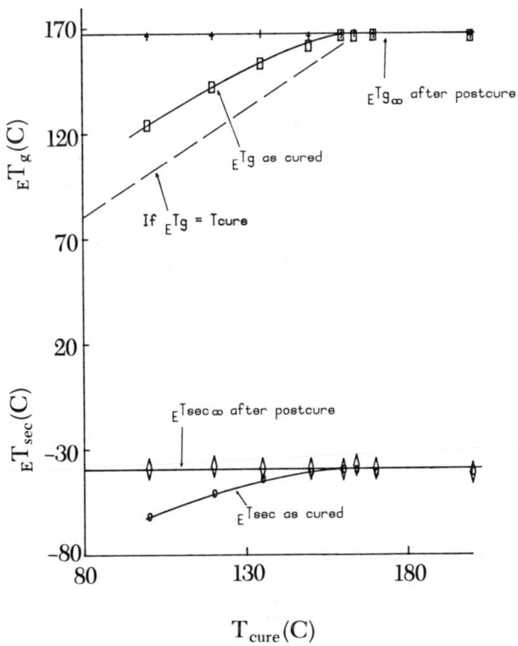

Figure 4. $_{E}T_g$ (□), $_{E}T_{g\infty}$ (+), $_{E}T_{sec}$ (○), *and* $_{E}T_{sec\infty}$ (◇) *vs.* T_{cure} *for the neat system, DER 331–TMAB.*

(*19*, *28*). A common cause for these interrelated phenomena (*28*) is the increasing free volume at RT with increasing extent of cure (*29*).

Prereacted Carboxyl-Terminated Rubber-Modified Epoxy System (DTK–293). The cure chemistry of the DTK–293 and neat systems are expected to be similar because the reactive functional end groups are the same. Representative TBA isothermal-cure spectra from low to high temperatures of cure (100–200 °C) are presented in Figure 5. As for the neat system, the isothermal-cure spectra of the DTK–293 system displayed gelation and vitrification peaks; peak times are summarized in the form of a TTT isothermal-cure diagram (Figure 6). The initial ${}_{E}T_{g}$ values of the DTK–293 and the neat systems were the same (i.e., ${}_{E}T_{g_0}$ = 2 °C). Because the lowest temperature of cure was 100 °C, only the upper part of the TTT diagram is shown. For comparison, the corresponding part of the TTT diagram for the neat system is also included. In general, the TTT diagram for the DTK–293 system follows the same pattern as for the neat system. The times to gelation and the activation energy of gelation (ΔE), obtained by plotting the gelation times in an Arrhenius manner, are also similar for the two systems. The activation energy of gelation for the DTK–293 system was found to be 14.2 kcal/mol compared to 14.1 kcal/mol for the neat system. However, the values of ${}_{E}T_{g\infty}$ for the DTK–293 system are not constant and are

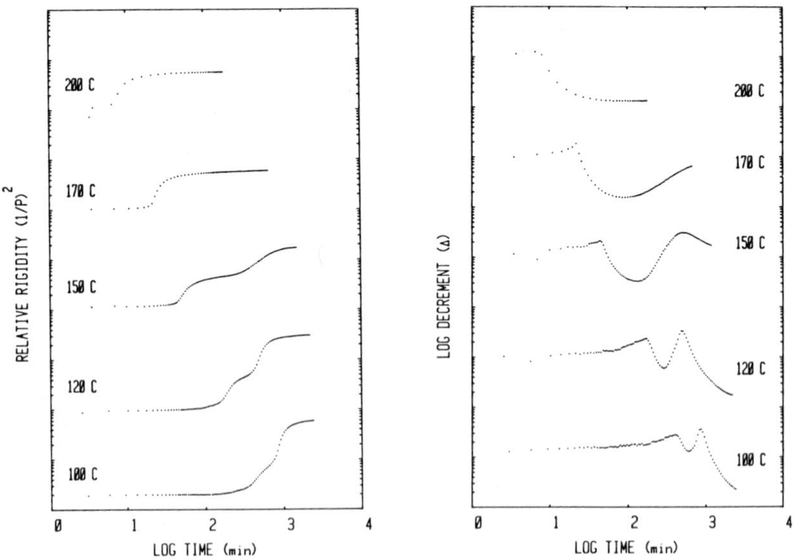

Figure 5. Representative TBA isothermal-cure spectra for the rubber-modified DTK–293 system.

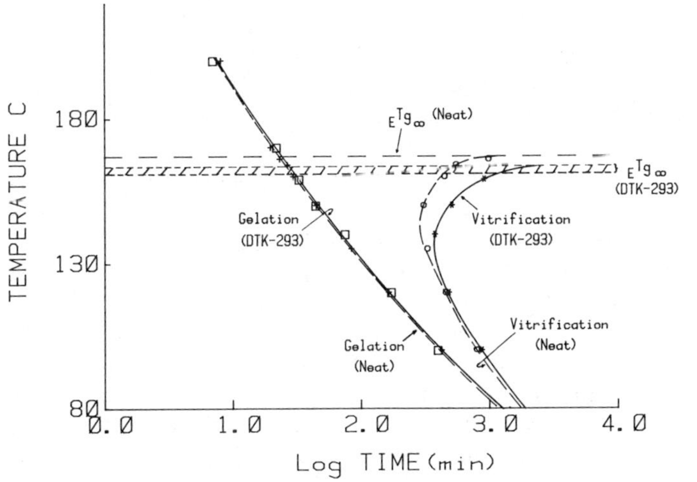

*Figure 6. TTT isothermal-cure diagram for the rubber-modified DTK–293 system vs. the neat system. Key: □, gelation (DTK–293); *, vitrification (DTK–293); +, gelation (neat); and ○, vitrification (neat).*

represented by a shaded band in the TTT diagram. The width of the band represents the range of ${}_{E}T_{g\infty}$ resulting from different cure conditions.

The light transmission results are included in the TTT isothermal-cure diagram (Figure 7). The time for onset of phase separation (cloud point) decreases with an increase in T_{cure}. A line de-

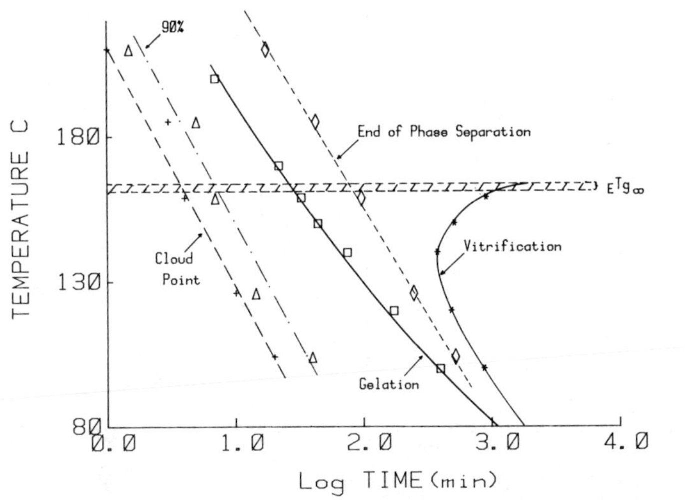

*Figure 7. TTT isothermal-cure diagram including phase separation for the rubber-modified DTK–293 system. Key: □, gelation; *, vitrification; +, cloud point; △, 90% decrease of light intensity; and ◇, end of phase separation.*

noting 90% decrease of the initial light intensity is also included to provide an estimate of how rapidly the morphology develops. Most of the phase separation of the dispersed phase occurred well before gelation, although some changes were observed afterwards, as has been reported (*30, 31*).

Representative thermomechanical spectra (Figure 8) show four relaxation processes in the logarithmic decrement plots: a high temperature relaxation associated with T_g of the epoxy matrix (${}_ET_g$), a broad relaxation at about −40 °C, a relaxation at about −50 °C associated with the T_g of the rubbery phase (${}_RT_g$), and a relaxation below ${}_RT_g$. Correlation of the volume fraction of the rubbery phase with the intensity of the relaxation associated with ${}_RT_g$ as in other reports (*4, 5, 32*) is difficult because of the overlapping of ${}_RT_g$ with other processes. The ${}_RT_g$ loss peaks were found to be located at or below the T_g of the unreacted liquid rubber even though the rubber in the dispersed phase was chemically combined with the epoxy resin. The suppression of ${}_RT_g$ is presumably a consequence of the triaxial stresses induced on the rubbery domains resulting from differences of thermal expansion coefficients of the two phases (*5*).

Values of ${}_ET_g$ and ${}_ET_{g\infty}$ versus T_{cure} for the DTK−293 and neat systems are shown in Figure 9 and are included in Table II. Values of ${}_ET_{g\infty}$ for the rubber-modified system were lower than that for the neat system by 4–6 °C. The decrease of ${}_ET_{g\infty}$ for cured rubber-mod-

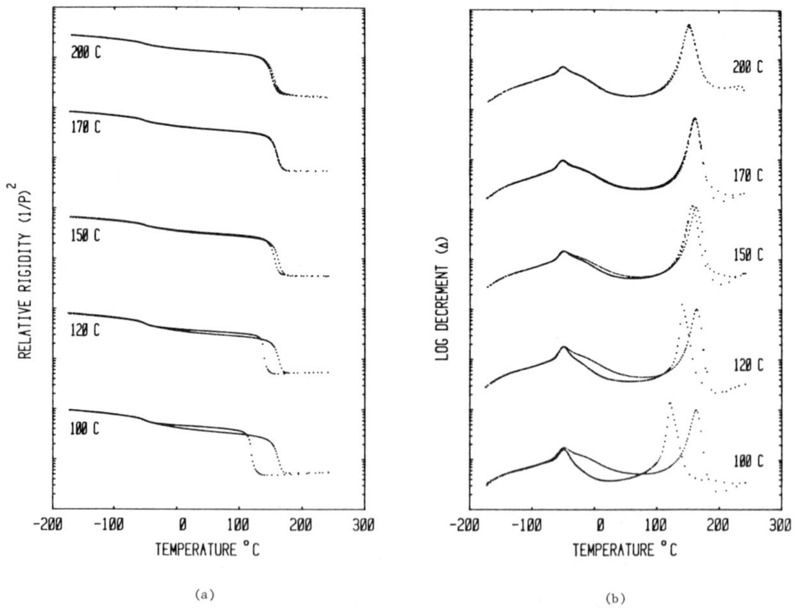

Figure 8. Representative TBA thermomechanical spectra after isothermal cure for the rubber-modified DTK–293 system. Key: (a) relative rigidity and (b) logarithmic decrement, both vs. temperature.

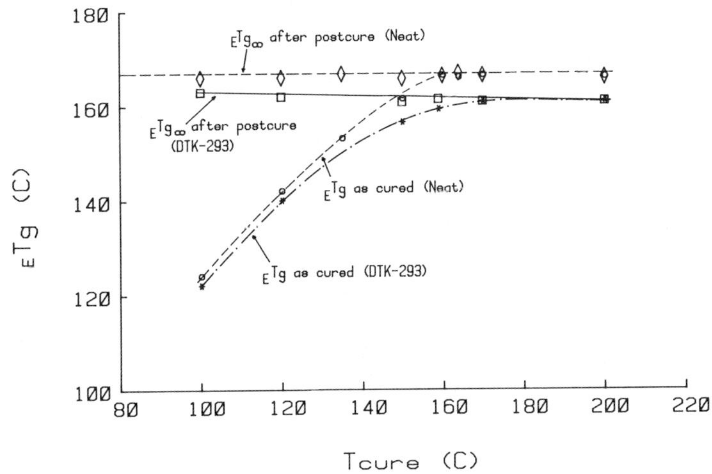

Figure 9. $_{E}T_{g}$ *and* $_{E}T_{g\infty}$ *vs.* T_{cure} *for the rubber-modified DTK–293 system and the neat system. Key:* *, $_{E}T_{g}$ *(DTK–293);* □, $_{E}T_{g\infty}$ *(DTK–293);* ○, $_{E}T_{g}$ *(neat); and* ◇, $_{E}T_{g\infty}$ *(neat).*

ified epoxy systems is considered to arise from incomplete phase separation; the dissolved rubber plasticizes the epoxy matrix. In principle, the amount of dissolved rubber versus phase-separated rubber can be determined from the values of $_{E}T_{g\infty}$ of the rubber-modified and the neat systems and the T_g of the rubber, if we assume that the chemical structures of the epoxy matrices for the two systems are similar. In the present work, the Fox equation (33) was used to estimate the amount of dissolved rubber:

$$\frac{1}{T_{gm}} = \frac{W_1}{T_{g1}} + \frac{W_2}{T_{g2}}$$

where $T_{gm} = {}_{E}T_{g\infty}$ of the rubber-modified epoxy matrix; $T_{g1} = {}_{E}T_{g\infty}$ of the neat system; $T_{g2} = T_g$ of the unreacted rubber (220 K = −50 °C); and W_1 and W_2 = weight fractions of epoxy and rubber in the rubber-modified epoxy matrix, respectively.

The weight fraction of the phase-separated rubber was obtained from the mass balance of W_2 calculated from this equation and the rubber used in the formulation (W_o = 9.68%). The weight fraction was converted to volume fraction of phase-separated rubber from the densities of the rubber (0.948 g/mL) and the fully cured neat epoxy specimen (1.218 g/mL). The volume fraction of the dispersed phase (V_f*) was then obtained from the calculated phase-separated rubber by assuming each rubber molecule was capped with two epoxy monomers in the dispersed phase. The values of the volume fraction of the dispersed phase determined from the decrease of $_{E}T_{g\infty}$ (Table III) were similar to those determined from TEM micrographs (*see* later).

Values of $_{E}T_{g\infty}$ for the rubber-modified DTK–293 system ap-

Table II. Glass Transition Temperatures vs. Cure Temperature

T_{cure}/*Time* (°*C*/*h*)	*Neat*		*DTK–293*		*DTAx16*	
	$_{E}T_{g}$	$_{E}T_{g\infty}$	$_{E}T_{g}$	$_{E}T_{g\infty}$	$_{E}T_{g}$	$_{E}T_{g\infty}$
100/40	124 (1.9)	166 (1.8)	122 (2.1)	163 (2.1)	120 (1.9)	125 (1.8)
120/36	142 (2.2)	167 (2.0)	140 (2.0)	162 (2.1)	135 (1.9)	136 (2.0)
150/24	161 (1.6)	166 (1.7)	157 (2.0)	161 (2.2)	145 (1.7)	145 (1.8)
170/10	167 (1.8)	167 (1.7)	161 (1.9)	161 (2.0)	146 (1.8)	146 (1.8)
200/3	166 (1.6)	166 (1.7)	161 (1.9)	161 (1.9)	148 (1.6)	148 (1.6)

NOTE: $_{E}T_{g}$, glass transition temperature as cured, degrees Celsius; $_{E}T_{g\infty}$, maximum glass transition temperature after postcure, degrees Celsius. The associated frequencies, from TBA experiments, are given (in Hertz) in parentheses under each transition temperature.

Table III. Effect of Cure Conditions on Morphology

T_{cure}/*Time* (°C/h)	*DTK–293*			*DTAx16*		
	$\bar{d}$[a] (μm)	V_f[b]	V_f^*[c]	$\bar{d}$[a] (μm)	V_f[b]	V_f^*[c]
100/40 and 170/5	0.6 (0.6)	0.11 (0.11)	0.13 —	3.1 (2.7)	0.34 (0.38)	0 —
200/3	1.5 (2.9)	0.13 (0.14)	0.13 —	1.8 (2.7)	0.15 (0.17)	0.08 —

[a] Mean diameter of particles determined by the Schwartz–Saltykov diameter method (the Spektor chord method values are in parentheses).
[b] Volume fraction of dispersed phase determined by Schwartz–Saltykov diameter method (Spektor chord method values are in parentheses).
[c] Volume fraction of dispersed phase determined with the Fox equation (*see* text).

peared to be only slightly dependent on cure conditions (Figure 9) and varied from 163 °C when cured at 100 °C to 161 °C when cured at 200 °C. The narrow range of $_ET_{g\infty}$ suggests that the extent of phase separation was insensitive to cure conditions. This observation is confirmed by the TEMs of the specimens cured at 100 (Figure 10a) and 200 °C (Figure 10b). The volume fractions of the dispersed phase (V_f) for the two specimens were approximately the same (V_f ~0.11 for 100 °C and V_f ~0.13 for 200 °C) according to the Schwartz–Saltykov diameter method (22) and the Spektor chord method (22) (Table III), although the size distributions for the two specimens were different. According to both methods, specimens displayed a unimodal distribution of particle size; and the mean diameter ($\bar{d}$) of the particles for the specimen cured at 100 °C was smaller than for the specimen cured at 200 °C (Table III).

The size distributions of the two specimens obtained by using the Schwartz–Saltykov diameter method are shown in Figure 11. The variation of domain sizes with cure conditions is presumably caused by the effect of temperature on the rates of nucleation and growth of the dispersed phase (*5*, *15*). At low T_{cure}, high nucleation rate and high viscosity (low growth rate) will probably produce a system with smaller but more numerous particles. On the other hand, low nucleation rate and low viscosity (high growth rate) at high T_{cure} will produce a system with larger but less numerous particles.

The apparently higher value of $\bar{d}$ of the particles for the 200 °C cured specimen determined by the Spektor method (Table III) may be a consequence of the underestimation of the percentage of smaller particles; smaller sections are less likely to be intersected than larger sections by straight lines drawn across the TEMs. The problem is more acute for the 200 °C cured specimen because of its wide range of particle sizes (Figure 11).

Amino-Terminated Rubber-Modified Epoxy System (DTAx16). The ATBN rubber used in this system contains a residual amount of

a

b

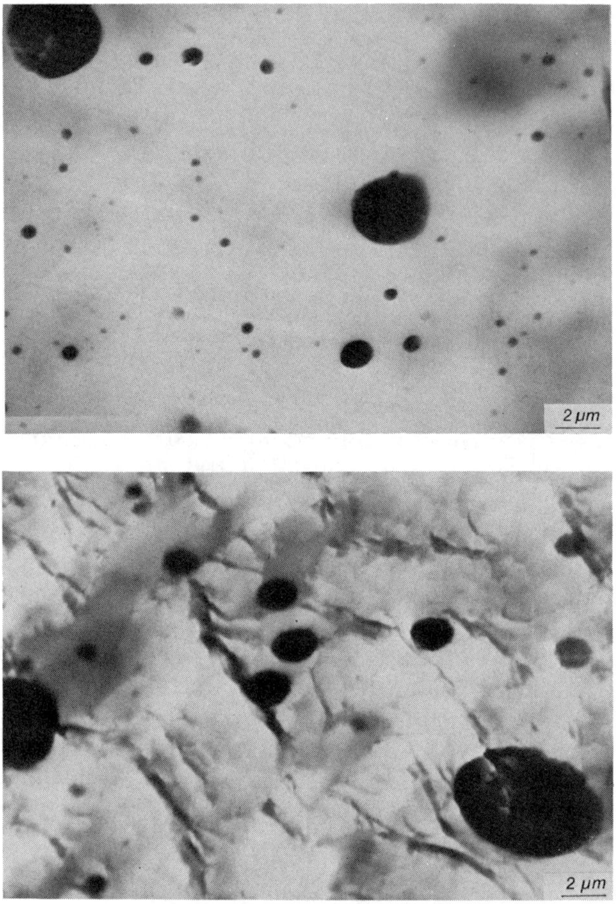

Figure 10. TEMs for cured rubber-modified DTK–293 specimens. Key: (a) T_{cure} *= 100 °C and (b)* T_{cure} *= 200 °C.*

AEP because it is made by reacting a CTBN rubber with AEP (*13*). The AEP content was estimated to be about 3% by weight. This value was obtained from information provided by the manufacturer on the NH equivalent weights of the residue-free ATBN molecule (~1900 g/NH equiv wt), AEP (~43 g/NH equiv wt), and the ATBN rubber with its residual AEP (~850 g/NH equiv wt).

Representative TBA isothermal-cure spectra from low to high temperatures (100–200 °C) for the DTAx16 system are presented in Figure 12. The times to gelation and vitrification for the DTAx16 system and the neat system are summarized in Figure 13 in the form of a TTT isothermal-cure diagram. The ${}_{E}T_{g_o}$ of the DTAx16 system was 2 °C, the same value as for the neat and DTK-293 systems. In contrast to the DTK–293 system, the times to gelation and the ac-

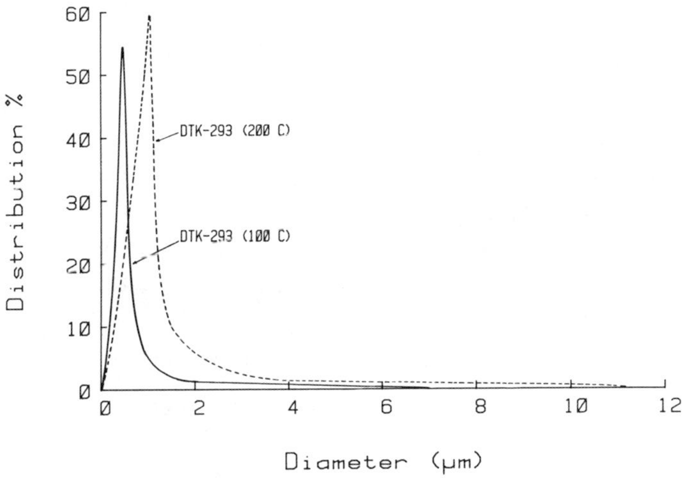

Figure 11. Size distributions of the rubber-modified DTK–293 specimens cured at 100 and 200 °C from analysis of TEMs by using the Schwartz–Saltykov diameter method.

tivation energy (ΔE) of gelation for the DTAx16 system differed significantly from those for the neat system. The times to gelation were decreased by the presence of the ATBN rubber, especially at low temperatures. The activation energy of gelation for the DTAx16 system was 10.2 kcal/mol compared to 14.1 kcal/mol for the neat

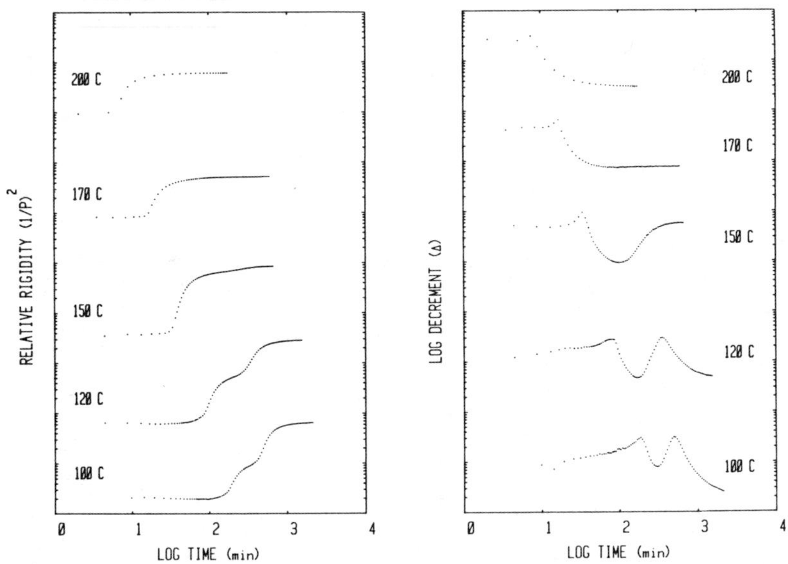

Figure 12. Representative TBA isothermal-cure spectra for the rubber-modified DTAx16 system.

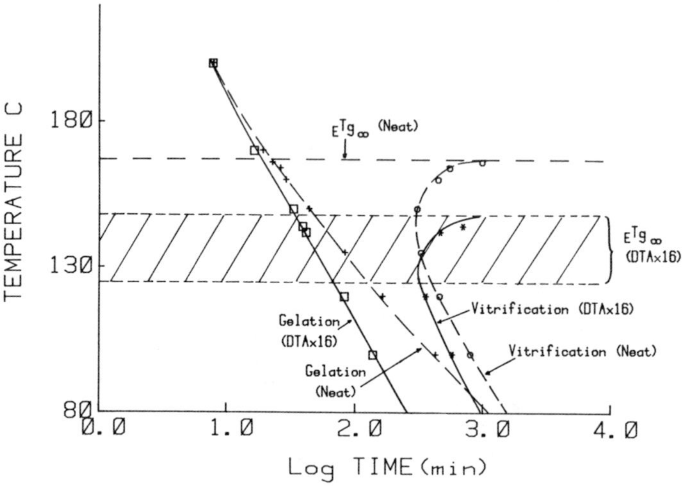

*Figure 13. TTT isothermal-cure diagram for the rubber-modified DTAx16 system vs. the neat system. Key: □, gelation (DTAx16); *, vitrification (DTAx16); +, gelation (neat); and ○, vitrification (neat).*

system and 14.2 kcal/mol for the DTK–293 system. Furthermore, values of $_ET_{g\infty}$ for the DTAx16 system were sensitive to the conditions of cure (indicated by the wide shaded band in the TTT diagram). The results of light transmission experiments in relation to the times for gelation and vitrification for the DTAx16 system are shown in Figure

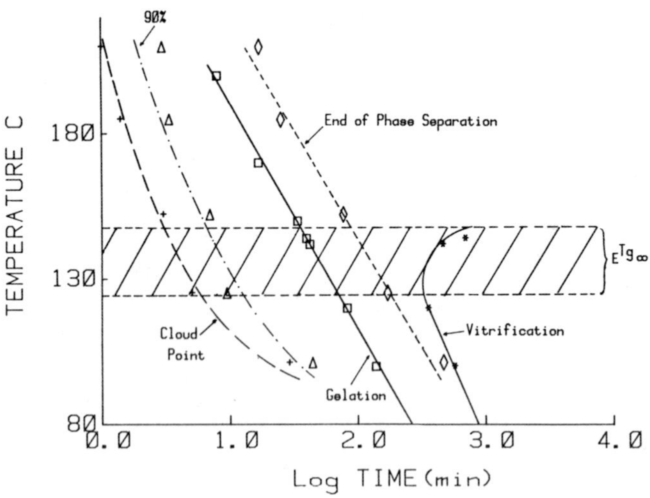

*Figure 14. TTT isothermal-cure diagram including phase separation for the rubber-modified DTAx16 system. Key: □, gelation; *, vitrification; +, cloud point; △, 90% decrease of light intensity; and ◇, end of phase separation.*

14. Most of the phase separation occurred well before gelation but, as was observed for the DTK–293 system, did not cease until after gelation.

The thermomechanical spectra obtained after prolonged cure at different isothermal temperatures (Figure 15) show four relaxation processes in the logarithmic decrement plots: a high temperature relaxation associated with ${}_ET_g$, a broad relaxation at about −40 °C, a relaxation at about −50 °C associated with ${}_RT_g$, and a relaxation below ${}_RT_g$. The values of ${}_ET_g$ and ${}_ET_{g\infty}$ versus T_{cure} are included in Figure 16 and Table II, together with data for the neat system. The values of ${}_ET_{g\infty}$ ranged from 125 °C (when cured at 100 °C) to 148 °C (when cured at 200 °C). In comparison, the value of ${}_ET_{g\infty}$ for the neat system was 167 °C and was independent of the isothermal temperature of cure. To cause a reduction of ${}_ET_{g\infty}$ equivalent to 42 °C (from 167 to 125 °C) for the specimen cured at 100 °C, the epoxy matrix must contain 10% rubber by weight as estimated by the Fox equation (*33*). However, the initial formulation for the DTAx16 system only contained 9.7% rubber by weight and yet the volume fraction of phase-separated rubber appeared to be very high (*see* Table III). Thus, the large decrease in ${}_ET_{g\infty}$ for the DTAx16 system could not be caused by the dissolved rubber alone. TEMs of a specimen cured at 100 °C (Figure 17a) and a specimen cured at 200 °C (Figure 17b) show that the 100 °C specimen had a much higher V_f than the 200 °C specimen

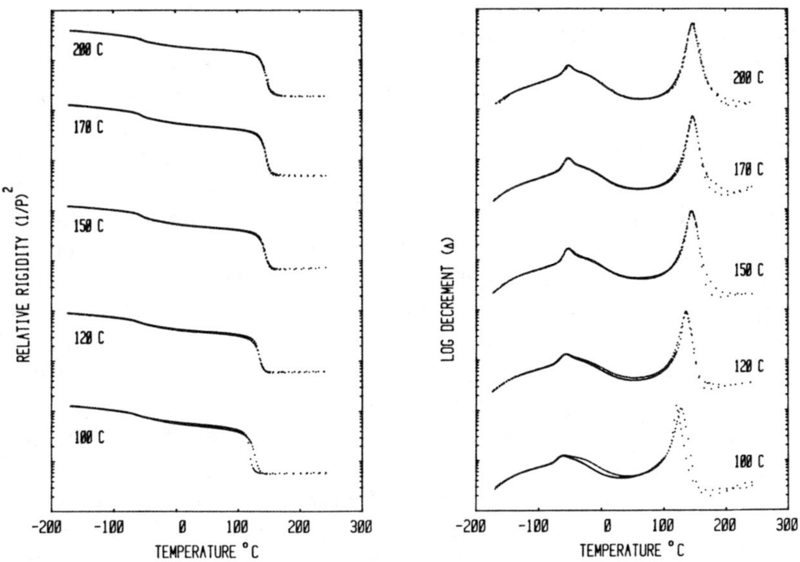

Figure 15. Representative TBA thermomechanical spectra after isothermal cure for the rubber-modified DTAx16 system.

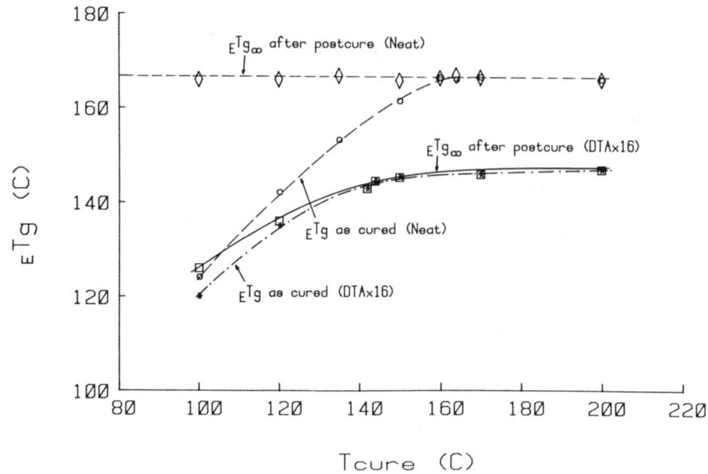

Figure 16. $_ET_g$ *and* $_ET_{g\infty}$ *vs.* T_{cure} *for the rubber-modified DTAx16 system and the neat system. Key:* *, $_ET_g$ *(DTAx16);* □, $_ET_{g\infty}$ *(DTAx16);* ○, $_ET_g$ *(neat); and* ◇, $_ET_{g\infty}$ *(neat).*

(*see* Table III), although the $_ET_{g\infty}$ for the 100 °C specimen was much lower. Furthermore, the V_f values of the DTAx16 specimens (both 100 and 200 °C) were higher than the V_f values for the DTK–293 specimens, although the values of $_ET_{g\infty}$ for the DTAx16 specimens were much lower than for the DTK–293 specimens.

Certain aliphatic amines promote homopolymerization of epoxy resins (*34*). In fact, DGEBA-type epoxy resin can be cured at room temperature with commercial ATBN rubber if more than 50 phr of ATBN rubber is used (*13*). To study the level of reactivity of ATBN rubber in the DTAx16 system in the absence of TMAB, 15 phr of the commercially available ATBN rubber was mixed with 100 phr of DER 331 at 50 °C, and the mixture was heated at 200 °C for 2 h. The T_g of the mixture increased by 62 °C (from −12 to 50 °C) as determined from TBA thermomechanical spectra obtained before and after cure. A sample of ATBN rubber with negligible residual AEP content (NH equivalent weight ~ 1900 g, The BFGoodrich Chemical Company) was also examined. The T_g of the mixture of 100 phr DER 331–15 phr residue-free ATBN rubber increased by 23 °C (from −12 to 11 °C) after cure at 200 °C for 2 h. The chemistry of cure for the DTAx16 system is therefore more complex than for the neat system. The latter is considered to involve a stoichiometric reaction between epoxy and NH end groups. In contrast, homopolymerization of DER 331 and reactions between the NH end groups of the ATBN and DER 331, AEP and DER 331, and TMAB and DER 331 can all occur competitively for the DTAx16 system. The differences in cure chemistry between the DTAx16 and neat systems were reflected in the reaction kinetics by the times of gelation (Figure 13) and the acti-

a

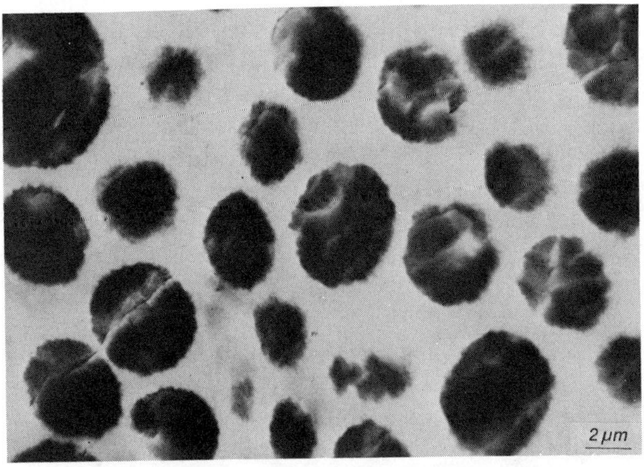

b

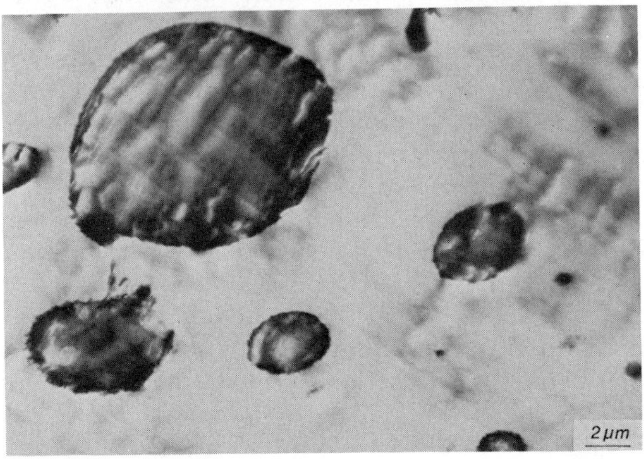

Figure 17. TEMs for cured rubber-modified DTAx16 specimens. Key: (a) T_{cure} = 100 °C and (b) T_{cure} = 200 °C.

vation energies of gelation for the two systems. Epoxy networks formed by stoichiometric reaction of epoxy resin and tetrafunctional aromatic diamine are highly cross-linked; for example, the DER 331–TMAB system gives a ${}_ET_{g\infty}$ of 167 °C. In contrast, cure of epoxy resins through homopolymerization usually produces more loosely cross-linked epoxy networks with low values of ${}_ET_{g\infty}$, for example, ${}_ET_{g\infty} \approx$ 100 °C for a (5 phr) piperidine-cured DGEBA-type epoxy system (*5*, *34*). Therefore, the large decrease of ${}_ET_{g\infty}$ for DTAx16 system may reflect the extent of homopolymerization rather than the amount of dissolved rubber. To isolate the influence of homopolymerization on ${}_ET_{g\infty}$ in the absence of rubber, the values of ${}_ET_{g\infty}$ versus T_{cure} for a formulation of DER 331–TMAB–0.76 phr of AEP (Aldrich Chemical Company) were obtained. In the formulation, the NH contribution from the AEP was made equivalent to the NH contribution from the commercial ATBN rubber (including its residual AEP) in the DTAx16 system. The values of ${}_ET_{g\infty}$ for the DER 331–TMAB–AEP formulation obtained from TBA experiments are presented in Figure 18; the data for the DTAx16 and neat systems are included for comparison. As was found for the DTAx16 system, values of ${}_ET_{g\infty}$ for the DER 331–TMAB–AEP formulation were also dependent on T_{cure} and ranged from 137 °C when cured at 100 °C to 161 °C when cured at 200 °C. At low T_{cure}, the reactivity of TMAB is low; therefore, significant amounts of DER 331 epoxy could have homopolymerized before reacting with TMAB to produce a low ${}_ET_{g\infty}$. At high T_{cure}, TMAB is more reactive; therefore, the one-to-one reaction of epoxy with amine hydrogen would be expected to occur more competitively. A consequence of homopolymerization of epoxy in the DTAx16 system is the exclusion of amine hydrogen of TMAB from reaction; this exclusion will have some deleterious effects on the physical properties.

The variation of ${}_ET_{g\infty}$ with T_{cure} for the DTAx16 system may have been caused by the vaporization of AEP from the TBA specimens, which increases with increasing temperature. However, values of ${}_ET_{g\infty}$ obtained from specimens (*16*) that had been machined from the large casting prepared in a sealed mold were the same as those obtained from the TBA experiments. Thus, the dependence of ${}_ET_{g\infty}$ on T_{cure} for the DTAx16 system results from the competing reactions during cure. A study of a DGEBA-type epoxy resin cured with a stoichiometric amount of AEP also concluded that the cure mechanism varied with cure conditions (*35*).

TEMs of the specimens cured at 100 and 200 °C (Figure 17)

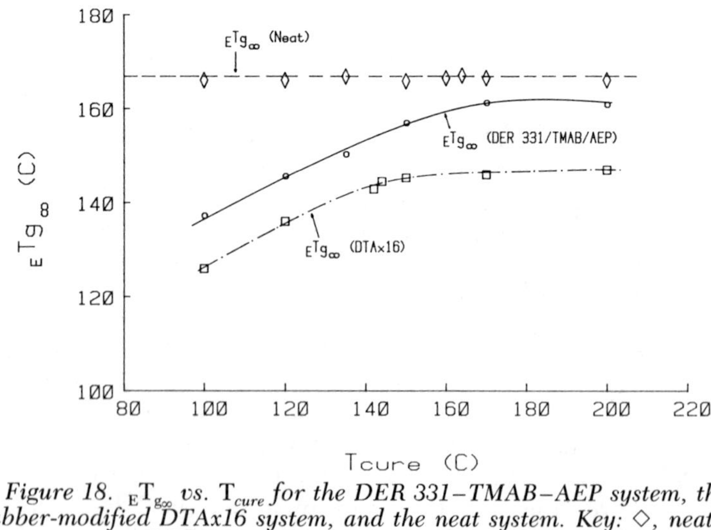

Figure 18. ${}_ET_{g\infty}$ *vs.* T_{cure} *for the DER 331–TMAB–AEP system, the rubber-modified DTAx16 system, and the neat system. Key: ◇, neat; ○, DER 331–TMAB–AEP; and □, DTAx16.*

show that the boundaries of the dispersed phase of the DTAx16 specimens were not as well-defined as in the DTK–293 specimens. Similar results have also been reported elsewhere (*36*). The values of V_f of dispersed phase for the two specimens determined by the Schwartz–Saltykov diameter method (22) and the Spektor chord method (22) were similar (Table III); the value of V_f for the 100 °C specimen (V_f ~0.34) was much higher than that for the 200 °C specimen (V_f ~0.15). According to both methods, the two specimens displayed a unimodal distribution of particle size and the 200 °C specimen has a higher percentage of smaller particles but a wider range of particle sizes. The size distributions of the two specimens obtained by using the Schwartz–Saltykov diameter method are shown in Figure 19. The apparent differences of the $\overline{d}$ values of the particles for the 200 °C specimen determined by the two methods (Table III) may be a consequence of the underestimation of the percentage of smaller particles in this specimen when using the Spektor method (as discussed previously for the DTK–293 200 °C specimen).

The high V_f of dispersed phase for the 100 °C specimen (V_f ~0.34) indicates that a large amount of epoxy was present in the dispersed phase; the initial V_f of rubber added was approximately 0.11. The inclusion of epoxy in the dispersed phase is probably a result of initiation of epoxy homopolymerization by the rubber. The 100 °C specimen is a translucent casting whereas the 200 °C specimen is an opaque solid resembling the rubber-modified DTK–293 specimens. The transparency of the 100 °C specimen is probably a consequence of the large amount of epoxy in the dispersed phase; the inclusions have a refractive index that is similar to the surrounding epoxy ma-

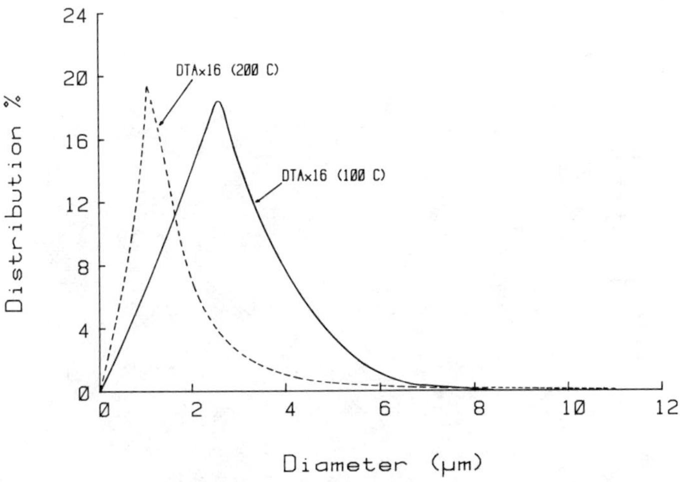

Figure 19. Size distributions of the rubber-modified DTAx16 specimens cured at 100 and 200 °C from analysis of TEMs by using the Schwartz–Saltykov diameter method.

trix. Because the increase in the relative amounts of epoxy in the domains is not accompanied by an increase in the temperature of ${}_{R}T_{g}$ (Figure 15), epoxy and rubber are not in solution in the domains. An SEM of the fracture surface of the 100 °C specimen (Figure 20) shows a characteristic texture inside the boundaries of the domains. In contrast, SEMs of the fracture surfaces of the 200 °C specimen and the DTK–293 (100 and 200 °C) specimens (*16*), as well as for other similar rubber-modified epoxy specimens (*6*, *7*), generally show "holes" for the sites of the dispersed phase (*7*).

Conclusions

The effect of cure conditions on the transitions and morphology of two rubber-modified epoxy systems has been investigated by using TBA, light transmission, and electron microscopy. TTT isothermal-cure diagrams and plots of the subsequent ${}_{E}T_{g}$ values versus T_{cure} were used as a basis for comparing the cure behavior and structure–property relationships for the neat and the two rubber-modified systems. During cure, most of the rubber phase separated well before gelation. Of the two rubber-modified systems, the V_f of the dispersed phase and the values of ${}_{E}T_{g\infty}$ for the ATBN-modified system were found to be more sensitive than the prereacted CTBN-modified system to cure conditions. The values of ${}_{E}T_{g\infty}$ for the systems were in the following order: neat (167 °C) > prereacted CTBN modified (161–163 °C) > ATBN modified (125–148 °C). Unlike the ATBN-modified system, the chemistry of cure for the prereacted CTBN-modified and neat systems are essentially the same. This similarity accounts for the similarity of the reaction kinetics as measured by the times to gelation and the activation energies of gelation. The relatively small decrease of ${}_{E}T_{g\infty}$ for the prereacted CTBN-modified

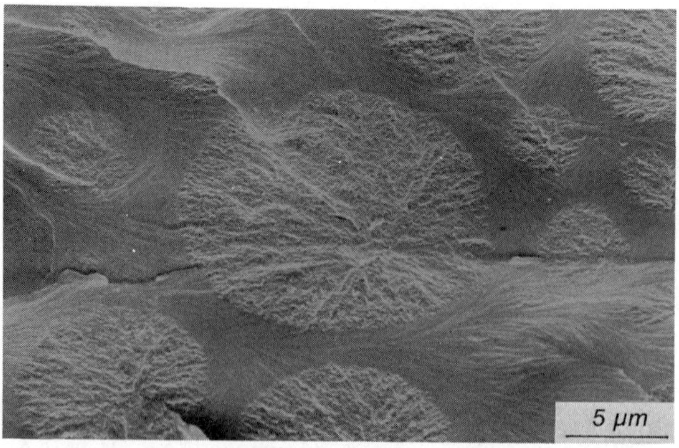

Figure 20. SEM of a rubber-modified DTAx16 specimen cured at 100 °C (fracture temperature = −20 °C).

system was attributed to dissolved rubber. In contrast, dissolved rubber cannot account for the large decrease of ${}_{E}T_{g\infty}$ for the ATBN-modified system, which also has the highest V_f of dispersed phase. The anomaly for the ATBN-modified system was attributed to the complexity of the cure chemistry introduced by using the rubber.

Acknowledgments

The authors thank C. K. Riew and A. R. Siebert (The BFGoodrich Chemical Company) and T. J. Deguchi and B. S. H. Royce (Princeton University) for their technical advice. Financial support by the Office of Naval Research and The BFGoodrich Chemical Company is also appreciated.

Literature Cited

1. McGarry F. J.; Willner, A. M. "Toughening of an Epoxy Resin by an Elastomeric Second Phase"; Massachusetts Institute of Technology: Cambridge, 1968; Research Report R68-8.
2. Sultan, T. N.; McGarry, F. J. *Polym. Eng. Sci.* **1973,** *13,* 29.
3. Riew, C. K.; Rowe, E. H.; Siebert, A. R. In "Toughness and Brittleness of Plastics"; Deanin, R. D.; Crugnola, A. M., Eds.; ADVANCES IN CHEMISTRY SERIES No. 154, American Chemical Society: Washington, D.C., 1976; p. 326.
4. Bucknall, C. B.; Yoshii, T., *Br. Polym. J.* **1978,** *10,* 53.
5. Manzione, L. T.; Gillham, J. K.; McPherson, C. A. *J. Appl. Polym. Sci.* **1981,** *26,* 889.
6. Manzione, L. T.; Gillham, J. K.; McPherson, C. A. *J. Appl. Polym. Sci.* **1981,** *26,* 907.
7. Kinloch, A. J.; Shaw, S. J.; Tod, D. A.; Hunston, D. L. *Polymer* **1983,** *24,* 1341.
8. Kinloch, A. J.; Shaw, S. J.; Hunston, D. L. *Polymer* **1983,** *24,* 1355.
9. Kunz-Douglass, S.; Beaumont, P. W. R.; Ashby, M. F. *J. Mater. Sci.* **1980,** *15,* 1109.
10. Sayre, J. A.; Kunz, S. C.; Assink, R. A. *Polymer Prepr., Am. Chem. Soc., Div. Polym. Mater.: Sci. Eng.* **1983,** *49,* 442.
11. Bascom, W. D.; Cottington, R. L.; Jones, R. L., Peyser, P. *J. Appl. Polym. Sci.* **1975,** *19,* 2545.
12. Laible, R. C.; McGarry, F. J. *Polym. Plast. Tech. Eng.* **1976,** *7,* 27.
13. Riew, C. K. *Rubber Chem. Technol.* **1981,** *54,* 374.
14. Gillham, J. K., In "The Role of the Polymeric Matrix in the Processing and Structural Properties of Composite Materials"; Seferis, J. C.; Nicolais, L., Eds.; Plenum: New York, **1983;** p. 127.
15. Williams, R. J. J.; Borrajo, J.; Adabbo, H. E.; Rojas, A. J. *Polym. Prepr., Am. Chem. Soc., Div. Polym. Mater.: Sci. Eng.* **1983,** *49,* 432.
16. Chan, L. C.; Gillham, J. K.; Kinloch, A. J.; Shaw, S. J. Chap. 16 this volume.
17. Chan, L. C.; Gillham, J. K. *Polym. Prepr., Am. Chem. Soc., Div. Polym. Mater.: Sci. Eng.* **1983,** *48,* 571.
18. "Mold Preparation for Araldite Resins, Instruction Manual", Ciba-Geigy Chem. Co., **1972.**
19. Gillham, J. K., In "Developments in Polymer Characterisation-3"; Dawkins, J. V., Ed.; Applied Science Publishers: London, **1982;** p. 159.
20. Enns, J. B.; Gillham, J. K. In "Computer Applications in Applied Polymer Science"; Provder, T., Ed.; ACS SYMPOSIUM SERIES No. 197, American Chemical Society: Washington, D.C., **1982;** p. 329.
21. Deguchi, T. J. M.Sci. Thesis, Princeton Univ., N.J., **1983.**

22. Underwood, E. E., In "Quantitative Microscopy", DeHoff, R. T.; Rhines, F. N., Eds.; McGraw-Hill: New York, **1968;** Chap. 6.
23. Enns, J. B.; Gillham, J. K. *J. Appl. Polym. Sci.* **1983,** *28,* 2567.
24. "High-Performance, Low-Energy-Curing Resins", National Materials Advisory Board, National Research Council, Publication NMAB-412, National Academy Press: Washington, D.C., **1984.**
25. Pogany, G. A. *Polymer* **1970,** *11,* 66.
26. Naé, H. N.; Gillham, J. K. *Polym. Prepr., Am. Chem. Soc., Div. Org. Coat. Plast. Chem.,* **1983,** *48,* 566.
27. Enns, J. B.; Gillham, J. K. In "Polymer Characterization: Spectroscopic, Chromatographic, and Physical Instrumental Methods"; Craver, C. D., Ed.; ADVANCES IN CHEMISTRY SERIES No. 203, American Chemical Society: Washington, D.C., **1983;** p. 27.
28. Enns, J. B.; Gillham, J. K. *J. Appl. Polym. Sci.* **1983,** *28,* 2831.
29. Shimazaki, A. *J. Polym. Sci., Part C* **1968,** *23,* 555.
30. Visconti, S.; Marchessault, R. H. *Macromolecules* **1974,** *7,* 913.
31. Wang, T. T.; Zupko, H. M. *J. Appl. Polym. Sci.* **1981,** *26,* 2391.
32. Keskkula, H.; Turley, S. G.; Boyer, R. F. *J. Appl. Polym. Sci.* **1971,** *15,* 351.
33. Fox, T. G. *Bull. Am. Phys. Soc.* **1956,** *1,* 123.
34. Lee, H.; Neville, K. "Handbook of Epoxy Resins"; McGraw-Hill: New York, **1967.**
35. Osinski, J. S.; Manzione, L. T. In "Epoxy Resin Chemisry II"; Bauer, R. S., Ed.; ACS SYMPOSIUM SERIES No. 221, American Chemical Society: Washington, D.C., **1983;** p. 263.
36. Kunz, S. C.; Sayre, J. A.; Assink, R. A. *Polymer* **1982,** *23,* 1897.

RECEIVED for review November 18, 1983. ACCEPTED May 29, 1984.

16

Rubber-Modified Epoxies

Morphology, Transitions, and Mechanical Properties

L. C. CHAN[1,3], J. K. GILLHAM[1], A. J. KINLOCH[2,4], and S. J. SHAW[2]

[1] Polymer Materials Program, Department of Chemical Engineering, Princeton University, Princeton, NJ 08544

[2] Ministry of Defence, Royal Armament Research and Development, Establishment (Waltham Abbey), Essex EN9 1BP, United Kingdom

The mechanical properties of two fully cured epoxy systems, one modified by an amino-terminated rubber and the other modified by a prereacted carboxyl-terminated rubber, were investigated in relation to variations in temperature (−90 to 140 °C). Various morphologies and maximum glass transition temperatures ($_ET_{g\infty}$) were developed by using different cure conditions; the neat system ($_ET_{g\infty}$ = 167 °C) was used as a control. The mechanical properties of the amino-terminated system were more sensitive to cure history than those of the prereacted carboxyl-terminated system. The improvement of fracture energy at a low strain rate for a rubber-modified epoxy system depends on the volume fraction of the dispersed phase and the inherent ductility of the matrix. The ductility is related to $_ET_{g\infty}$.

ADDITION OF REACTIVE LIQUID RUBBER to uncured epoxy formulations can improve the crack resistance of cured epoxy materials (*1*, *2*). A cured rubber-modified material usually exhibits a two-phase structure consisting of finely dispersed rubber-rich domains (~0.1–5 μm) bonded to the epoxy matrix. The mechanical properties are dependent on the relative amounts of dissolved and phase-separated rubber, the domain size and size distribution of the dispersed phase, and the chemical and physical compositions of the matrix and of the dispersed phase (*3–8*).

The cure process and its relationship to the development of morphology and transitions for two rubber-modified epoxy systems have been studied (*9*). A two-step cure process was used to develop a fully

[3] Current address: Bell Laboratories, Whippany, NJ 07981

[4] Current Address: Department of Mechanical Engineering, Imperial College, London University, London SW7 2BX United Kingdom

0065-2393/84/0208-0261$06.00/0

cured but distinct cure-dependent morphology. First, the resin was cured isothermally at different temperatures (T_{cure}) until reactions ceased; and second, the cured resin was postcured by heating above the maximum glass transition temperature ($_{E}T_{g\infty}$) of the system to complete the reactions of the matrix. An aromatic, tetrafunctional, diamine-cured diglycidyl ether of bisphenol A (DGEBA) type epoxy resin was selected as the neat system because of its high $_{E}T_{g\infty}$ (167 °C). The two rubber-modified systems, each containing 15 parts per hundred parts resin (phr) of rubber, were obtained by modifying the neat system with a commercial, prereacted carboxyl-terminated rubber and a commercial, amino-terminated rubber, separately. Both rubbers had been made from the same copolymer of butadiene and acrylonitrile (AN).

The main objective of this research was to investigate the mechanical properties of fully cured specimens of the two rubber-modified epoxy systems. Each system had varying morphologies and $_{E}T_{g\infty}$ values produced by using different cure conditions. The neat system was the control. Dynamic mechanical properties in torsion, uniaxial compressive modulus, yield stress and strain behavior, and tensile fracture behavior at a low strain rate were obtained over a range of test temperatures (−90 to 140 °C).

Experimental

Materials. The chemical structures of the materials used for the neat and the two rubber-modified systems have been described (9). The neat system was a DGEBA-type epoxy resin (DER 331, Dow Chemical Company) cured with propyl 1,3-bis(4-aminobenzoate) (trimethylene glycol di-*p*-aminobenzoate, TMAB, Polacure 740M, Polaroid Corporation). The first system, denoted DTK–293, was modified with a commercial, prereacted carboxyl-terminated butadiene–acrylonitrile (CTBN) copolymer (K–293, Spencer Kellog Corporation). The K–293 rubber had been made by reacting a carboxyl-terminated rubber containing 17% AN (CTBN × 8, The BFGoodrich Chemical Company) with an excess of DGEBA resin. The second system, denoted DTA × 16, was modified with a commercial, amino-terminated butadiene–acrylonitrile (ATBN) copolymer also containing 17% AN (ATBN × 16, The BFGoodrich Chemical Company). The commercial ATBN rubber contained a residual amount (~3% by weight) of *N*-(2-aminoethyl)piperazine (AEP) from its synthesis (*10*). The formulations for the neat and the two rubber-modified systems, each containing 15 phr of the butadiene–acrylonitrile copolymer, are included in Table I. The neat epoxy system was also modified with 15 phr of non-prereacted CTBN × 8 in a preliminary attempt to investigate the effect of poor interfacial bonding on fracture behavior.

Specimen Preparation. The mixture of liquid epoxy resin and liquid rubber (for the modified systems) was heated to 120 °C in an open beaker. The solid curing agent was then added and dissolved with the aid of mechanical stirring for 5 min. The solution was degassed at 100 °C for 25 min in a preheated vacuum oven at a pressure of about 1 torr and was then poured into two preheated (at T_{cure}) molds [precoated with a release agent (QZ13, Ciba-Geigy Chemical Company) (*11*)]. The systems were then cured according to the chosen cure conditions (Table II) to form a large casting for the compact-tension and film

Table I. Chemical Formulations for the Neat, DTK–293, and DTA×16 Systems

Component	*Neat*	*DTK–293*	*DTA×16*
DER 331	100.0	100.0	100.0
TMAB	41.0	51.0	39.9
K–293[a]	—	42.0	—
ATBN×16	—	—	15.0

NOTE: Neat system (no rubber)—1 epoxy end group/1 amine hydrogen. Rubber-modified systems—15 phr of rubber/100 phr of unreacted epoxy and 1 free epoxy end group/1 amine hydrogen. [It is assumed that all epoxy end groups in DTK–293 and in DER 331 react with TMAB and that all NH in the ATBN rubber (including the NH in the residual AEP) and the TMAB react with the epoxy.]

[a] 1 mol of K–293 contains 0.44 mol of rubber and 0.56 mol of epoxy.

specimens, and small rods for the compression specimens. The large casting (220 × 220 × 6 mm) was prepared in an air oven; the open end of the mold was sealed with a plug made of silicone rubber (RTV660, General Electric Company) to minimize exposure to air during cure. The rods (length ≅ 60 mm, diameter = 7 mm) were prepared under nitrogen in an oven. After cure, the molds were allowed to cool freely inside the ovens to room temperature (~1 °C/min). The casting and rods were then removed from their molds, and specimens were machined to the specified dimensions for mechanical testing. Residual stresses were removed by an annealing process in nitrogen that involved heating the specimens from room temperature to 200 °C before freely cooling them to room temperature.

Transmission Electron Microscopy. Morphologies of cured specimens were examined by using transmission electron microscopy (TEM) (9). Specimens (from previously fractured compact-tension specimens) were stained with osmium tetroxide and microtomed at room temperature. The volume fraction and mean diameter of the dispersed phase were determined by using the Schwartz–Saltykov diameter method (*12*) and the Spektor chord method (*12*). Approximately 20–30 TEMs were examined for each cure condition. The volume fractions of the dispersed phase obtained by the two methods were in good agreement (9).

Dynamic Mechanical Analysis. The use of an automated torsional braid analysis (TBA) instrument to study cure behavior and its relationship to the properties of the cured state involved obtaining the dynamic mechanical properties of composite specimens; each specimen was formed by impregnating a braid substrate with a reactive epoxy resin (9). The TBA instrument (Plastics Analysis Instruments, Inc.) was used as a conventional, freely decaying torsional pendulum (TP) (*13*) to obtain the quantitative values of shear modulus (G') and logarithmic decrement (Δ) of specimens (~0.8 × 3.0 × 60 mm) that had been machined from the large casting. Dynamic mechanical spectra of the fully cured specimens were obtained, after heating from room temperature to 200 °C, by cooling to −170 °C at a rate of 1.5 °C/min. Transitions were identified by the temperatures of maxima (and the associated frequencies, ~1 Hz) in the logarithmic decrement. The shear modulus was determined from the natural period (P) and Δ with the following equation (*14*):

$$G' = \frac{4\pi^2 Il}{NP^2}\left(1 + \frac{\Delta^2}{4\pi^2}\right) - \frac{mgb^2}{12N} \quad (1)$$

Table II. Effect of Cure Conditions on Transitions and Morphology

	Neat		*DTK–293*				*DTA×16*			
T_{cure}/*Time* (°C/h)	${}_{E}T_{g_\infty}$ (*frequency*)	${}_{E}T_{sec_\infty}$ (*frequency*)	${}_{E}T_{g_\infty}$ (*frequency*)	${}_{R}T_{g}$ (*frequency*)	$\bar{d}^a$ (μm)	V_f^b	${}_{E}T_{g_\infty}$ (*frequency*)	${}_{R}T_{g}$ (*frequency*)	$\bar{d}^a$ (μm)	V_f^b
100/40 and 170/5	167 (1.3)	−35 (0.3)	163 (2.4)	−47 (0.3)	0.6 [0.6]	0.11 [0.11]	125 (1.6)	−52 (0.3)	3.1 [2.7]	0.34 [0.38]
200/3	167 (2.8)	−35 (0.3)	161 (1.6)	−50 (0.3)	1.5 [2.9]	0.13 [0.14]	148 (1.2)	−52 (0.3)	1.8 [2.7]	0.15 [0.17]

NOTE: Transitions are designated by temperature in degrees Celsius and frequency in Hertz (parentheses).
[a] Mean diameter by the Schwartz–Saltykov diameter method (Spektor chord method values in square brackets).
[b] Volume fraction of dispersed phase by the Schwartz–Saltykov diameter method (Spektor chord method values in square brackets).

in which the form factor $N = [(a^3b)/3]\,[1-(0.63)(a/b)]$, and where a is thickness, b is width ($a < b/3$), and l is length of the specimen; I is the moment of inertia of the oscillating parts of the pendulum; m is the mass supported by the specimen (32 g); and Δ is $\ln(A_i/A_{i+1})$ (A_i is the amplitude of the ith oscillation of a freely damped wave).

Uniaxial Compression Experiments. The true yield stress and yield strain and the true compressive modulus were determined from uniaxial compression experiments because highly cross-linked epoxy materials fracture prior to plastic yielding when tested under uniaxial tension (*6, 7*). Compression rods (l/d = 1.8; length = 12.6 mm, diameter = 7 mm) were machined from rods ($l \sim 60$ mm). After annealing, the rods were deformed in a compression cage (which had been lubricated with molybdenum disulfide grease) between lightly polished steel plates, at a constant crosshead displacement rate (1 mm/min) over a range of temperatures (−90 to 140 °C). The true stress, σ, was determined from the following equation (*7*):

$$\sigma = \frac{P}{A_o}(1 - e) \tag{2}$$

where P = load from the load-displacement record; A_o = initial cross-sectional area of specimen; and e = nominal strain from the crosshead displacement after correction for machine deflection with a steel specimen.

Fracture Experiments. The fracture behavior of the cured epoxy materials was examined by using compact-tension specimens (100 × 96 × 6 mm) (Figure 1) that had been machined from the large casting. After annealing, a slot was made along the centerline of a specimen with a handsaw; a sharp crack was then formed at a base of the slot by carefully tapping a fresh razor blade in the base and, thus, causing a natural crack to grow for a short distance ahead of the blade. The specimen was mounted in a tensile testing machine (Instron); a constant displacement rate (1 mm/min) was applied; and the associated load, P, versus displacement was recorded (*6, 7*). Experiments were conducted over a range of temperatures (−90 to 140 °C). The value of the fracture energy, $\mathcal{G}_{Ic}$ (kJ/m²), was determined (*15*) from Equations 3 and 4:

$$\mathcal{G}_{Ic} = \frac{K_{Ic}^{\,2}}{E}(1 - \nu^2) \tag{3}$$

where K_{Ic} = stress intensity factor (MNm$^{-3/2}$); E = Young's modulus (GPa) (≡ initial compressive modulus); and ν = Poisson's ratio (=0.40); and

$$K_{Ic} = P_cQ/HW^{\frac{1}{2}} \tag{4}$$

where P_c = load at crack initiation; Q = geometric factor; H = thickness of specimen; W = width of specimen as defined in Figure 1; and a = crack length. [$Q = 29.6\,(a/W)^{\frac{1}{2}} - 185.5\,(a/W)^{\frac{3}{2}} + 655.7\,(a/W)^{\frac{5}{2}} - 1017\,(a/W)^{\frac{7}{2}} + 638.9\,(a/W)^{\frac{9}{2}}$].

All measurements of K_{Ic} are valid according to the procedure and limits described in ASTM E399–72, except at high temperatures (starting at approximately 50 °C below ${}_ET_{g\infty}$ for each system) that were beyond the recommended limits (*16*).

For examination by scanning electron microscopy (SEM) after fracturing at different temperatures, fracture surfaces of the compact-tension specimens were coated with a thin layer (~600 Å) of gold by using a high vacuum sputterer.

Results and Discussion

Cure, Transitions, and Morphology. The cure process and its relationship to the development of morphology and transitions for the two rubber-modified epoxy systems have been discussed (*9*). Of

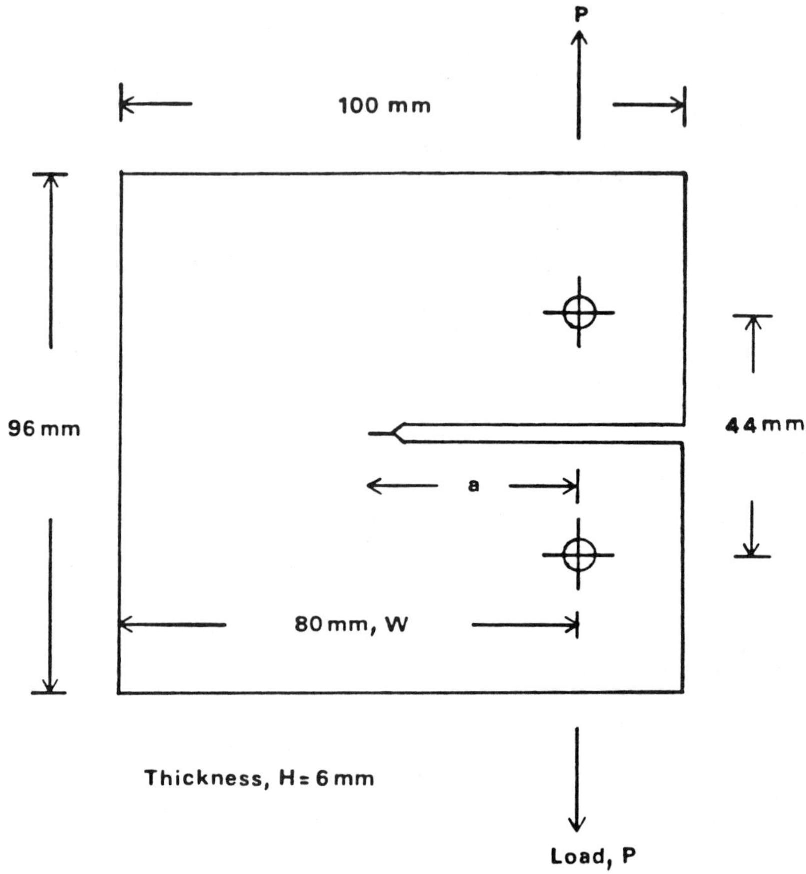

Figure 1. Dimensions of compact-tension specimen used for fracture experiments.

the two rubber-modified epoxy systems, the volume fraction of the dispersed phase and values of ${}_ET_{g\infty}$ for the DTA×16 system were more sensitive to cure conditions. The values of ${}_ET_{g\infty}$ for the systems studied were in the following order: neat (167 °C) > DTK–293 (161–163 °C) > DTA×16 (125–148 °C). A decrease in ${}_ET_{g\infty}$ can be a consequence of dissolved rubber. However, dissolved rubber cannot account for the large decrease in ${}_ET_{g\infty}$ for the present DTA×16 system, which also has the highest volume fraction of dispersed phase. The anomaly for the DTA×16 system has been attributed to the complexity of the cure chemistry that is introduced by using the ATBN rubber (9). In contrast, the cure chemistry for the neat and DTK–293 systems is essentially the same because the reactive end groups in each are identical.

In order to investigate the effect of morphology and ${}_ET_{g\infty}$ on the mechanical properties, two extreme cure conditions (100 °C/40 h +

170 °C/5 h; and 200 °C/3 h) were selected to provide specimens with the widest variation of properties. The cure conditions, transitions, and details of morphology for these rubber-modified DTK–293 and DTA × 16 specimens are contained in Table II, which also includes data for the neat system. The volume fractions of the dispersed phase ranged from about 0.11 for the DTK–293 specimen cured at 100 °C to about 0.34 for the DTA × 16 specimen cured at 100 °C; the latter had about three times the volume fraction of rubber added initially. The mean diameter of the dispersed phase ranged from about 0.6 μm for the DTK–293 specimen to about 3 μm for the DTA × 16 specimen (both specimens were cured at 100 °C). The values of ${}_ET_{g\infty}$ ranged from 125 °C for the DTA × 16 specimen to 163 °C for the DTK–293 specimen.

The transitions listed in Table II were obtained from TP logarithmic decrement data (Figure 2) with fully cured specimens. For the neat epoxy specimens, three relaxations were observed: a high-temperature relaxation associated with ${}_ET_{g\infty}$, a relaxation at approximately −35 °C (${}_ET_{sec\infty}$), and a weak relaxation below ${}_ET_{sec\infty}$. For

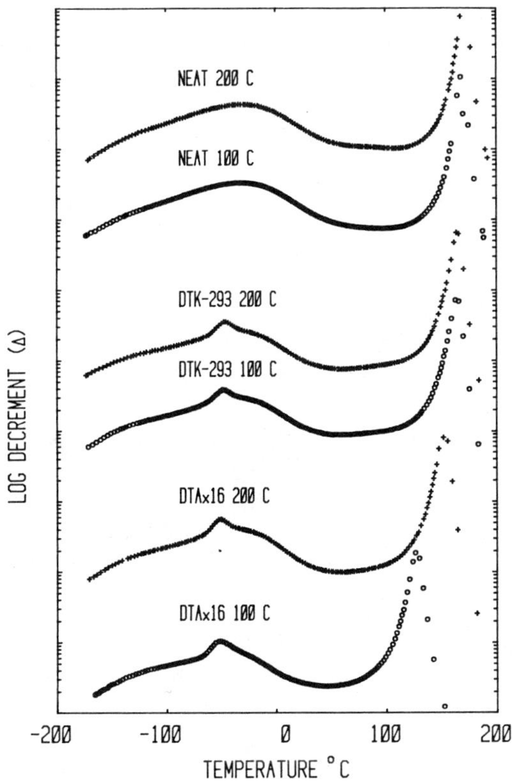

Figure 2. Torsional pendulum (TP) thermomechanical spectra for specimens.

both of the rubber-modified epoxy specimens, four relaxations were observed: a high-temperature relaxation associated with ${}_ET_{g\infty}$, a relaxation at approximately −35 °C, a relaxation at about −50 °C associated with the rubber glass transition (${}_RT_g$); and a weak relaxation below ${}_RT_g$. The values of ${}_ET_{g\infty}$ obtained with specimens were the same as those obtained with TBA specimens (9). However, the values of ${}_ET_{\text{sec}\infty}$ for the neat specimens and values of ${}_RT_g$ for the rubber-modified specimens obtained in the TP mode appeared to be slightly higher than those obtained by using the TBA mode (by ~4 °C).

Shear and Compressive Moduli. G' versus test temperature for the DTK–293 and the DTA×16 specimens, as well as data for the neat system, are shown in Figures 3 and 4, respectively. The corresponding compressive modulus versus test temperature plots are shown in Figures 5 (DTK–293) and 6 (DTA×16). [The compressive modulus data at the lowest temperature may be unreliable because of severe frictional effects (*17*).] For elastic isotropic materials, G' is related to Young's modulus (E) (≡ compressive modulus) by (*18*):

$$E = 2G' (1 + \nu) \tag{5}$$

where ν = Poisson's ratio. The ratio of the compressive moduli to shear moduli was 2.8; consequently, ν was 0.4, a value similar to reported values for other glassy polymeric materials (*18*).

The values of the modulus for all of the systems are expected to

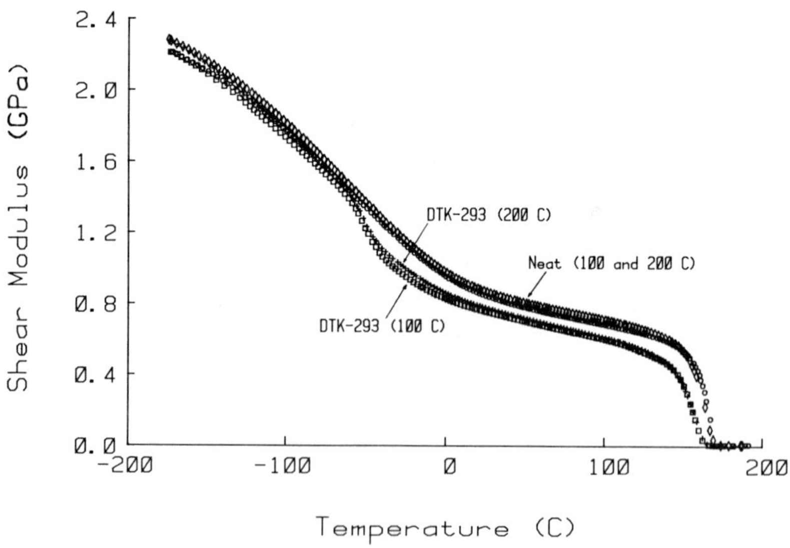

Figure 3. Shear modulus vs. test temperature for the DTK–293 and neat systems.

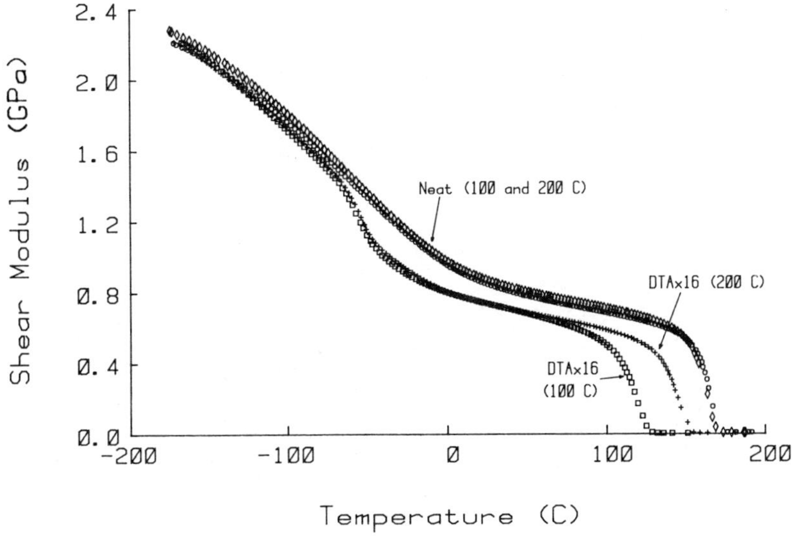

Figure 4. Shear modulus vs. test temperature for the DTA × 16 and neat systems.

be similar below ${}_RT_g$, but, at higher temperatures, to reflect the extent of phase separation and the values of ${}_ET_{g\infty}$. In general, between ${}_RT_g$ and ${}_ET_{g\infty}$, the moduli of the rubber-modified specimens were lower than those for the neat specimens. The moduli for the neat specimens cured at 100 and 200 °C were about the same throughout

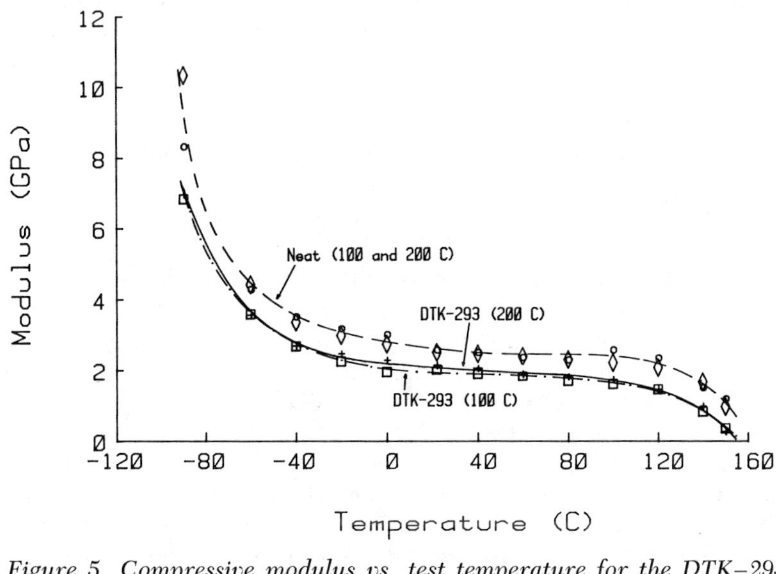

Figure 5. Compressive modulus vs. test temperature for the DTK–293 and neat systems.

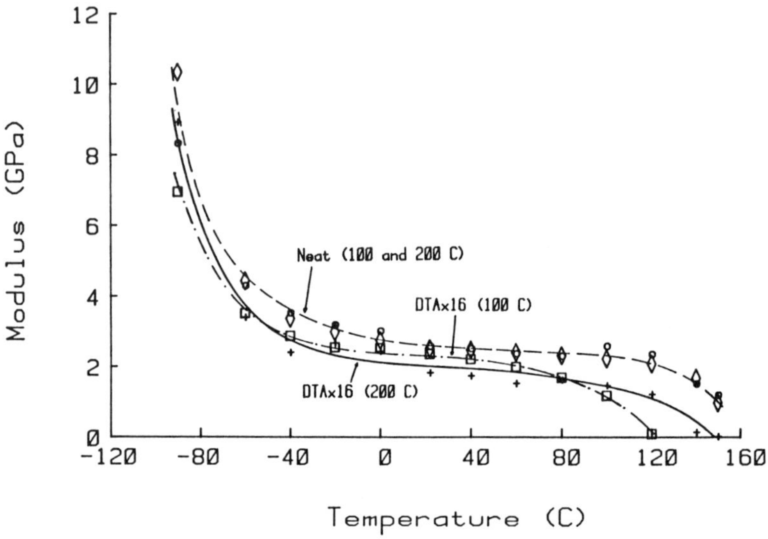

Figure 6. Compressive modulus vs. test temperature for the DTA × 16 and neat systems.

the test temperature range because of the identical values of ${}_ET_{g\infty}$ for the two specimens. Similarly, the values of modulus for the DTK–293 specimens cured at 100 and 200 °C were about the same as a result of the insensitivity of ${}_ET_{g\infty}$ and the volume fraction of the rubber-rich domains to cure conditions. In contrast, different values of shear moduli and compressive moduli were observed at test temperatures above 80 °C for the DTA × 16 specimens cured at 100 and 200 °C because of the very different values of ${}_ET_{g\infty}$ for these two specimens.

Yield Stress and Strain. The true uniaxial compressive yield stress versus test temperature for the DTK–293 system and DTA × 16 system, as well as data for the neat system, are presented in Figures 7 and 8, respectively. The corresponding yield–strain data are shown in Figures 9 (DTK–293) and 10 (DTA × 16).

The values of yield stress for all the systems decreased as the test temperature increased because of increasing ductility of the matrix. The yield stress and strain behavior for the neat and DTK–293 specimens was insensitive to differences in their previous cure history. In contrast, differences were observed at high temperatures between the DTA × 16 100 and 200 °C specimens. In general, the values of yield stress decreased in the order: neat; DTK–293 100 °C ≅ 200 °C; DTA × 16 200 °C; and DTA × 16 100 °C. This is the order of decreasing ${}_ET_{g\infty}$ and increasing volume fraction of the dispersed phase. The values of yield strain decreased with increasing temperature and, above 0 °C, appeared to follow the order DTK–293 > neat > DTA × 16.

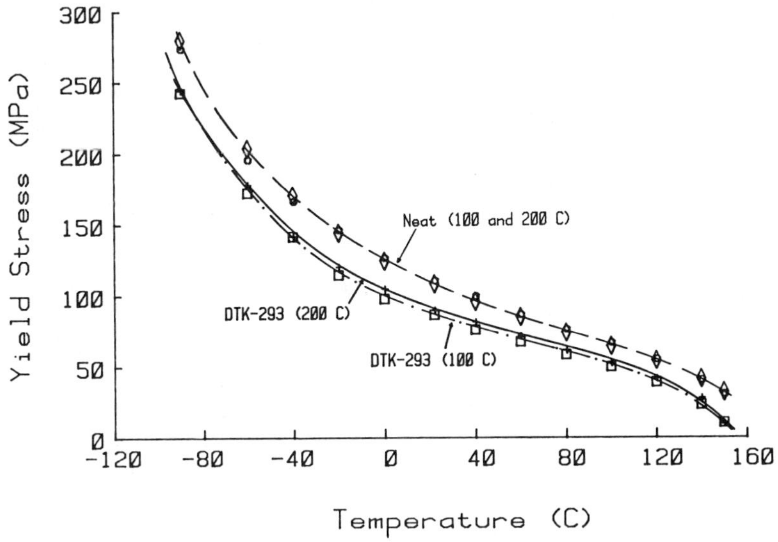

Figure 7. Yield stress vs. test temperature for the DTK-293 and neat systems.

Fracture Behavior. As found for other epoxy systems (*6*), three basic types of crack growth could be identified from the load-deflection curves: brittle, stable crack growth at low test temperatures; brittle, unstable crack growth (where crack propagation occurred in-

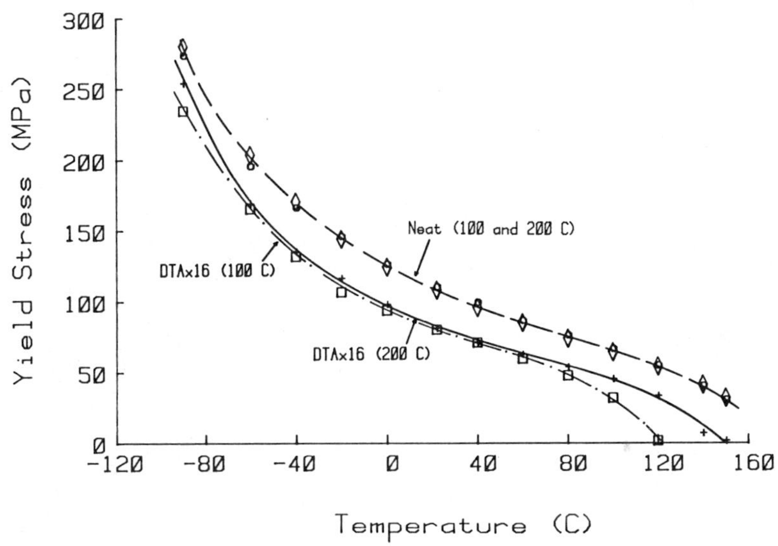

Figure 8. Yield stress vs. test temperature for the DTA×16 and neat systems.

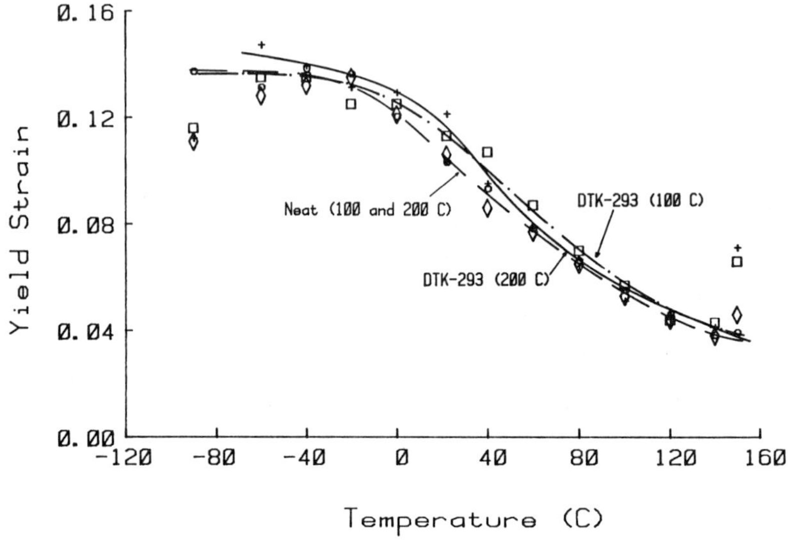

Figure 9. Yield strain vs. test temperature for the DTK–293 and neat systems.

termittently in a "stick–slip" manner) at intermediate temperatures; and ductile, stable crack growth at higher temperatures.

$\mathcal{G}_{Ic}$ versus test temperature for the systems is shown in Figure 11. For the neat system, the properties of the specimens fully cured

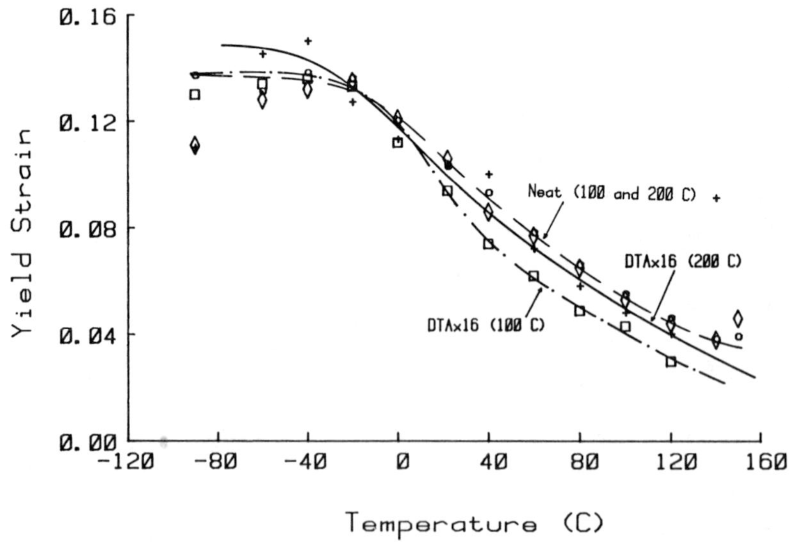

Figure 10. Yield strain vs. test temperature for the DTA×16 and neat systems.

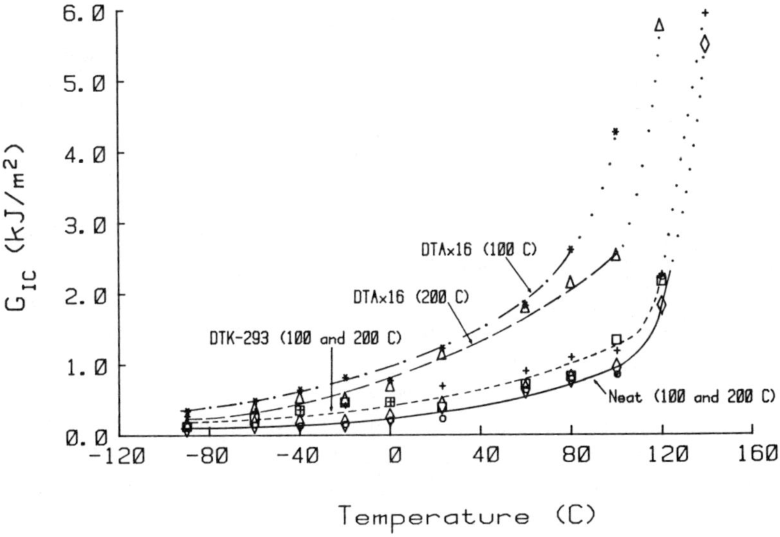

Figure 11. Fracture energy ($\mathcal{G}_{Ic}$) vs. test temperature for the DTK–293, DTA × 16, and neat systems.

at 100 and 200 °C should be independent of their cure histories because ${}_ET_{g\infty}$ was not affected by the time–temperature reaction path of cure. Therefore, the values of fracture energies for the neat 100 and 200 °C specimens were the same. The fracture energies for the two DTK–293 specimens (100 and 200 °C) were also insensitive to their cure histories because of the similar values of ${}_ET_{g\infty}$ and volume fractions of dispersed phase for these specimens. In contrast, the fracture energies for the DTA × 16 specimens (100 and 200 °C) were dependent on their cure histories; the cure chemistry and volume fraction of dispersed phase of this system is dependent on the time–temperature path of cure (9). The fracture energy of the DTA × 16 specimen cured at 100 °C, which has a higher volume fraction of dispersed phase and lower ${}_ET_{g\infty}$, was higher than that of the 200 °C specimen at all temperatures.

A small improvement in fracture energy above that for the neat specimens was observed for the DTK–293 specimens throughout the test temperature range. On the other hand, a larger improvement in fracture energy was observed for the DTA × 16 specimens, especially at high temperatures. The fracture energies of the rubber-modified specimens decreased in the order DTA × 16 100 °C; DTA × 16 200 °C; and DTK–293 100 °C ≅ 200 °C. This is the order of increasing values of ${}_ET_{g\infty}$ and decreasing volume fraction of the dispersed phase. The ratios of fracture energies of the rubber-modified specimens to those of the neat specimens at room temperature varied from 1.7 for the DTK–293 (100 and 200 °C) specimens, to 3.5 for the DTA×16 200 °C specimen, and to 3.7 for the DTA×16 100 °C specimen. Fracture energy has been related to the volume fraction of the dispersed phase by other researchers (*5, 15, 19, 20*). The similar values of fracture energies for the two DTK–293 specimens suggest that the fracture energy is less sensitive to the domain size (within the range of domain size attained); the mean diameters of the rubbery domains for these two specimens were different although the volume fractions were about the same (Table II). The distribution of particle size in these materials was unimodal (*9*). The fracture energy may be dependent on the size distribution of rubber-rich inclusions for bimodal distributions (*1, 2, 20*).

The improvement of fracture energy for the rubber-modified epoxy materials is a result of the increased extent of energy-dissipating deformations occurring in the vicinity of the crack tip during loading (*6, 7*). This fact is demonstrated by comparing the fractographs of a neat specimen cured at 200 °C (Figure 12) and a DTK–293 specimen cured at 200 °C (Figure 13). Both specimens were fractured at 23 °C. The lesser extent of shear deformation in the neat specimen results in a lower fracture energy because shear yielding is a principal source of energy dissipation. For the rubber-modified specimen, the deformation processes can involve multiple plastic shear yielding in the epoxy matrix, void formation (cavitation) either in the domain particles or at the particle–matrix interface, and tear of domain material.

The difference in coefficients of thermal expansion between the rubber and the matrix causes the rubber to be in hydrostatic tension on cooling after cure, so the particle contracts after it fails. Cavitation of the rubber produces the appearance of many hollow deep holes in fractographs; some of these holes (Figure 13) appear to have been

Figure 12. SEM of the fracture surface of a neat specimen cured at 200 °C. (Fracture temperature = 23 °C.) Crack-growth direction is from right to left, and the vertical line is the crack arrest–crack reinitiation line (6, 15).

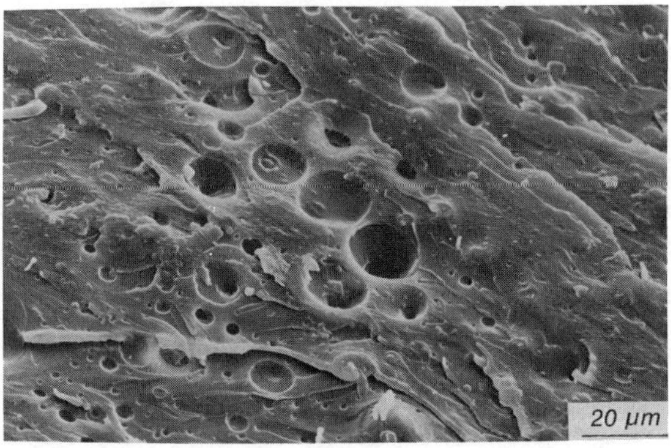

Figure 13. SEM of the fracture surface of a DTK–293 specimen cured at 200 °C. (Fracture temperature = 23 °C.)

enlarged by the deformation of the matrix that occurs on fracture. The holes become "hillocks" after the rubber is swollen with a solvent; therefore, most of the rubber is still in the holes as a lining in the cavities (*6*). In contrast, fracture surfaces of the DTA × 16 100 °C specimen appear to show fracture of the inclusions without their disappearance; this observation is attributed to the presence of large amounts of epoxy in the domains (*9*). Change in the composition of the inclusions may affect fracture-energy values (*21*).

The rubber-modified epoxies (i.e., DTK–293, DTA × 16, and nearly all of those reported in the literature) have well-bonded particles as a result of the chemical reactivity of the rubber. A preliminary attempt was made to investigate the effects of poor bonding by modifying the neat system with the non-prereacted CTBN. Poor interfacial bonding is revealed in the fractograph for the non-prereacted CTBN-modified specimen (Figure 14). The volume fractions for the non-prereacted, CTBN-modified specimen and the prereacted CTBN-modified DTK–293 specimens were about the same, although the particles for the former specimen were larger. The ratio of fracture energies for the non-prereacted CTBN-modified specimen to that for the neat specimens at room temperature was 1.2, compared to 1.7 for the prereacted CTBN-modified DTK–293 specimens. This difference is an indication of the importance of attaining adequate interfacial bonding. The adverse effects of poor interfacial bonding on the fracture energy would have been displayed more convincingly by using a pre-reacted rubber-modified system that produces a significant improvement of fracture energy above that of the neat system.

For all of the systems, the fracture energy increased at an accelerating rate with test temperature because of increasing ductility of the matrix. At low temperatures, the crack is relatively sharp and the fracture energy is low because the yield stress is high (Figures 7 and 8). Thus, the extent of plastic deformation and associated crack-

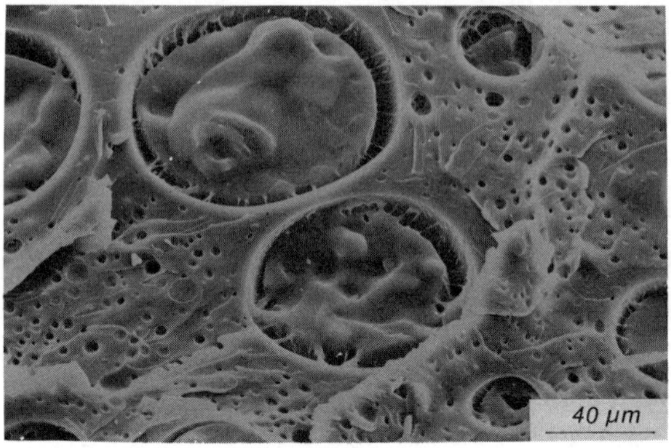

Figure 14. SEM of the fracture surface of a non-prereacted CTBN-modified specimen cured at 200 °C for 3 h. (Fracture temperature = 130 °C.)

tip blunting is relatively limited. As the temperature increases, the yield stress decreases so more crack-tip yielding is possible; as a result, the crack becomes blunter. This blunting results in higher failure loads and higher fracture energies at higher temperatures. The increase of ductility of the matrix with temperature is apparent in the fractographs of the DTA×16 100 °C specimens fractured at −60, 0, and 120 °C, shown in Figures 15, 16, and 17, respectively.

The effect of decreasing ${}_ET_{g\infty}$ is equivalent to the effect of increasing temperature; therefore, the fracture energy of the DTA×16 100 °C specimen should be the highest of all the specimens because of its low ${}_ET_{g\infty}$. However, a second factor is the volume fraction of dispersed phase, and this specimen also has the highest volume fraction (Table II). The relatively poor improvement in fracture energy for the DTK–293 specimens results from the similar values of ductility of the neat (${}_ET_{g\infty}$ = 167 °C) and the rubber-modified matrices (${}_ET_{g\infty}$ = 161 and 163 °C) and/or from the relatively low

Figure 15. SEM of the fracture surface of a DTA×16 specimen cured at 200 °C. (Fracture temperature = −60 °C.)

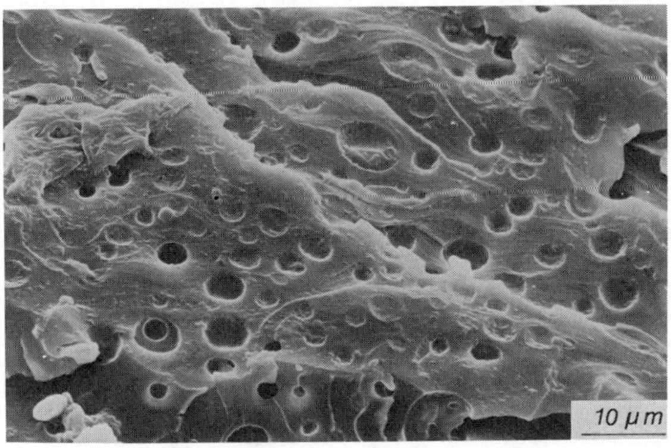

Figure 16. SEM of the fracture surface of a DTA × 16 specimen cured at 200 °C. (Fracture temperature = 0 °C.)

volume fraction of dispersed phase obtained with this rubber. To further investigate the effect of ${}_ET_{g\infty}$ on the fracture energy, the fracture energies were compared after normalizing the test temperature (T) relative to each system's ${}_ET_{g\infty}$. The normalized results of $\mathcal{G}_{Ic}$ versus $[-({}_ET_{g\infty} - T)]$ (Figure 18) clearly show that some of the improvement in $\mathcal{G}_{Ic}$ for the DTA × 16 specimens is due to the increased ductility of the matrix. The increased ductility in these specimens arises from their lower ${}_ET_{g\infty}$ values (compare Figures 11 and 18). However, as would be expected, the volume fraction of the dispersed phase also plays a role in improving the fracture energy (compare DTK–293 and DTA × 16 specimens, Figure 18).

As a caveat, it should be noted that the yield behavior of the matrix is controlled by the details of molecular structure of the matrix. The value of ${}_ET_{g\infty}$ is just a convenient "first-order" parameter to reflect the ductility of the matrix.

For high T_g matrix materials, the improvement of fracture en-

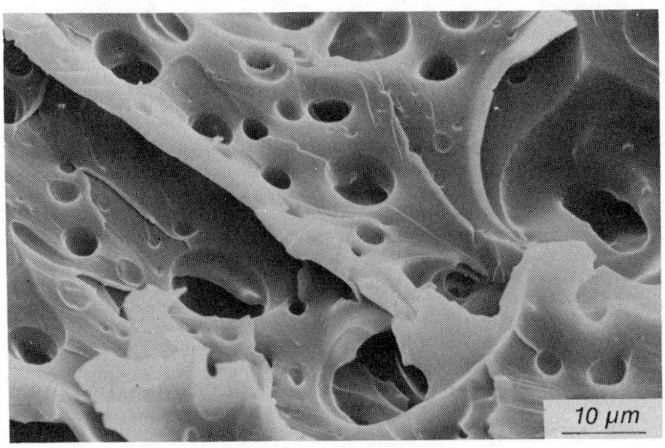

Figure 17. SEM of the fracture surface of a DTA × 16 specimen cured at 200 °C. (Fracture temperature = 120 °C.)

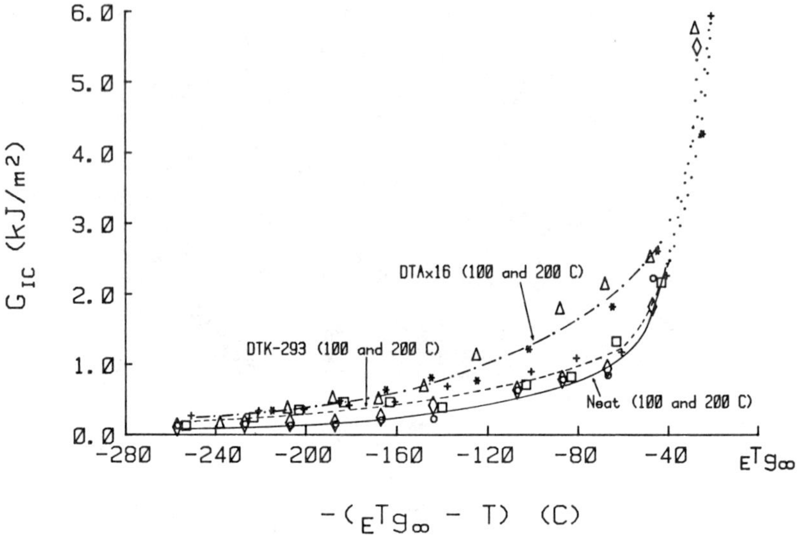

Figure 18. Fracture energy vs. test temperature normalized to $_{E}T_{g\infty}$ for the DTK–293, DTA×16, and neat systems.

ergy at room temperature by rubber modification, with the inclusions produced in the present work, is not substantial. For example, in the systems studied, at 100 °C below $_{E}T_{g\infty}$ the maximum improvement in fracture energy over the neat system was about threefold.

Conclusions

The mechanical properties of the two fully cured rubber-modified epoxy systems have been investigated versus test temperature as a function of different cure conditions by using the neat system as control. The mechanical properties included shear and compressive moduli, uniaxial yield stress and strain in compression, and the fracture energy ($\mathcal{G}_{Ic}$) at a low strain rate. Of the two rubber-modified systems, the mechanical properties of the ATBN-modified system were more sensitive to cure history than the prereacted CTBN-modified system. This observation was a result of the higher sensitivity of the volume fraction of the dispersed phase and the $_{E}T_{g\infty}$ of the matrix to cure conditions for the ATBN modified system. (Additional cure reactions are introduced with the amino-terminated rubber.) The fracture energy increased with temperature (−90 to 140 °C) for all of the systems as a result of increasing ductility of the matrix. In general, the fracture energy followed the order: amino-terminated rubber > prereacted carboxyl-terminated rubber > neat system. The fracture energies of the different systems were also compared after normalizing the test temperature to the $_{E}T_{g\infty}$ of each system because of the large variations of $_{E}T_{g\infty}$ for the systems [i.e., neat (167 °C) >

prereacted carboxyl-terminated rubber (161–163 °C) > amino-terminated rubber (125–148 °C)]. The fracture energy was dependent on both the volume fraction of the dispersed phase and the ductility of the matrix; the ductility was related to ${}_{E}T_{g\infty}$.

Acknowledgments

We thank M. Bennett (Waltham Abbey, United Kingdom) and A. Davala (Drexel University) for their assistance. Financial support by the Office of Naval Research is acknowledged.

Literature Cited

1. Sultan, T. N.; McGarry, F. J. *Polym. Eng. Sci.* **1973,** *13*, 29.
2. Riew, C. K.; Rowe, E. H.,; Siebert, A. R. In "Toughness and Brittleness of Plastics"; Deanin, R. D.; Crugnola, A. M., Eds.; ADVANCES IN CHEMISTRY SERIES No. 154, American Chemical Society: Washington, D.C., 1976; p. 326.
3. Bucknall, C. B.; Yoshii, T. *Br. Polym. J.* **1978,** *10,* 53.
4. Manzione, L. T.; Gillham, J. K.; McPherson, C. A. *J. Appl. Polym. Sci.* **1981,** *26*, 889.
5. Manzione, L. T.; Gillham, J. K.; McPhearson, C. A. *J. Appl. Polym. Sci.* **1981,** *26*, 907.
6. Kinloch, A. J.; Shaw, S. J.; Tod, D. A.; Hunston, D. L. *Polymer* **1983,** *24*, 1341.
7. Kinloch, A. J.; Shaw, S. J.; Hunston, D. L. *Polymer* **1983,** *24,* 1355.
8. Kunz-Douglass, S.; Beaumont, P. W. R.: Ashby, M. F. *J. Mater. Sci.*, **1980,** *15*, 1109.
9. Chan, L. C.; Gillham, J. K.; Kinloch, A. J.; Shaw, S. J. Chap 15, this volume.
10. Riew, C. K., *Rubber Chem. Technol.* **1981,** *54*, 374.
11. "Mold Preparation for Araldite Resins, Instruction Manual," Ciba-Geigy Chem. Co., 1972.
12. Underwood, E. E. In "Quantitative Microscopy"; DeHoff, R. T.; Rhines, F. N., Eds.; McGraw-Hill; New York, **1968;** Chap. 6.
13. Enns, J. B.; Gillham, J. K. In "Computer Applications in Applied Polymer Science"; Provder, T., Ed.; ACS SYMPOSIUM SERIES No. 197, American Chemical Society: Washington, D.C., 1983; p. 329.
14. McCrum, N. G.: Read, B. F.; Williams, G. "Anelastic and Dielectric Effects in Polymeric Solids"; John Wiley and Sons: London, 1967.
15. Kinloch, A. J.; Young, R. J. "Fracture Behavior of Polymers"; Applied Science: London, 1983.
16. "Plane-Strain Fracture Toughness of Metallic Materials", American Society for Testing and Materials, Philadelphia, 1972; E-399.
17. Hayden, H. W.: Moffatt, W. G.; Wulff, J. "The Structure and Properties of Materials, Volume III, Mechanical Behavior"; John Wiley and Sons: New York, 1975; p. 10.
18. Nielsen, L. E. "Mechanical Properties of Polymers"; Reinhold: New York, **1962.**
19. Pearson, R. A.; Yee, A. F. *Polym. Prepr., Am. Chem. Soc., Div. Polym. Mat.: Sci. Engr.* **1983,** *49*, 316.
20. Hunston, D. L.; Kinloch, A. J.; Shaw, J.; Wang, S. S. In "International Symposium on Adhesive Joints"; American Chemical Society, Kansas City, Sept. 1982, (to be published by Plenum Press, New York, 1984; Mittal K. L., Ed.).
21. Sayre, J. A.; Kunz, S. C.; Assink, R. A. *Polymer Prepr., Am. Chem. Soc., Div. Polym. Mat.: Sci. Engr.* **1983,** *49*, 442.

RECEIVED for review November 18, 1983. ACCEPTED May 29, 1984.

17

Rubber-Modified, Flame-Retardant, High Glass Transition Temperature Epoxy Resins

H. N. NAE[1], S. REICH[1], and Z. NIR[2,3]

[1] Department of Plastics Research, The Weizmann Institute of Science, Rehovot, Israel 76100

[2] National Air and Space Administration, Ames Research Center, Moffett Field, CA 94035

Addition of brominated polymeric additive (BPA) and rubber to high glass transition temperature epoxy systems has been studied. Thermochemical properties, glass transition temperatures, and morphology are discussed. The cure of the modified systems is similar to that of the unmodified systems. Gelation and vitrification times are about the same as in the unmodified systems with similar apparent activation energies. Addition of BPA does not significantly change the thermochemical properties of the material. Rubber particles phase separate to form spheres homogeneously dispersed in the epoxy matrix. Above a critical concentration the rubber spheres aggregate to form cells.

Addition of various amounts of carboxyl-terminated butadiene–acrylonitrile (CTBN) rubber to polyfunctional epoxy resins improves crack resistance and impact strength of the cure system. This improvement has been attributed to the in situ formation of rubbery domains of a definite size and shape during cure (*1–4*). The rubber particles apparently introduce energy dissipation mechanisms. Phase separation occurs when the rubber and epoxy become incompatible and rubber-rich domains precipitate in the epoxy-rich matrix. Gillham et al. (*5–7*) found that phases begin to separate well before gelation of the cured system. The size and distribution of the rubbery domains were dependent on the temperature and time of cure. Longer gel times may promote separation; quick gelation, such as in reactive materials or at high cure temperature, may result in smaller,

[3] Permanent address: Makhteshim Chemical Works, P.O. Box 60, Beer–Sheva, Israel 84100.

0065-2393/84/0208-0281$06.00/0

fewer domains. However, Romanchick et al. (*8*) noted that morphological changes occur after gelation in a rubber-modified epoxy system cured at 200 °C.

Thermosetting systems containing branched epoxy resins with relatively short distances between cross-linking points and cured with aromatic diamines have higher maximum glass transition temperatures ($T_{g\infty}$) than systems containing only difunctional epoxy resins. A linear correlation between cross-link density and $T_{g\infty}$ has been found (*9*). Brominated polymeric additives (BPA) of the diglycidyl type were added to epoxy systems to impart flame retardancy and were used for graphite-reinforced epoxy composites (*10*). Comparison of nonbrominated and brominated epoxy systems with different molecular weights of BPA has shown that the mechanical properties of the cured systems are a function of the Br content, the polymeric backbone length, and the rubber content as well as the nature of the epoxy resin. Addition of BPA also improves the durability in moist atmospheres and lowers the sorption capacity (*11*).

Our objective is to study the effect of the incorporation of brominated additives and rubber on the cure behavior, thermochemical properties, and morphology of polyfunctional epoxy resins.

Experimental

Materials. Triglycidyl ether of trishydroxyphenylmethane [TEN (Tris Epoxy Novolac), XD 7342.OOL, The Dow Chemical Company] and tetraglycidyl-4,4′-diaminodiphenylmethane (TGDDM, MY 720, Ciba-Geigy) were cured with 4,4′-sulfonylbisbenzamine [4,4′-diaminodiphenyl sulfone (DDS, Ciba-Geigy)]. A BPA containing 50% Br (w/w) with an epoxy equivalent weight (EEW) of 545 (F2001P, Makhteshim Chemical Works) was used as the bromine source. F2001P was prereacted with a carboxyl-terminated butadiene–acrylonitrile copolymer (CTBN, HYCAR 1300×13, The BFGoodrich Chemical Company) to give a diepoxide containing 50% (w/w) rubber.

Five formulations were prepared for each resin to obtain 19 wt% Br (to impart flame retardancy) and 0, 2, 4, 6, and 8 wt% rubber. DDS was added to form stoichiometric compositions (1 mol epoxy with 1 mol amine hydrogen). Each formulation was dissolved in methyl ethyl ketone (MEK), 1:1.5 w/v. Formulas of raw materials are shown on page 283.

Cure behavior was studied during isothermal cure. Specimens for morphology studies were cured according to the following cure cycle: 45 min under vacuum at 80 °C, 8 min at 125 °C, and 2 h at 177 °C. After cooling to room temperature, the specimens were further cured for 4 h at 192 °C.

Dynamic Mechanical Analysis. An automated torsional braid analyzer (TBA) (Plastics Analysis Instruments, Inc.) was used to monitor isothermal cure (*7*). Modulus and mechanical loss (logarithmic decrement) versus time at about 1 Hz were measured under helium flow. Temperature scans were run after isothermal cure at 1.5 °C/min.

Morphology. An optical microscope (Zeiss), a scanning electron microscope (SEM) (JEOL JSM 35C), and a scanning transmission electron microscope (STEM) (Philips) were used to study the morphology. SEM micrographs were obtained from gold-coated fractured surfaces of the fully cured specimens. STEM micrographs were obtained from samples that were cured by the same cure cycle directly on copper grids and stained with osmium tetroxide.

1. Tetraglycidyl-4,4′-Diaminodiphenylmethane (TGDDM)

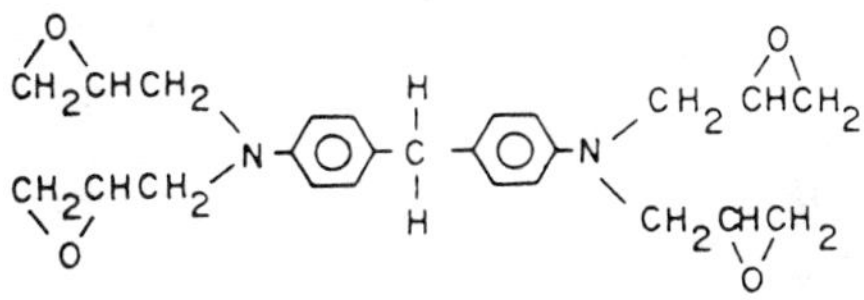

2. Tris(hydroxyphenyl)methane Triglycidyl Ether (TEN)

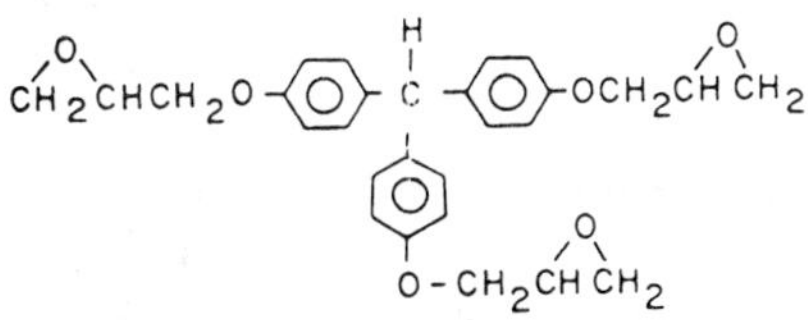

3. Diaminodiphenyl Sulfone (DDS)

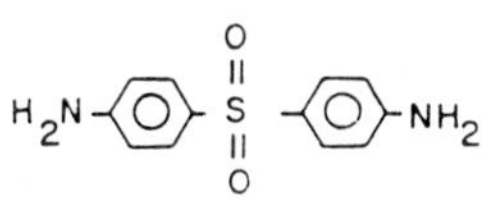

4. Brominated Polymeric Additive (BPA)

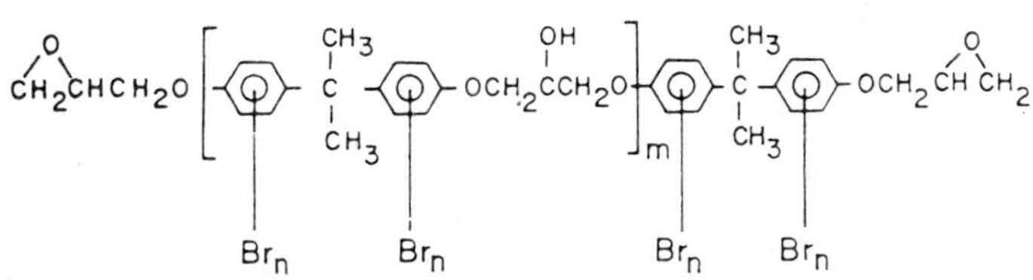

5. Carboxyl-Terminated Butadiene–Acrylonitrile Copolymer (CTBN)

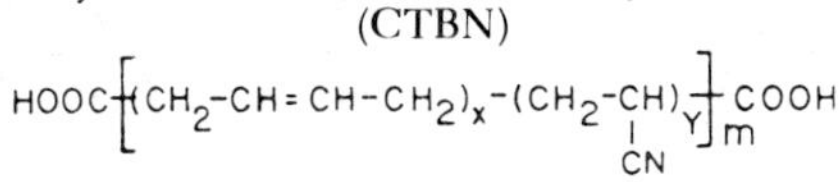

A laser light-scattering analyzer, similar to the apparatus used in the study of polymer blends (*12*) was used to identify phase separation and size of the rubbery domains. We used a classical diffraction arrangement. A 1-mW He–Ne laser was focused to a spot of ~30 μm onto a thin layer of the epoxy system. The direct beam emerging from the sample was blocked, and the far-field diffraction pattern was projected onto a translucent screen. The image of the scattered light on the screen was photographed. The average sizes of the phase-separated rubber particles were calculated by using the Bragg relationship, $2d \sin \theta/2 = \lambda$, where d is the dimension of the particle, θ is the ring angle, and λ is the light wavelength.

Results and Discussion

To avoid competing polymerization mechanisms such as reactions of the carboxyl groups of the rubber with hydroxyl groups formed during cure, and because carboxyl groups facilitate reactions between epoxy groups and the hydroxyl groups (*8*), we decided that the reacting composition will contain only epoxides that are cured by the curing agent, without a catalyst. The rubber was therefore end capped by the BPA to form a diepoxide. The concentration of epoxide groups is, therefore, the sum of the epoxide concentrations of the epoxy resin, the BPA, and the rubber. The BPA and the BPA–rubber serve as chain extenders that are linked to the major epoxy component via the curing agent through their epoxy end groups.

Representative isothermal TBA spectra for systems containing 19% Br and 2% rubber are shown in Figure 1. The three transition

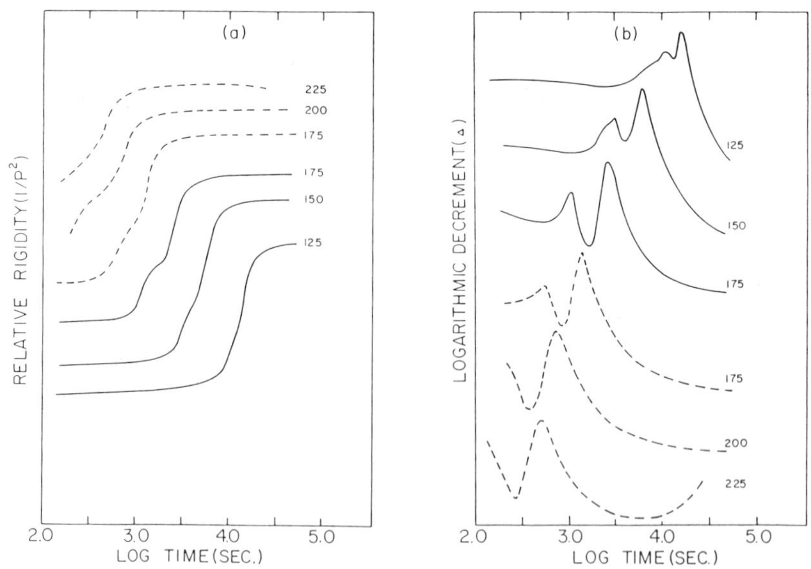

Figure 1. TBA spectra during isothermal cure of systems containing TGDDM (——) and TEN (----) epoxy resins. Key: a, relative rigidity; and b, logarithmic decrement. Note: the order of temperatures in (a) is reversed in (b).

events have been designated as pregel, gelation, and vitrification transitions (*7*). The gel point is the critical conversion at which branched molecules of infinite molecular weight are first formed. Vitrification is the formation of a glassy solid from the liquid or the rubbery state. In the systems investigated, the trifunctional epoxy system gels and vitrifies sooner than the system containing the tetrafunctional epoxy. This result seems contradictory to the calculated conversions from the functionality of the epoxy resin (*13*). However, because the chemical nature of the epoxy resins is different and the distances between branching points are different, this difference in transition times is possible. Another reason for the difference may be the different reactivities of the epoxide groups in the two resins. The reactive additive should slow the reaction, but a comparison to similar systems without Br and rubber (Table I) shows that, in an isothermal cure, gelation and vitrification times are about the same. Similar cure behavior has been observed with other rubber-modified systems (*6*). This behavior is a result of phase separation of the rubber segments before gelation so that gelation and vitrification depend mainly on the chemical nature of the epoxy resin and the curing agent. A fourth event, degradation, is observed at higher temperatures. At these temperatures, cure and degradation compete, and the material may degrade well before reaching maximum cure (*9*). Degradation, which occurs below $T_{g\infty}$, prevents such systems from being fully cured. The significance of this observation for structural purposes is obvious. Such materials must be cured below their $T_{g\infty}$; however, they may then age with time and change their mechanical and thermal properties. The degradation temperature serves, therefore, as the upper limit for practical applications. A value for $T_{g\infty}$ can only be calculated theoretically for such systems.

Apparent activation energies were calculated from log time versus $1/T$. A comparison of these activation energies (Table II) shows that they are about the same as those of systems without Br and rubber. This result also indicates that the cure reaction energy bar-

Table I. Gelation and Vitrification Times for the Trifunctional System

	Temperature (°C)					
Parameter	*175*		*200*		*225*	
System[a]	A	B	A	B	A	B
Gelation (min)	10.4	11.5	3.5	3.2	—	2.5
Vitrification (min)	25.7	29.0	13.3	15.1	6.7	9.9

[a] A = TEN epoxy cured with DDS (9); and B = TEN epoxy, 19% Br, and 2% rubber cured with DDS.

Table II. Apparent Activation Energies

System[a]	ΔE_a, *Gelation*	ΔE_a, *Vitrification*
A	13.3	11.3
B	14.0	9.6
C	16.0	12.3

NOTE: Activation energies in kcal/mol.
[a] A = TEN epoxy cured with DDS (9); B = TEN epoxy, 19% Br, and 2% rubber cured with DDS; and C = TGDDM, 19% Br, and 2% rubber cured with DDS.

riers depend mainly on the major component epoxy resin and the curing agent, even though the amount of additives is about 30% of the system.

Figure 2 shows temperature scans after isothermal cure. On subsequent heating above the cure temperature, T_{cure}, the material exhibits a characteristic glass transition temperature, ${}_{c}T_{g}$, and an ap-

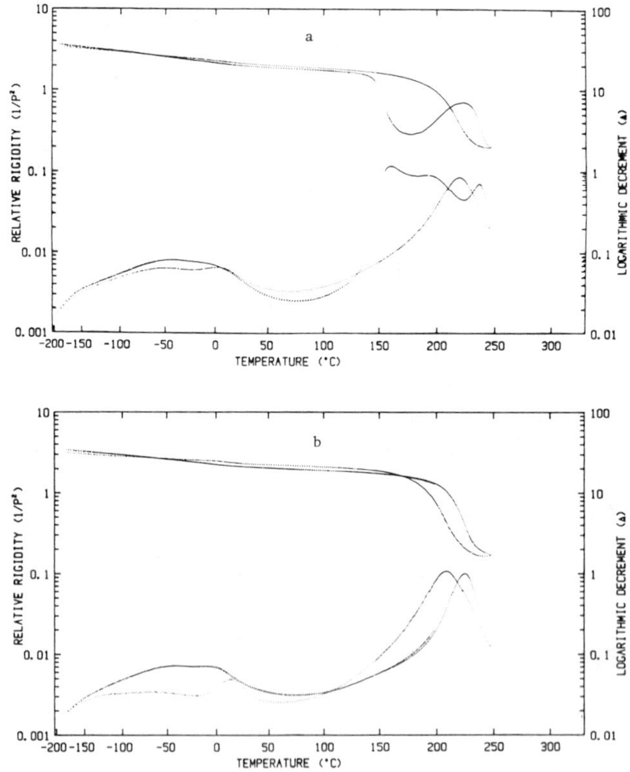

Figure 2. Temperature scans after isothermal cure of a system containing TGDDM–DDS, 19% Br, and 2% rubber. Cycle: a, 125 °C → −180 °C → 250 °C at 1.5 °C/min; and b, 200 °C → −180 °C → 250 °C at 1.5 °C/min.

Table III. Characteristic Glass Transition Temperatures

System[a]	T_{cure}	${}_{c}T_g$	${}_{app}T_g$	$T_{g\infty}$
A	175	238	315	d
	200	271	313	d
B	175	225	273	d
	200	258	266	d
	225	261	266	d
C	177	—	—	225
D	125	164	239	220
	150	194	241	223
	175	221	236	220
	200	226	226	210

NOTE: T_g values in degrees Celsius.
[a] A = TEN–DDS (calculated $T_{g\infty}$ = 334 °C) (*9*); B = TEN–DDS, 19% Br, and 2% rubber; C = TGDDM–DDS (*14*); D = TGDDM–DDS, 19% Br, and 2% rubber; and d = degrades.

parent glass transition temperature, ${}_{app}T_g$. The latter is a result of further cure of the system; unreacted end groups, which were trapped in the glassy state, are exposed as a result of the heat gained by the system. Characteristic T_g values are shown in Table III. Because of its chemical structure, the network formed by the trifunctional epoxy is denser than that formed by the tetrafunctional one. Therefore, $T_{g\infty}$ of the trifunctional system is higher. In comparison with the neat system, however, $T_{g\infty}$ is lower because of the additives. The distance between cross-links and the T_g of the initial reactants, T_{go}, are different from those of the neat systems. Because $T_{g\infty}$ cannot be reached in the trifunctional system, the best experimental approximation of $T_{g\infty}$ is the highest T_g obtained during temperature scans (i.e., ${}_{app}T_g$). This T_g is in a region in which the material degrades and therefore is lowered in subsequent temperature scans.

A sub-zero relaxation corresponding to the T_g of the rubber is observed in temperature scans of the cured systems. This relaxation overlaps the broad secondary transition of the epoxy resin. The rubber T_g, ${}_{R}T_g$, of the 50% rubber–BPA concentrate is −9 °C. This transition is apparently the shoulder observed near 0 °C in temperature scans after isothermal cure (Figure 2). After heating above 250 °C, the material starts to degrade, and the rubber relaxation becomes more distinct. This result is apparently because of further phase separation that occurs upon further curing during the temperature scan. The rubber particles are chemically bonded to the epoxy matrix; therefore, further cure increases the energy needed for the glass transition of the rubber because the rubber particles are now embedded in a tighter environment. Thus, ${}_{R}T_g$ moves to higher temperatures.

Thermogravimetric analysis (TGA) of all systems (*10*) shows a catastrophic weight loss at the same temperature, 310–350 °C. The char yields at 800 °C are 35–45% in nitrogen and 0–5% in air. Because the $T_{g\infty}$ of the BPA–rubber formulations is about 240 °C for the TGDDM–DDS system and about 275 °C for the TEN–DDS system, the thermostability of these systems is not a limiting factor when compared with the unmodified systems. However, prolonged exposure above 250 °C may result in degradation well before the catastrophic failure. A Br content of 19% was chosen as sufficient to impart flame retardancy (UL 94 V = 0) to the cured systems.

The sub-zero relaxation during temperature scans indicated changes in phase separation of the rubber–BPA system during cure. Therefore, the morphologies of systems containing 0, 2, 4, 6, and 8% rubber (w/w) were studied. An epoxy system containing BPA without rubber is transparent and does not show phase separation. Although the system seems to form an homogeneous matrix with the epoxy resin, SEMs of the system containing TEN epoxy resin show some inhomogeneities. These areas have an average diameter of 0.6 μm and resemble the two-phase system observed by etching styrene and bromostyrene cross-linked polyesters (*15*). Upon addition of rubber to the BPA–epoxy system, distinct rubber particles are observed. These particles have the form of a sphere and core. Figure 3 shows homogeneously dispersed rubber particles in a brittle epoxy matrix. Although the the outer boundaries of the spheres are linked to the epoxy matrix, the inside core is totally incompatible. The spheres have an average diameter of 2–4 μm, and the core diameter is 1–2 μm. The rubber particles are about the same size in systems con-

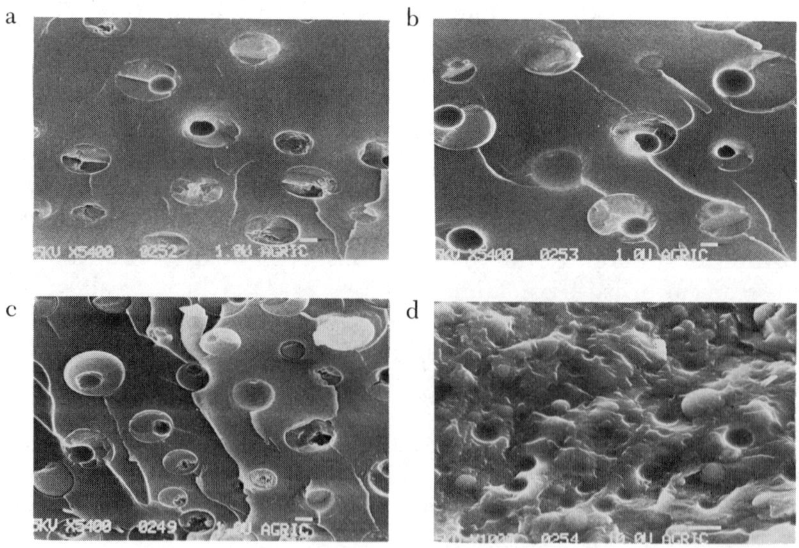

Figure 3. SEMS of TEN–DDS, 19% Br, and 2% (a), 4% (b), 6% (c), and 8% (d) rubber.

taining 2, 4, and 6% rubber. This size is apparently determined by the cure conditions prior to gelation (5). In systems containing 8% rubber, the rubbery domains are larger and have a sphere diameter of about 8 μm and a core diameter of 3 μm. Particle diameters and distributions are shown in Table IV. Because all samples were cured with the same curing cycle, this result indicates that, above a certain percentage of rubber, the rubber particles aggregate to form larger agglomerates. STEMs of samples stained with osmium tetroxide (Figure 4) also show rubber particles with an average diameter of about 1 μm. These samples were cured directly on copper grids; therefore, their thickness limits the growth of larger spheres. This limit results in the larger domains shown in the figure. The agglomeration of small rubber particles inside a cell has also been observed in the STEM samples for the TEN–DDS and the TGDDM–DDS systems containing 8% rubber.

The mechanism for the formation of the rubber particles is as follows: The rubber is chemically bonded to the epoxy matrix through the BPA. The BPA–rubber concentrate arranges itself in the epoxy matrix so that the rubbery part is aligned toward the inside and the BPA part is aligned toward the outside. As long as the chains are growing, these arrangements are only local. When the system becomes denser (i.e., as branches and cross-links start to form) these local arrangements form "droplets." The droplets have a higher concentration of rubber in the center; this rubber is apparently the core of the sphere shown in the SEMs. The sphere itself is bonded chemically to the matrix through the BPA moiety but is different from the matrix because of the polarity of the Br atoms. The rubber itself is also polar as a result of the acrylonitrile segments. This polarity has been cited as the reason for better compatibility of the rubber in the epoxy matrix in the early stages of the polymerization. When the concentration of the rubber is low, the droplets form spheres with distinct cores. When the concentration of the rubber is above a critical concentration, many cores are locked together within a BPA matrix that forms a cell containing a number of rubber cores (*16*). These

Table IV. SEM Morphology of Rubber-Modified TEN Epoxy

Rubber (%)	d[a] *Sphere*	d[a] *Core*	n[b]
0	0.6	—	15.1
2	2.8	1.4	4.5
4	3.8	1.6	3.3
6	2.7	1.3	7.1
8	8.3	3.4	7.1

[a] Average diameter of particle in μm.
[b] Number average of particles/100 μm^2.

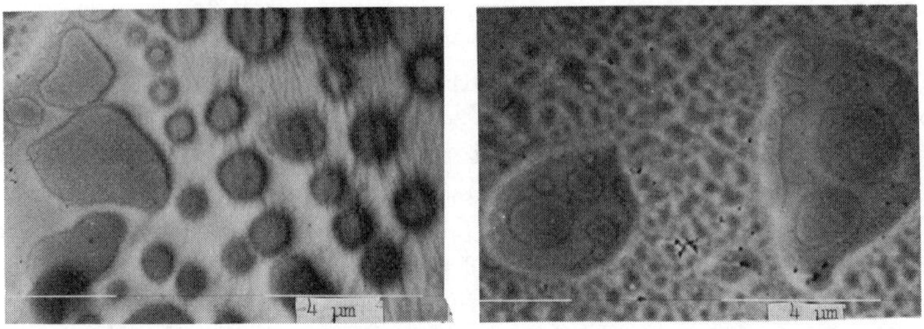

Figure 4. STEMS of TEN–DDS, 19% Br, and 8% rubber.

cells, which are larger in size, are observed in thin layers of the epoxy systems that were cured on glass slides (Figure 5). These cells have an average diameter of 20–180 µm and an envelope with an average thickness of 10–20 µm. The inner rubber particles are, however, still 2–4 µm in diameter. The envelope may have been formed by phase inversion of the rubbery phase, which is depleted on both sides by the epoxy moiety. Phase separation in the TGDDM–DDS system is hardly seen in the SEMs, probably because the rubber is more com-

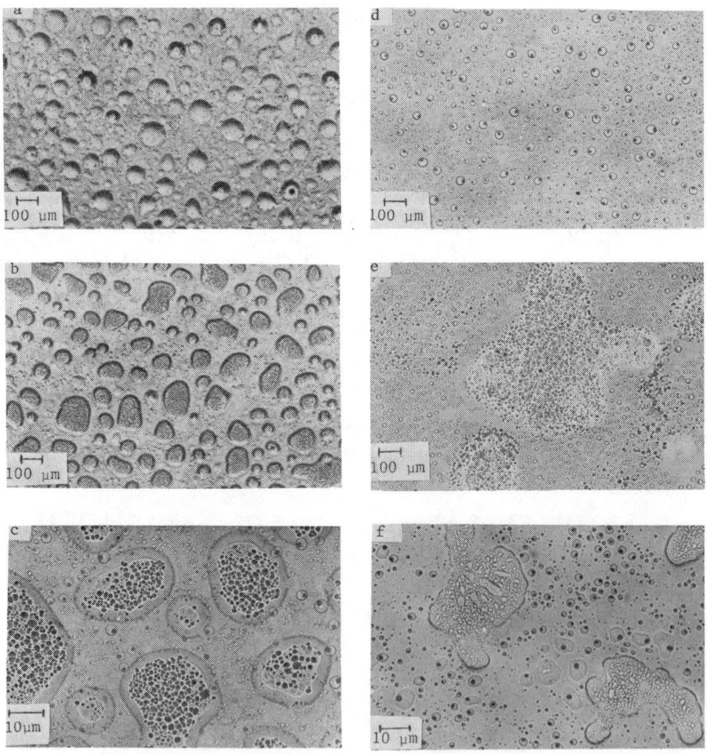

Figure 5. Optical micrographs of TEN–DDS with 19% Br and 4% (a), 8% (b), and 8% (c) rubber, and of TGDDM–DDS with 19% Br and 4% (d), 8% (e), and 8% (f) rubber.

patible with the epoxy matrix. However, thin-layer optical micrographs of the rubber-modified TGDDM–DDS system show that rubber particles are phase separated but form smaller domains. In this system, the aggregates formed in the 8% rubber concentration are less pronounced and do not have the envelopes seen in the TEN–DDS system. Although aggregates are formed above a critical rubber concentration, individual spheres are still dispersed in the matrix.

Crack propagation is mainly unidirectional and stops at the rubber particles (Figure 3). This phenomenon has been cited as the energy dissipation mechanism that improves toughness (3).

The laser light-scattering experiment gave us data on the packing of the phase-separated rubber particles and their average size. The diffraction pattern produced by a random array of N identical spherical particles is the same as that produced by one particle alone, except that it is N times brighter. Any deviation from randomness would produce enhancement of scattered light intensity in certain directions. An arrangement in a regular array, for example, would produce in certain directions an intensity N^2 times the intensity produced by a single particle. In our diffraction pattern (Figure 6), we see a deviation from a perfect ring to an hexagonal shape with an increase in the scattered-light intensity at the corners. This result indicates partial short-range packing of the rubber particles. Indeed, the cluster-forming tendency of the rubber particles in this sample is evident from Figure 5. The intensity of the scattered light is much higher in the TEN–DDS system; this result indicates that, as a result of the denser matrix (higher cross-link density), the rubber phase separates to form a more "ordered" system. From the average scattering angle for the ring shown in Figure 6, the average particle size was calculated as $d = 2.6\ \mu m$.

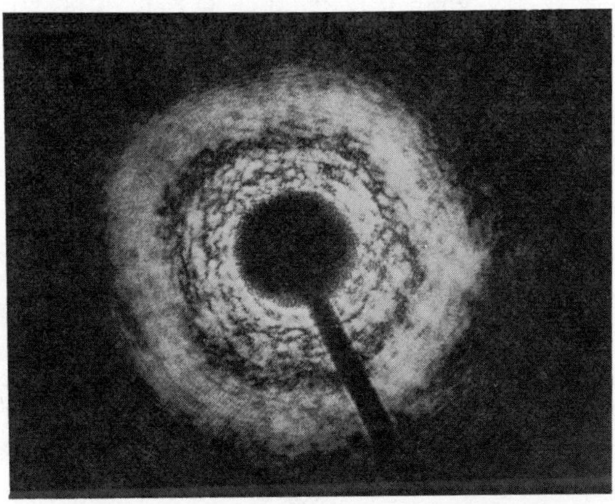

Figure 6. Laser light scattering from TEN–DDS, 19% Br, and 4% rubber.

Conclusions

Systems containing epoxy resins with a brominated polymeric additive and rubber have high T_g values, and their cure behavior is similar to that of the unmodified systems. The T_g of these systems is lowered as a result of the additives. These resins combine the advantages of the use of BPA as a flame retardant and the use of the rubber for improved fracture toughness. Rubber particles phase separate to form homogeneously dispersed spheres in the epoxy matrix. These spheres are not observed in the systems containing the BPA without rubber. Above a critical concentration, the rubber spheres aggregate to form cells. This phenomenon may limit the use of rubber in these systems to concentrations lower than the critical concentration for cell formation.

Acknowledgment

We are grateful to J. K. Gillham, Princeton University, for the use of the TBA and many helpful discussions. Discussions with J. A. Parker, NASA Ames Research Center, are highly appreciated.

Literature Cited

1. Rowe, E. H.; Siebert, A. R.; Drake, R. S. *Mod. Plast.* **1970,** *47*, 110.
2. Riew, C. K.; Smith, R. W. *J. Polym. Sci.* **1971,** A-1, *9*, 2739.
3. Sultan, J. N.; McGarry, F. J. *Polym. Eng. Sci.* **1973,** *13*, 29.
4. Riew, C. K.; Rowe, E. H.; Siebert, A. R. In "Toughness and Brittleness of Plastics"; Deanin, R. D.; Crugnola, A. M., Eds.; ADVANCES IN CHEMISTRY SERIES NO. 154, American Chemical Society: Washington, D.C., 1976; p. 326.
5. Manzione, L. T.; Gillham, J. K.; McPherson, C. A. *J. Appl. Polym. Sci.* **1981,** *26*, 889; Ibid., **1981,** *26*, 907.
6. Gillham, J. K.; Chan, L. C. *Am. Chem. Soc. Div. Polym. Mater. Sci. Eng., Prep.* **1983,** *49*, 455.
7. Gillham, J. K. In "Development in Polymer Characterization—3"; Dawkins, J. V., Ed.; Applied Science: London, 1982; Chap. 5.
8. Romanchick, W. A.; Sohn, J. E.; Geibel, J. F. In "Epoxy Resin Chemistry II"; Bauer, R. S., Ed.; ACS SYMPOSIUM SERIES No. 221, American Chemical Society: Washington, D.C., 1983; p. 85.
9. Nae, H. N.; Gillham, J. K. *Am. Chem. Soc. Div. Org. Coat. Plast. Chem., Prep.* 1983, *48*, 566.
10. Nir, Z.; Gilwee, W. J.; Kourtides, D. A.; Parker, J. A. *SPE ANTEC*, **1983,** May 2–5.
11. Goobich, J.; Marom, G. *Polym. Eng. Sci.* **1982,** *22*, 1052.
12. Snyder, H. L.; Meakin, P.; Reich, S. *Macromolecules* **1983,** *16*, 757.
13. Flory, P. J. "Principles of Polymer Chemistry"; Cornell University: Ithaca, 1953.
14. Morgan, R. J.; O'Neal, J. E.; Muller, D. B. *J. Mater. Sci.* **1979,** *14*, 109.
15. Cohn, D.; Marom, G. *Polymer* **1983,** *24*, 223.
16. Schmitt, B. J. *Angew. Chem. Int. Ed. Engl.* **1979,** *18*, 273.

RECEIVED for review November 18, 1983. ACCEPTED May 16, 1984.

18

Fluoroelastomer-Modified Thermoset Resins

JOVAN MIJOVIĆ, ELI M. PEARCE, and C.-C. FOUN

Polymer Science and Engineering Program, Departments of Chemical Engineering and Chemistry and the Polymer Research Institute, Polytechnic Institute of New York, Brooklyn, NY 11201

Toughening of epoxy and polyimide matrices with fluorocarbon elastomers was investigated. The introduction of polar groups into the fluorocarbon backbone gave rise to compatible blends of elastomer and partially cured epoxy matrix. During postcure, however, the elastomer phase separated out. Through a careful control of postcure conditions, production of "tailor-made" morphology and, consequently, physical–mechanical properties may be possible. In epoxy blends, a bimodal distribution of the elastomer particle size and a concomitant fourfold increase in the critical strain energy release rate were found. Blends containing the polyimide matrix were incompatible at all stages of cure. The effect of addition of elastomer on thermal stability of the blend was more pronounced with the polyimide than the epoxy matrix.

THE MODIFICATION OF THERMOSET RESINS by incorporation of elastomers continues to be an active area of research. The inherent brittleness of thermosets restricts the range of their use, particularly as the matrix material in composites. Hence, considerable research has been conducted on the toughening of thermosets. Advances in studies of the most common toughening route (i.e., the addition of elastomers) previously were compiled (*1*) and are further discussed in this book.

In general, toughened thermosets are characterized by an increase in impact strength and, unfortunately, a simultaneous decrease in the glass transition temperature (T_g) and modulus. Various elastomers have been used to toughen thermosets. For instance, modification of epoxy resins with carboxyl-terminated copolymers of butadiene and acrylonitrite (CTBN) significantly improved the toughness (*2, 3*). Many factors affect the nature of the elastomer phase, such as its chemical composition and concentration, the type of curing agent, and the kinetics of cure. Some workers have concluded that

0065-2393/84/0208-0293$06.00/0

the improved toughness was related to the size and distribution of rubber domains (*4*, *5*), whereas others have expressed the view that the resin–rubber compatibility determined the toughness (*6*). In spite of the reported results, however, the fundamental relationships between processing (cure), morphology, and properties of elastomer-modified thermosets remain incompletely understood.

We have undertaken an innovative approach in controlling the formation of rubbery phase. A fluorocarbon elastomer, initially incompatible with resin, was chosen as toughening agent. Our first hypothesis was that a chemical modification of the elastomer (e.g., introduction of polar groups) would result in compatibility with partially cured resin network through various intermolecular interactions. The latter, however, would be disrupted at high temperature (postcure) leading to the separation of the rubbery phase. A careful control of postcure conditions (time and temperature) would then allow one to control the morphology of the blends; the morphology, in turn, determines their physical–mechanical properties.

We have recently embarked on a research program of investigating the synthesis–morphology–property relationship in elastomer-modified thermosets. The objectives of this part of that program were to investigate the mechanism of formation of rubbery phase in epoxy and polyimide matrices during cure and to correlate the morphology of cured systems to their physical–thermomechanical properties.

Experimental

Epoxy Systems. In the first set of experiments, a diglycidyl ether of bisphenol A (DGEBA) liquid epoxy resin (Epon 828, Shell Chemical Corporation) was cured with 4,4′-sulfonylbisbenzamine [4,4′-diaminodiphenyl sulfone (DDS)]. Both compounds were used without further purification.

Polyimide Systems. For the second set of experiments, a polyimide resin was synthesized according to Scheme 1.

The phosphine oxide component (**I**), used as the starting material in the synthesis of polyimide resin, was made from triphenylphosphine oxide by nitration and a subsequent reduction of nitro groups to amino groups. The product was characterized by IR spectroscopy and had a melting point of 260 ± 2 °C. The imide monomer (**II**) was also identified by IR spectroscopy, and the characteristic absorption bands resulting from the presence of imide rings were recorded at 1710 and 1780 cm^{-1}, as seen in Figure 1.

Fluoroelastomer Modifier. A fluorocarbon elastomer (Viton GF, Du Pont) was used as a modifier. Because of its incompatibility with Epon 828, Viton GF was further modified, first by the introduction of double bonds into the main chain and then by the addition of various polar groups on the double bond. The entire sequence of chemical reactions used to modify Viton GF is shown in Scheme 2. The modified Viton GF used in this study in blends with epoxy matrix is shown by Structure C in Scheme 2 and is hereafter referred to as the polar Viton GF. Both double-bond containing Viton GF and polar Viton GF were used in blends with polyimide matrix.

Preparation of Polyblends. The polyblends were prepared by the solution cast method described in Table I.

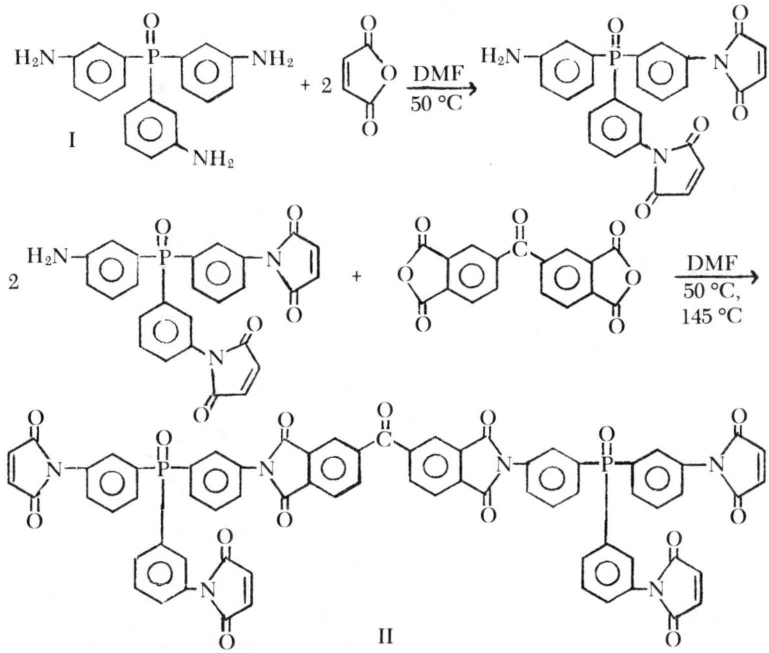

Scheme 1

Techniques. After processing, all samples were allowed to cool to 20 °C and were then kept in closed vials in a desiccator at 20 °C for 1 week before testing.

An IR spectrometer (Digilab–FTS–20B) was used for the analysis of chem-

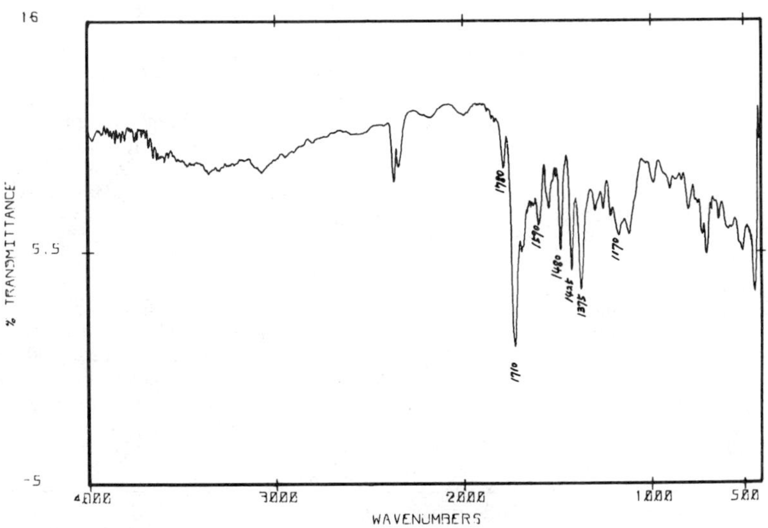

Figure 1. FTIR spectrum of trisimide monomer. Note the absorbance of imide rings at 1710 and 1780 cm^{-1}.

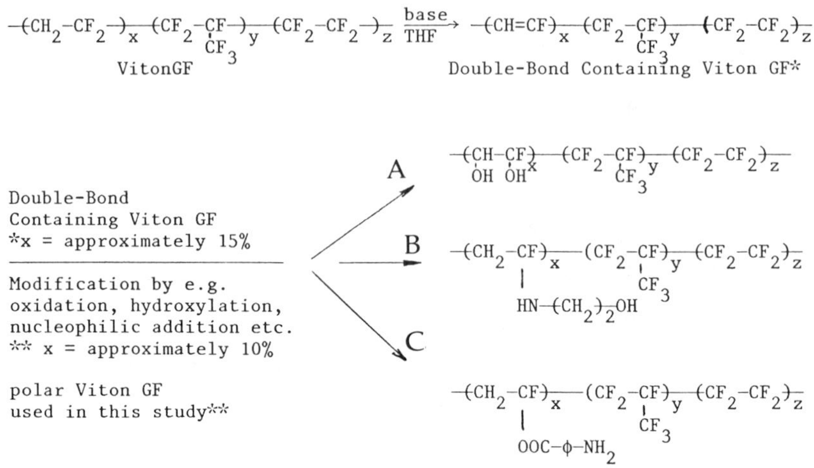

Scheme 2

ical composition. The specimens of modified Viton GF for Fourier transform IR (FTIR) analysis were first cast onto a glass plate at room temperature. Then, dried films (1 × 2 × 0.005 cm) were cut. Imide samples and various blends were prepared and analyzed in the form of KBr pellets.

Dynamic mechanical measurements were performed in the Du Pont 981 model DMA connected to the 1090 thermal analyzer. Tests were run at a heating rate of 10 °C/min and at an oscillation amplitude of 0.2 mm peak-to-peak.

Linear elastic fracture mechanics (LEFM) analysis was applied to calculate

Table I. Various Polyblends and Their Preparation

Formulation	*Variable*	*Cure Schedule*
Epoxy		
Epon 828 + DDS (40 phr) + Viton GF	Viton GF (0–100 phr)	Components dissolved in acetone and mixed at room temperature for 10 min. Solvent removed by heating for 24 h at 40 °C and 12 h at 60 °C. Cured in silicon molds for 24 h at 125 °C and 5-h postcure at 200 °C.
Epon 828 + DDS (40 phr) + polar Viton GF	Polar Viton GF (0, 5, 10, and 15 phr)	Same as above
Polyimide		
Imide/double-bond containing Viton GF	Double-bond containing Viton GF (0–100 phr)	Components dissolved in DMF–acetone (1:1) and mixed at room temperature for 30 min. Solvent removed in a vacuum oven at 60 °C overnight. Cured at 300 °C for 30 min.
Imide/polar Viton GF	Polar Viton GF (0–100 phr)	Same as above

NOTE: phr = parts per hundred parts resin.

the fracture energy of samples prepared in the form of adhesive joints. The use of height-tapered double cantilever beam (HTDCB) specimens for fracture energy measurements of similar systems has been described elsewhere (7). The preparation of aluminum beam surfaces has also been reported (8). The HTDCB samples were fractured in a tensile tester (Instron) at 20 °C/50% relative humidity (RH) and a crosshead speed of 0.05 in./min.

Thermogravimetric analysis (TGA) was performed in the Du Pont 951 model TGA connected to the 1090 thermal analyzer. Tests were run at a heating rate of 10 °C/min.

Differential scanning calorimetry (DSC) was performed in the Du Pont model DSC connected to the 1090 thermal analyzer. Tests were run at a heating rate of 10 °C/min.

Cast specimens were immersed in liquid nitrogen, removed, and immediately fractured in air. A scanning electron microscope (SEM) (AMR/1200) was used to study fracture surfaces. Prior to SEM, all samples were coated with a thin layer of gold. Transmission electron micrographs (TEM) (with a JEOL 300D model TEM) were obtained on thin sections prepared with an ultramicrotome (LKB 3800 III) equipped with a glass knife.

Results and Discussion

Modification of Viton GF. Viton GF was first modified by dehydrofluorination in order to produce the double-bond containing Viton GF. This reaction was carried out in the presence of dehydrofluorination catalysts [e.g., NH_4OH, $(C_2H_5)_2NH$, $(C_2H_5)_3N$, and $(CH_2)_3N$].

Polar Viton GF (Scheme 2, Structure A) was obtained by reaction of osmium tetroxide (OsO_4) and double-bond containing Viton GF. Osmium tetroxide adds slowly and forms an osmic ester (*see* Structure 1). Upon subsequent hydroxylation, –OH groups are produced.

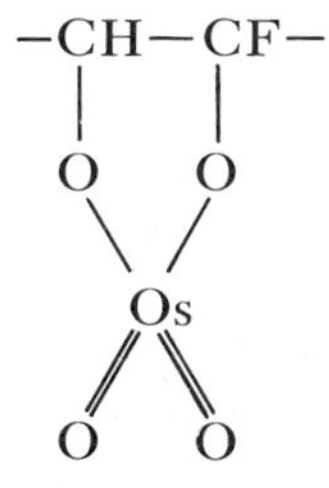

I

Ethanolamine was added to the double-bond containing Viton GF in order to produce polar Viton GF (Scheme 2, Structure B). Similarly, *p*-aminobenzoic acid reacted with double-bond containing Viton GF to form the molecule shown in Scheme 2, Structure C. The Fourier transform IR (FTIR) spectra of double-bond containing Viton GF and polar Viton GF are shown in Figures 2 and 3, respectively. Note the appearance of characteristic C=C bond absorbance in Figure 2 at 1700 cm^{-1} and the amino group absorbances at 3350 and 3450 cm^{-1} in Figure 3.

Compatibility of Blends. In the next set of experiments, a series of blends of the epoxy formulation and Viton GF were prepared. All blends were incompatible as judged by the appearance of two separate endothermic events in DSC thermograms (Figure 4). (In blends containing 20% or less of either phase, the glass transition endotherm of that phase was very subtle.) The lower endotherm at −7 °C is representative of the T_g of Viton GF, and the endotherm at approximately 193 °C corresponds to the T_g of the epoxy resin. This finding was not surprising because of the low likelihood of compatibility of fluorocarbon polymers with epoxy resins. The calculation of solubility parameters by the group contribution method yielded values of 23 and 12.3 $(J/cm^3)^{\frac{1}{2}}$ for Epon 828 and Viton GF, respectively. We then postulated that the introduction of polar groups into the main chain of Viton GF would enhance its compatibility with the epoxy resin. Our hypothesis was corroborated by the calculation of solubility parameters of Viton GF containing a hydroxyl group (21.1) and a *p*-aminobenzoic acid group (18.0). Both numbers were much closer to the solubility parameter for Epon 828. We then prepared and investigated a series of blends of epoxy formulation with different amounts of polar Viton GF (5,10, and 15 phr). Blends that were not exposed to the final postcure of 5 h at 200 °C showed an increase in T_g upon the addition of polar Viton GF (Table II). The T_g was taken as the peak value in the loss modulus trace in the dynamic mechanical spectrum. This peak value is also clearly seen in the dynamic mechanical spectra of these blends shown in Figure 5. Hereafter, the glass transition of all samples is defined by the location of the α peak in the loss modulus curve. The single value of glass transition, observed in dynamic mechanical spectra, indicates that the precured blends of Epon 828 and polar Viton GF are compatible. Analogous results were obtained from DSC measurements and are shown in Figure 6. In general, however, the sensitivity of DSC results was inferior to that of dynamic mechanical measurements; hence, the latter were preferred. The observed compatibility is, we believe, a direct consequence of the modification of Viton GF. The apparent increase in T_g as a function of the concentration of polar Viton GF clearly suggests the existence of interactions between the resin network and polar Viton GF. The most pronounced effect upon T_g is obtained by the addition of up to 5% polar Viton GF. The low-temperature β transition (T_β), which is believed to be caused by various relaxations within the epoxy network, is also affected by the presence

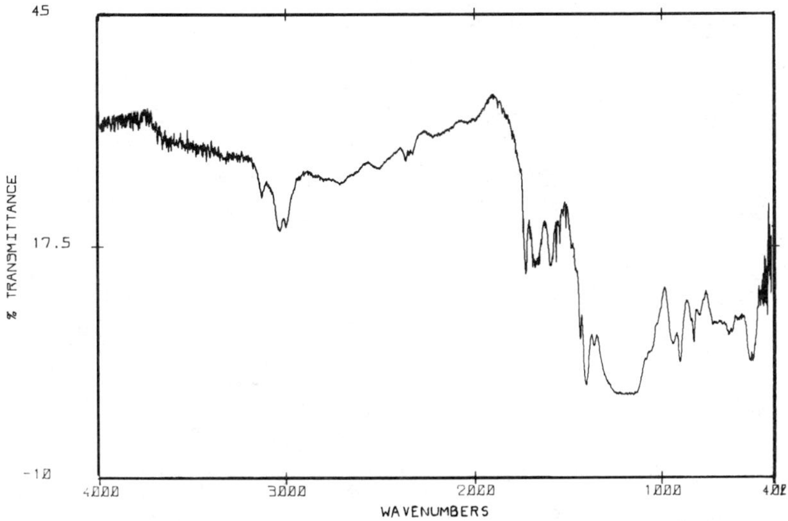

Figure 2. FTIR spectrum of double-bond containing Viton GF. Note the absorbance of C=C bond at 1700 cm^{-1}.

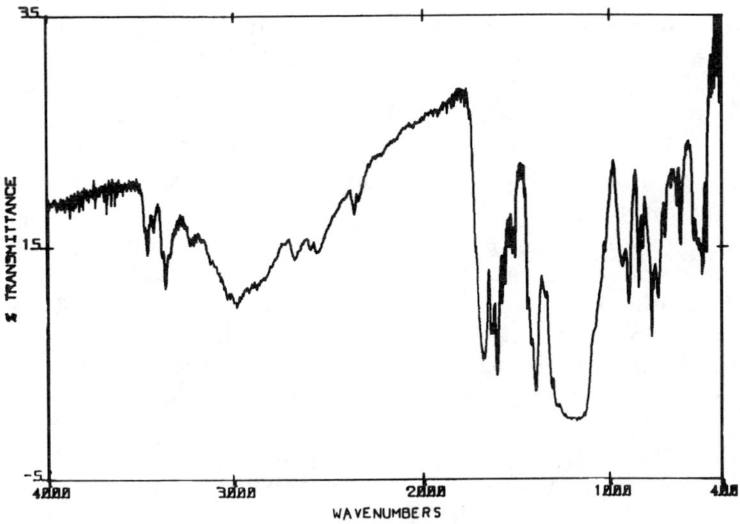

Figure 3. FTIR spectrum of polar Viton GF. Note the disappearance of C=C bond and appearance of NH_2 groups at 3350 and 3450 cm^{-1}.

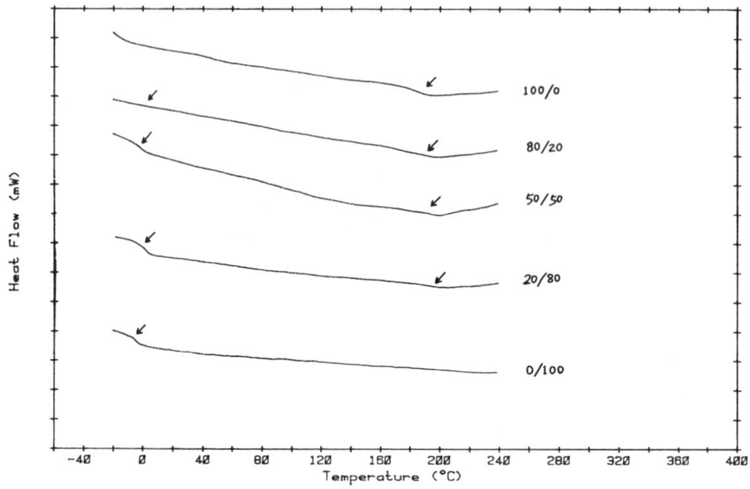

Figure 4. DSC thermograms of a series of Epon 828–Viton GF blends. The curves were shifted along the y axis for clarity.

of polar Viton GF. A small increase in T_β also suggests overall stiffening of the network.

Continuing with our hypothesis, we tested the permanence of the interactions between the epoxy network and the polar Viton GF. Next in our experimental procedure, the samples were subjected to the 5-h postcure at 200 °C (Formulation 2, Table I). As a result of this exposure to high temperature, the intermolecular interactions between the epoxy network and the elastomer were lost, and the rubbery phase separated. Simultaneously, the T_g of all blends increased, as shown in Figure 7. Analogous results were obtained from DSC measurements.

This increase in T_g results from additional cross-linking reactions within the epoxy network. However, with an increased concentration of polar Viton GF in cured blends, T_g decreased. Not all of the elastomer molecules are located within the discrete inclusions (*see* Figures 9 and 10). Instead, some remain scattered throughout the epoxy network and exert an apparent plasticizing effect. The changes

Table II. Glass Transition Temperatures of a Series of Epoxy–Polar Viton GF Blends Before and After Postcure

Concentration of Polar Viton GF (%)	T_g[a] (°C)	T_g[b] (°C)
0	120	193
5	160	191
10	163	189
15	164	176

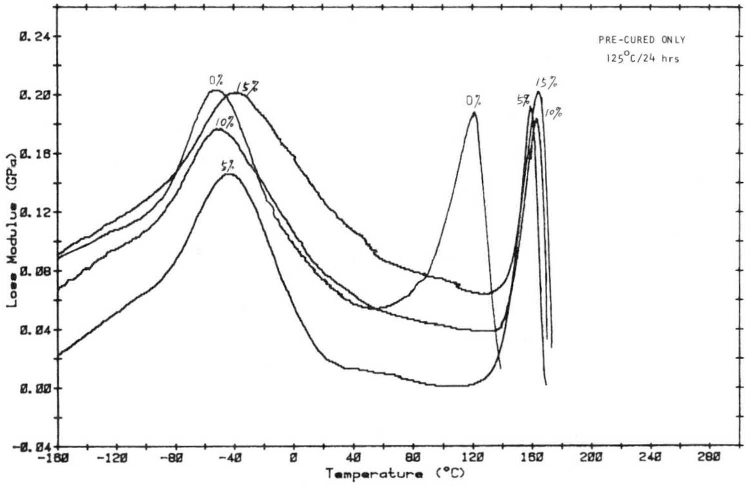

Figure 5. Loss modulus as a function of temperature with concentration of polar Viton GF as a parameter for precured epoxy blends.

in T_g of various blends after the initial cure and the postcure are summarized in Table II.

The analogous approach has been undertaken for blends of polyimides with double-bond containing Viton GF and polar Viton GF. Thus far, however, we have not been able to produce initially compatible blends. For instance, blends of polyimide double-bond containing Viton GF are incompatible, as seen in Figure 8. The T_g of each component in the blend shows no change as a function of com-

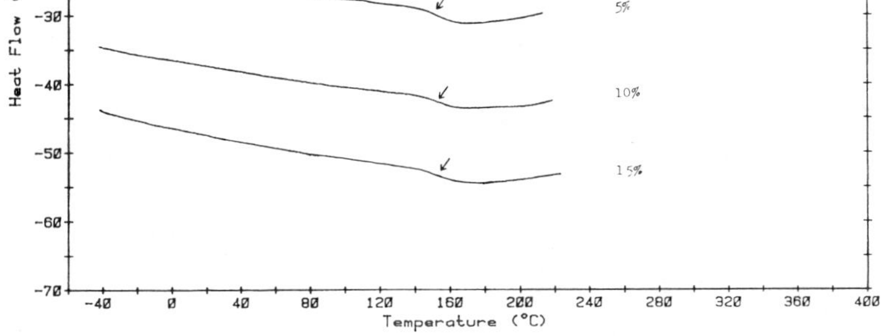

Figure 6. DSC thermograms of a series of precured Epon 828–polar Viton GF blends. The curves were shifted along the y axis for clarity.

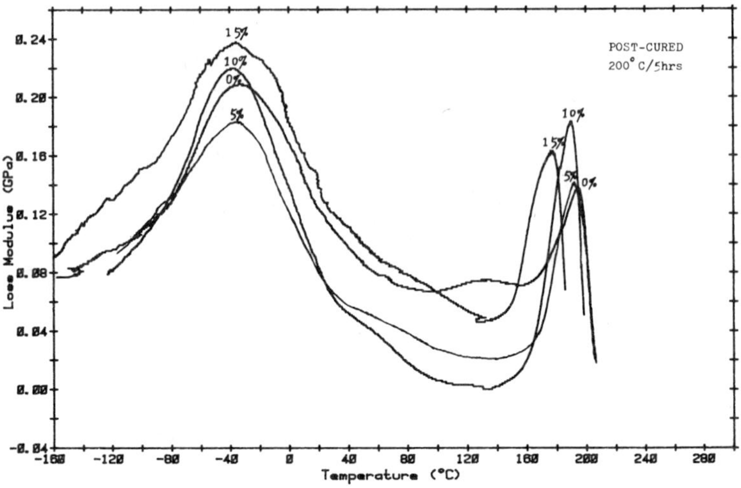

Figure 7. Loss modulus as a function of temperature with concentration of polar Viton GF as a parameter for cured epoxy blends.

position. Analogous results were observed with blends containing polar Viton GF.

Morphology of Blends. SEM and TEM were used to investigate the size and distribution of the rubbery phase in the epoxy matrix. The fracture surfaces of precured blends were featureless, an observation indicating that those blends were compatible. On the other hand, in the postcured blends of epoxy resin and 15% polar Viton GF, a bimodal distribution of rubber particle sizes was found. The diameter of smaller rubber particles ranged from 0.15 to 0.35

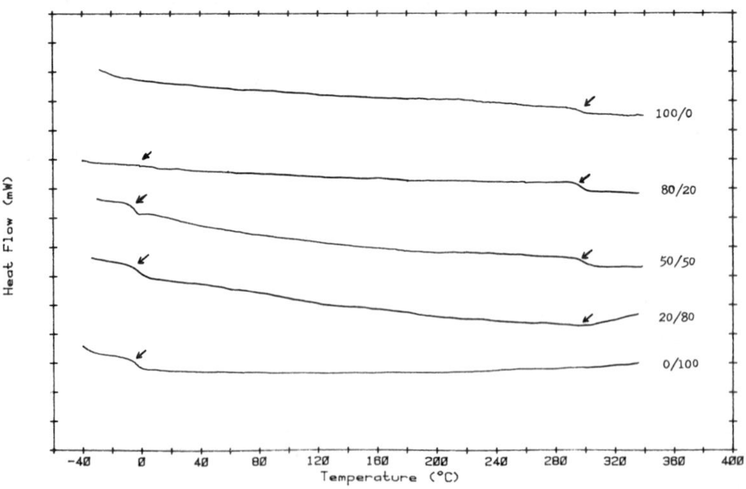

Figure 8. DSC thermograms of a series of polyimide–double-bond containing Viton GF blends.

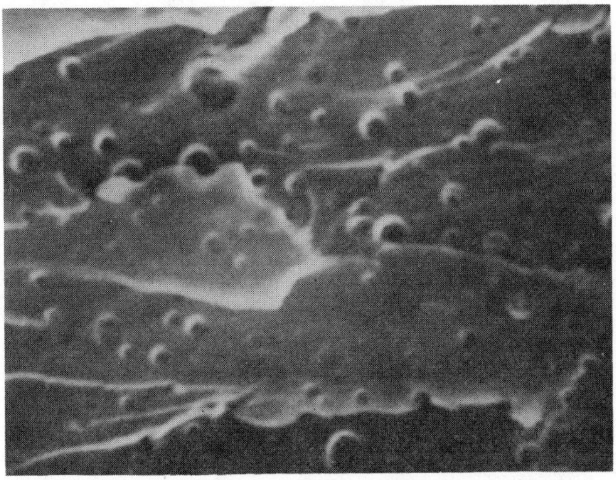

Figure 9. SEM of a fracture surface of an Epon 828–15% polar Viton GF blend. Magnification, 6,800×.

μm. The size and distribution of rubber particles were determined from the electron microscopy. Examples of an SEM and a TEM are shown in Figures 9 and 10, respectively. The transformation of compatible blends (before postcure; Figure 5) into incompatible ones (Figure 7) apparently occurs as a function of time and temperature of postcure. An important question is how those variables influence the size and distribution (morphology) of rubbery particles. Control of the morphology is of paramount significance, for it could allow the tailoring of blends with desired properties.

In the 15% polar Viton GF–epoxy blends, the rubbery phase comprises approximately 25% of the surface. The latter number was

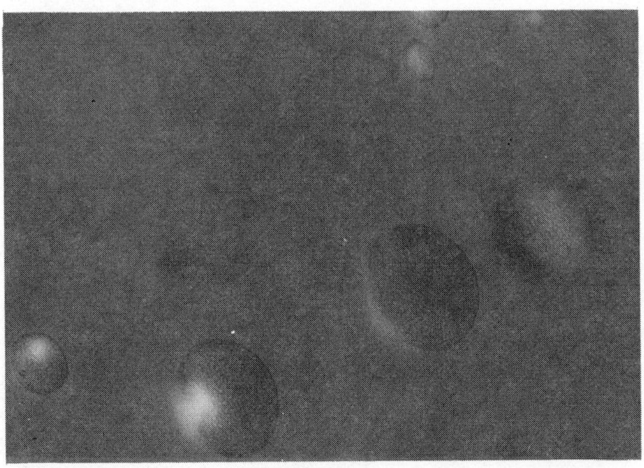

Figure 10. TEM of a thin section of an Epon 828–15% polar Viton GF blend. Magnification, 25,000×.

obtained from electron micrographs of the fracture surface. An apparent "excess" of rubbery material indicates the existence of epoxy matrix within the rubbery inclusions. A two-phase-like appearance of the rubbery phase can be seen in Figure 10. The domain size did not increase in direct proportion with the increase in concentration of polar Viton GF. Apparently, molecules of elastomer exist within the epoxy network, particularly in the blend containing 15% of polar Viton GF. The presence of elastomer, in turn, explains the observed depression of the glass transition (Table II).

Fracture Energy Measurements. Changes in the toughness of a blend upon the addition of an elastomer can be measured by various tests of ultimate mechanical properties. In this study, the value of critical strain energy release rate in mode I ($\mathcal{G}_{Ic}$) was calculated. Samples were cured as outlined in Table I. The exact manner of the application of adhesive onto aluminum beams has been described elsewhere (*7, 8*). An apparent fourfold increase in $\mathcal{G}_{Ic}$ was recorded as a result of the addition of 15% polar Viton GF to the epoxy network. The $\mathcal{G}_{Ic}$ values for the neat epoxy and the epoxy–15% polar Viton GF formulations were 273 and 1069 J/m^2, respectively. Only blends containing 15% polar Viton GF were tested because of the relatively large decrease in T_g in these blends (*see* Table II).

Thermal Stability of Blends. In addition to the effect of elastomers on physical–mechanical properties of modified thermosets, one is also concerned with the effect on thermal stability. A series of TGA thermograms, shown in Figure 11, indicates a small decrease in thermal stability of epoxy blends upon the addition of polar Viton GF. Thermal stability is reduced as the result of substitution of C–F

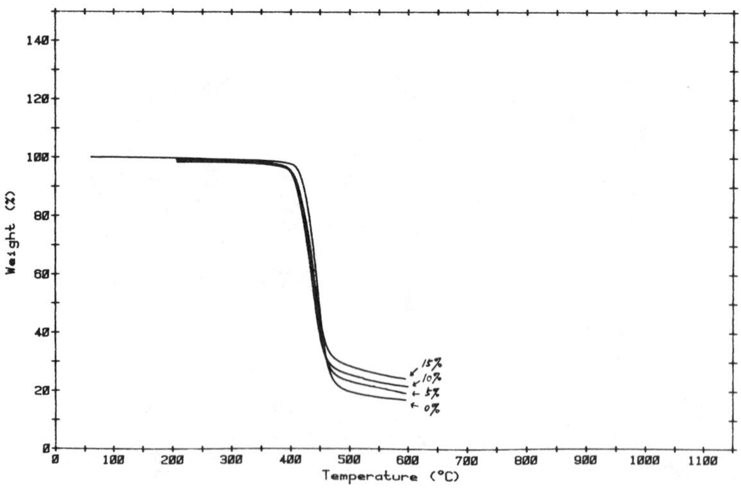

Figure 11. TGA thermograms for a series of Epon 828–polar Viton GF blends with the concentration of the polar Viton GF as a parameter.

bonds with weaker C–N and/or C–O bonds. A decrease in decomposition temperature was also noted; the largest drop observed was between the neat resin and the formulation containing 5% of the modifier. On the other hand, the char yield increased in proportion with the amount of polar Viton GF in the blend (Figure 11).

TGA thermograms of a series of blends of polyimide and double-bond containing Viton GF are shown in Figure 12. The onset of weight loss in the neat polyimide resin was observed at approximately 510 °C, and the maximum rate of decomposition was recorded at 610 °C. The char yield of 70% was obtained at 800 °C in nitrogen atmosphere. The double-bond containing Viton GF decomposes rapidly at temperatures above 400 °C. The thermal stability of various polyimide–double-bond containing Viton GF blends depends on the ratio of the two components in the blend, as seen in Figure 12. Similar results were observed in polyimide–polar Viton GF blends as shown in Figure 13.

Conclusions

Several fluoroelastomer-modified thermoset resins were prepared and investigated. The nonmodified elastomer was incompatible with the epoxy and the polyimide matrix. Modification via the introduction of polar groups resulted in a compatible blend of polar elastomer and partially cured epoxy resin. During the subsequent postcure, however, the elastomer phase separated. We believe that a careful control of postcure conditions could produce the desired blend morphology.

For the polyimide matrix, compatibility was not observed, regardless of the type of modification on the fluorocarbon elastomer.

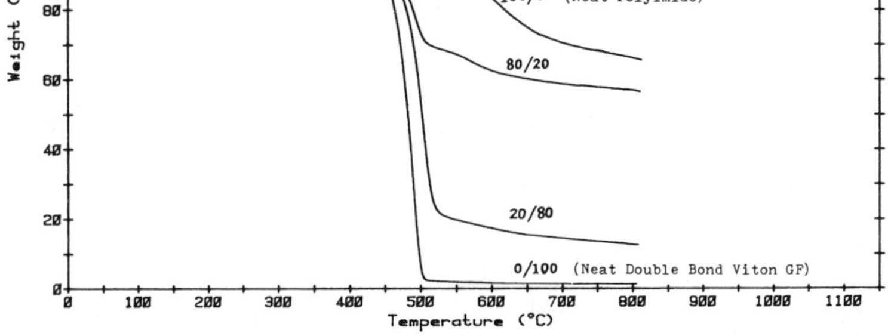

Figure 12. TGA thermograms for a series of polyimide–double-bond containing Viton GF blends with the concentration of the double-bond containing Viton GF as a parameter.

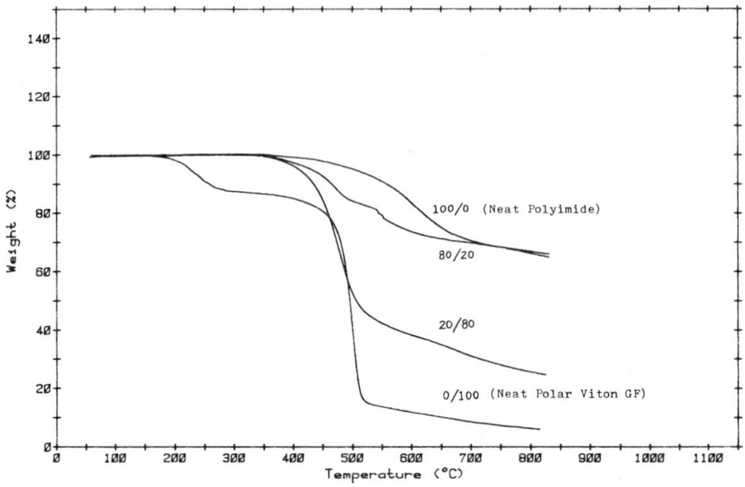

Figure 13. TGA thermograms for a series of polyimide–polar Viton GF blends with the concentration of the polar Viton GF as a parameter.

Blends containing 15% polar elastomer and epoxy matrix were characterized by the bimodal distribution of rubbery particles. Electron microscopy has shown that some epoxy resin exists within the elastomer inclusions. Interestingly, some elastomer molecules (particularly in blends containing 15% elastomer) remain within the epoxy network and cause an apparent decrease in T_g. Simultaneously, a fourfold increase in fracture energy was observed in comparison with the neat epoxy resin.

Finally, the polar fluorocarbon elastomer was found to be less thermally stable than the nonmodified one. The effect of addition of polar elastomer on the stability of blends was more pronounced with epoxies than polyimides.

Acknowledgment

We acknowledge the support of this research, in part, by the NASA-Ames Research Center and the encouragement of John A. Parker.

List of Symbols

E'' Loss modulus, Pa
$\mathcal{G}_{Ic}$ Critical strain energy release rate in mode I, J/m^2
T_g Glass transition temperature, °C

Literature Cited

1. Drake, R. S.; Siebert, A. R. *Am. Chem. Soc. Org. Coat. Appl. Polym. Sci., Proc. Prepr.* **1983,** *48,* 491.
2. McGarry, F. J. *Proc. R. Soc., London* **1970,** *A 319,* 59.
3. Bascom, W. D.; Cottington, R. L.; Timmons, C. O. *J. Appl. Polym. Sci., Appl. Polym. Symp.* **1977,** *32,* 165.

4. Riew, C. K.; Rowe, E. H.; Siebert, A. R. In "Toughness and Brittleness of Plastics"; Deanin, R. D.; Crugnola, A. M., Eds.; ADVANCES IN CHEMISTRY SERIES No. 154; American Chemical Society: Washintgon, D. C., **1976;** p. 326.
5. Sohn, J. E. *Am. Chem. Soc. Org. Coat. Plast. Chem. Prepr.* **1981,** *44*, 38.
6. Meeks, A. C. *Polymer* **1974,** *15*, 675.
7. Mijović, J. *J. Appl. Polym. Sci.* **1980,** *25*, 1179.
8. Mijović, J. *Ind. Eng. Chem. Prod. Res. Dev.* **1982,** *21* (2), 290.

RECEIVED for review December 23, 1983. ACCEPTED April 13, 1984.

APPLICATIONS

19

Reaction Injection Molding of Interpenetrating Polymer Networks

LUU T. NGUYEN and NAM P. SUH

MIT–Industry Polymer Processing Program, Massachusetts Institute of Technology, Cambridge, MA 02139

In the processing of interpenetrating polymer networks (IPNs) by reaction injection molding (RIM), the physical properties of the resulting alloy are highly dependent on the reaction kinetics and the degree of mixing. Improvements in heat sag and tensile strength were observed for a polyurethane–polyester IPN system. At high stream Reynolds numbers, dynamic mechanical measurements provided evidence of interphase mixing by shifts in the transition temperatures of the individual components.

An interpenetrating polymer network (IPN), in its simplest form, is an alloy of two polymers synthesized via two noninterfering reactions (e.g., a combination of a polycondensation and a polyaddition or a polycondensation and a chain polymerization). Current interest centers in this new class of materials for three main reasons:

1. The IPN polymerization route represents a new way of blending two polymers with a minimal degree of covalent bonding and phase separation between the networks (*1*, *2*).
2. The permanent chain entanglement in three-dimensional space provides a good illustration of the concept of topological isomerism (*3*). The resulting macromolecular catenation yields nonlinear elastic restoring forces deviating widely from ideal rubber elasticity (*4*).
3. Tailoring the compositions and adjusting the corresponding reaction kinetics can produce networks with a wide range of morphologies and synergistic behavior (*5–7*).

An IPN can be synthesized by two methods. With a sequential IPN, a mixture of monomer B together with its initiator and cross-linking agent is swollen into prereacted network A and polymerized

0065-2393/84/0208-0311$06.00/0

in situ. This process depends on the solubility of monomer B in network A, takes an inordinate amount of time, and is therefore unsuitable for mass production (*8*). On the other hand, a simultaneous IPN starts out as a homogeneous mixture of monomers, linear polymers, initiators, and cross-linkers. This mix can be prepolymerized until just short of the Flory gelation point and delivered into a mold with continued polymerization (*9*). For this reason, the latter reaction scheme is readily suitable to reaction injection molding (RIM) processing (*10*). With the current thrust in high-modulus RIM parts, IPN systems offer an attractive alternative to sheet molding compound (SMC) and reinforced reaction injection molding (RRIM) without the drawbacks associated with the processing of fiber–resin mixtures (*11–14*). For example, IPNs made up of elastomeric polyurethanes with glassy polymers such as epoxy and polyester can have stiffness and creep resistance equal to those of SMC and RRIM with low reinforcing filler loadings.

In the past, the impingement-type mixing technique could not be used to produce true IPNs because of the high pressure required for intimate molecular mixing. Tucker and Suh (*15*) showed that the scale of mixing impinging streams is determined by the turbulent eddy size, which is a function of the stream Reynolds number based on the nozzle diameter. In commonly used RIM machines, the Reynolds number for processing is less than 1000; this value results in a relatively coarse scale of mixing (around 10 μm). IPNs with molecular-level mixing cannot be created at these Reynolds numbers. Therefore, research has been undertaken to investigate if IPN systems can be synthesized by the impingement process at very high Reynolds numbers. The critical issues being investigated are the relationships between the Reynolds number and the resulting IPN morphology and properties.

Experimental

The materials used in this investigation are as follows: a polyoxypropylenetriol, endcapped with ethylene oxide and grafted with 10 wt% acrylonitrile (AN) (Niax D–337, Union Carbide); 4,4′-(carbonylamino)diphenylmethane modified with 30 wt% NCO (Isonate 191, Upjohn Company); an orthophthalic polyester with about 20% styrene monomer (Laminac 4123, USS Chemicals); dibutyltin dilaurate (T–12, M&T Chemicals); methyl ethyl ketone peroxide in plasticizer (Lupersol Delta X–9, Pennwalt Company); and cobalt naphthenate (cobalt promoter, USS Chemicals).

To remove any absorbed moisture, both polyol and polyester resins were dried at 70 °C under vacuum (2 mm Hg) for 12 h. The isocyanate, promoter, and catalysts were used as received.

Impingement mixing of the reactants was carried out in a specially constructed high-pressure RIM system. The system consists of positive displacement cylinders that provide an independent control of pressure and reactant metering ratios, means of setting the stream temperatures independently, a three stream impingement head, and a mixing chamber capable of withstanding an impingement pressure of up to 100,000 psi. A detailed description of the system is given elsewhere (*16*).

Experiments were carried out by impinging three streams. Reactants in the first stream consisted of catalysts (Lupersol Delta X-9 and T–12) blended with the demoisturized polyol. The second and third streams were polyester and Isonate 191, respectively. The viscosity of each stream was controlled by adjusting the stream temperatures individually. Similarly, the mold was heated to a preselected temperature that determined the gelation time of the resulting IPN mixture. Preparation for each shot also involved setting the desired blend ratio, stream velocities, and impingement pressure.

After the shot and upon gelation, the samples were postcured at 121 °C for 1 h in order to complete the polymerization process. These samples were then cut to the appropriate dimensions and tested without any further preparation. Heat-sag and tensile-strength experiments were carried out according to ASTM D3769–81 and ASTM D638–80, respectively. Glass transitions of the polyurethane–polyester IPN were obtained with a direct-reading dynamic viscoelastometer (Rheovibron Model DDV II).

Results and Discussion

Reynolds Number and Morphology. The chemical reaction occurring in RIM is mixing activated. Consequently, the physical properties of a part depend on the accurate metering and thorough mixing of the fluid reactants. This goal is accomplished in all RIM machines by delivering the reactants at relatively high pressure (1500–3000 psi) through nozzles into a small chamber where they mix by impingement. Such a design allows rapid reduction of the scale of segregation to a size sufficiently small for diffusion to take over. The time available for diffusion is limited by the gel time of the particular resin system under consideration. Various estimates of this required scale of segregation have been calculated by Suh and Tucker (*17*) and Richter and Macosko (*18*). All values are on the order of 10 μm. Therefore, for any resin system and combination of catalysts and processing temperature, a critical scale exists below which complete reaction will occur. The faster the reaction or the lower the molecular diffusion of the resins, the better the mixing needed.

Basic work done by Malguarnera and Suh (*19*) indicated that the Reynolds number (*Re*) was the governing processing parameter in impingement mixing when the effects of surface tension are negligible. Dimensional analysis indicated that the mix quality depends on $Re^{-9/4}$. Similarly, the small-scale features of the mixture vary according to the relationship $l_m/D \sim Re^{-3/4}$ where l_m denotes the characteristic length scale of the mixture and D is the nozzle diameter.

This mixing theory is based on the hypothesis that the microscale of segregation of the mixture depends on the relative size of the turbulent eddies once the stream *Re* exceeds the critical value for the laminar–turbulent transition ($Re \sim 50$). On the other hand, the macromixing of the fluid components that control the large-scale segregation always occurs at high *Re* values.

In order to predict the small-scale features of the mixing process on which the mixing theory is based, past experiments were conducted with relatively low viscosity systems such as water and glycerine (*20*) and liquid metals (*21, 22*). One metal processing technique (*23*) uses the eddy motion in the turbulent-fluid flow situation encountered commonly in RIM to mix liquid metals. The morphology of the resulting alloy is more affected by fluid mechanics and heat-transfer considerations than by the equilibrium thermodynamics or the kinetics of phase transformations inherent in the particular metallic system. General agreements were found between *Re* and the resulting eddy and particle size.

For typical RIM reactions currently practiced commercially, a *Re* of 200 is used; this value seems to provide good mixing (*24*). Although the corresponding eddy size is large (~ 30 μm), additional laminar mixing occurs as the liquids flow out of the impingement chamber, down a runner, through a gate, and into the mold cavity.

Reynolds Number and Physical Properties. ASTM D3769–81 is used to characterize the high-temperature thermal dimensional stability of microcellular urethanes (*25*). The method measures the magnitude of the deflection of a specimen subjected to a high-temperature heating cycle under gravity loading. Such a test gives a good qualitative comparison between the high-modulus materials used in RIM fascia. Figures 1a, 1b, and 1c show the heat-sag characteristics of the polyurethane–polyester IPNs generated at an impingement pressure of 18,000 psi. The IPN shown in Figure 1a has the slowest curing kinetics and that shown in Figure 1c has the fastest curing kinetics. The gelation time is defined as the induction time required to reach the maximum temperature in the exotherm generated. With this definition, gelation times of 738, 417, and 125 s were recorded for mold temperature settings at 25, 50, and 120 °C, respectively. For all three settings, the stream temperatures and velocities of the displacement pistons were adjusted to yield the following stream *Re* values: polyol, 2,170; polyester, 1,862; and Isonate 191, 12,744. Testing was done at 110 °C for 2 h in a forced-air oven.

For the given conditions of stream *Re*, the observed trend indicates better thermal dimensional stability with faster reaction kinetics. A 50:50 polyurethane–polyester composition, for instance, exhibits improvements in heat-sag properties slightly over 30% for a corresponding reduction in gelation time by a factor of six.

Also shown in Figures 1a–c are the upper and lower bounds delineated by the rule of mixtures and the inverse rule of mixtures, respectively. These simple rules can be expressed by the following equations, where P = property of interest (heat sag) and ϕ_i = volume fraction of component i.

Rule of mixtures $$P = P_1\phi_1 + P_2\phi_2 \quad (1)$$

Inverse rule of mixtures $$\frac{1}{P} = \frac{\phi_1}{P_1} + \frac{\phi_2}{P_2} \quad (2)$$

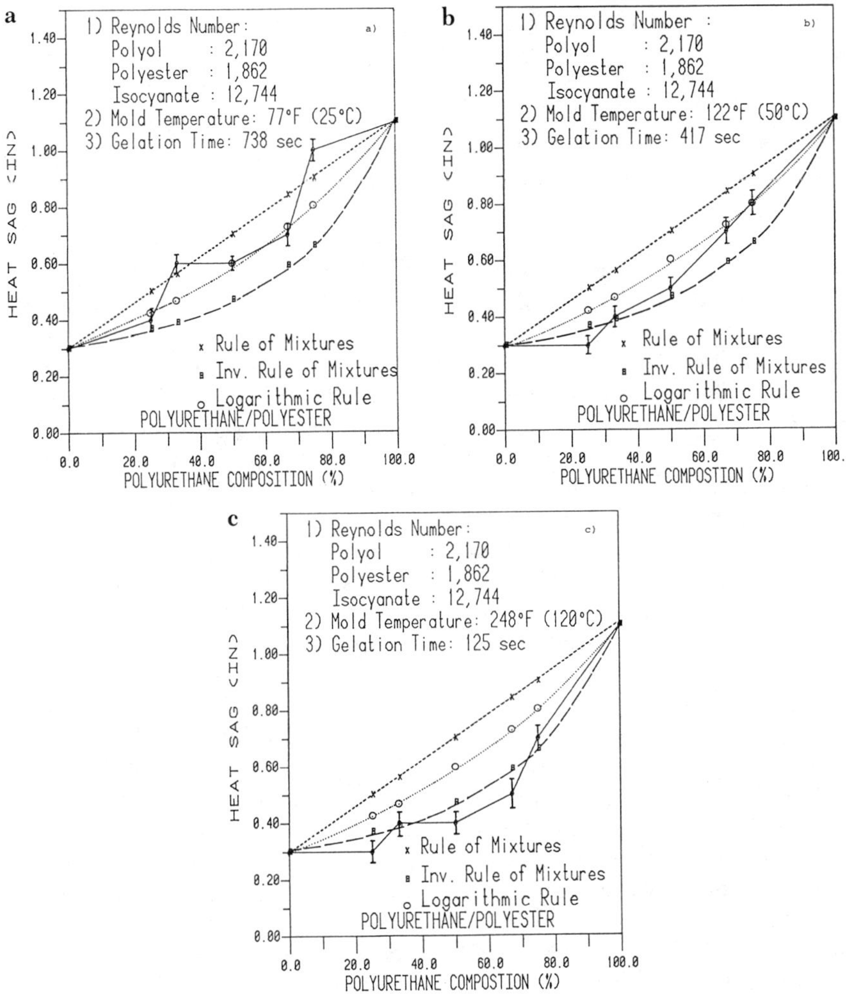

Figure 1. Heat sag of a polyurethane–polyester IPN mixed by impingement and reacted at various temperatures: 25 °C (a), 50 °C (b), and 120 °C (c).

The assumptions underlying both rules are that no interaction exists between components and that the property of the mixture is some linear combination of the properties of the components. In a majority of mixtures, these rules cannot predict accurately the properties of a binary mixture. For more accurate predictions, additional information on morphology is required. Mixture rules for two contiguous phases (*26*, *27*) and for one phase dispersed in a continuous phase have been derived theoretically (*28–30*). Specific examples of the former system include IPNs, open-celled foams, laminates of sheets glued together, and mats and felts. The corresponding properties can often be predicted by the general mixing equation:

$$P^n = \phi_1 P_1^n + \phi_2 P_2^n \quad (3)$$

When $n = 0$, the previous equation can be rewritten in the following form (also known as the logarithmic rule)

$$\log P = \phi_1 \log P_1 + \phi_2 \log P_2 \quad (4)$$

Equation 4 agrees fairly well with the experimental heat-sag data of IPNs molded at 25 °C (Figure 1a) except at those compositions (33 and 67% by volume of polyurethane) where phase inversion occurs. As the gelation time decreases, however, the morphology of the IPN changes. The change in morphology results in a poor fit of the data to the equation (Figures 1b and 1c). Morphology plays an important role when one phase is dispersed but becomes less important when both phases are continuous, such as in an IPN structure (*31*). In general, however, the morphology of systems exhibiting one continuous phase can be incorporated into Equation 3. Unfortunately, no theory relating the dependence of the exponent n in Equation 3 to the morphology has been developed yet.

The dramatic effect of high-speed impingement mixing can be observed in Figures 2 and 3. In a typical run, the *Re* of the polyol (Re_{polyol}) is approximately 10% higher than that of polyester ($Re_{polyester}$). Although the stream with the lowest *Re* controls the mixing scale and, consequently, the morphology of the resulting IPN, Re_{polyol} rather than $Re_{polyester}$ was plotted on the abscissa in order to provide a more familiar comparative measure with current RIM processing equipment. (Current commercial RIM machinery meters reactants at *Re* values of a few hundreds.) The smaller mixing scale obtained with higher *Re* values led to improved heat-sag and tensile properties. For a 50:50 polyurethane–polyester composition molded at 25 °C and 120 °C, thermal stability improves by 12 and 20%, respectively, as the *Re* was increased by an order of magnitude (Figure 2). Such enhancement can yield potential benefits, for example, during the paint application and baking operation of RIM fascia parts in an assembly process. Similarly, the same increase in *Re* led to approximately a 16% increase in tensile strength (Figure 3). Faster reaction kinetics can presumably "freeze" the finely dispersed microstructure obtained upon impingement before the occurrence of any significant phase separation resulting from surface-tension forces and thermodynamic incompatibility.

Figure 4 shows the phase relationships of the polyurethane–polyester IPNs determined by dynamic mechanical measurements, which have been described elsewhere (*33*). T_{g_1} and T_{g_2} are the glass transition temperatures of the soft and hard segments of the polyurethane, respectively, and T_{g_3} represents that of the polyester. The

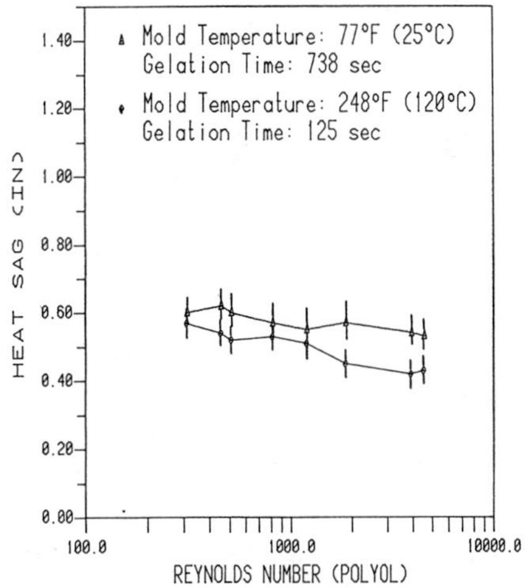

Figure 2. Effect of mixing on the heat sag of a 50:50 polyurethane–polyester IPN reacted at 25 and 120 °C.

observed trend indicates a shift to higher values for T_{g_1} and to lower values for both T_{g_2} and T_{g_3}. This inward shift is good qualitative evidence of interphase mixing. The extent of interpenetration and the nature of the interface certainly depend on the mixing scale and the polymerization kinetics. Under conditions of thermodynamic incompatibility, surface tension was also found to affect the formation of the two-component transition region (32). Viscoelastic and rheological data on polymer blends and IPNs suggest two possible types

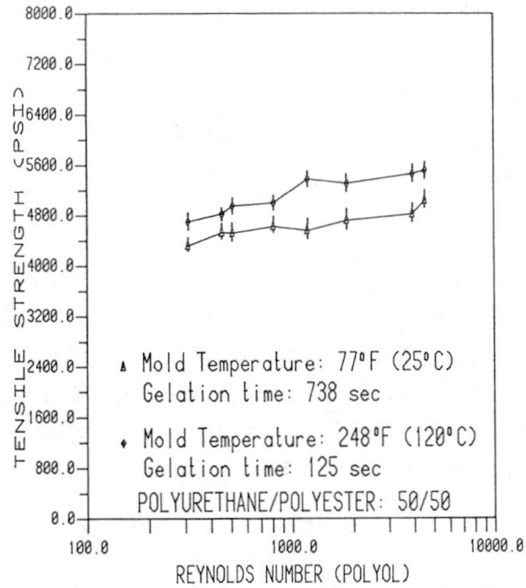

Figure 3. Effect of mixing on the tensile strength of 50:50 polyurethane–polyester IPN reacted at 25 and 120 °C.

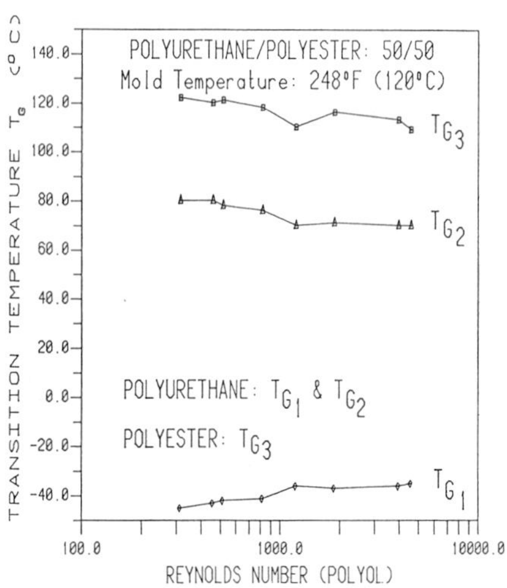

Figure 4. Effect of mixing on the phase relationships of a 50:50 polyurethane–polyester IPN reacted at 120 °C.

of contact regions. The first one exhibits two boundary layers delineated by a boundary surface and characterized by a constant composition and morphological distinction from the polymer in bulk. The second one presents an induced, formed emulsion of one polymer in the other and is characterized mainly by a variable composition.

Conclusions

Improvements in the thermal stability and tensile strength of a polyurethane–polyester IPN processed by RIM were observed. With higher *Re* values, dynamic mechanical analysis indicated a shift in the T_g values of both components. This qualitative evidence of interphase mixing seemed to depend on two factors, namely, the macroscale and microscale of mixing generated by the intense turbulence within the impingement chamber and the kinetics of gelation governing the rate of phase separation.

Acknowledgments

This investigation was supported by the MIT–Industry Polymer Processing Program. We are grateful to our friends at the following companies: AMP, Inc.; Boeing; Eastman Kodak; Du Pont; General Motors; Instrumentation Labs; ITT; Martin Marietta; and Xerox. L. T. Nguyen is also pleased to acknowledge the fellowship given by the Fannie and John Hertz Foundation.

List of Symbols

Re Reynolds number

P Property of interest in mixture rules (heat sag, tensile strength, impact strength, . . .)

P_i Property of component i (i = 1, 2)
ϕ_i Volume fraction of component i (i = 1, 2)
n Constant in general mixing rule ($-1 \leq n \leq 1$)
T_{g_i} Glass transition temperature of component i (°C)
l_m Characteristic length scale of the mixture (in.)
D Nozzle diameter (in.)

Literature Cited

1. Matsuo, M.; Kwei, T. K.; Klempner, D.; Frisch, H. L. *Polym. Eng. Sci.* **1970,** *10,* 327–331.
2. Frisch, K. C.; Klempner, D.; Migdal, S.; Frisch, H. L.; Ghiradella, H. *Polym. Eng. Sci.* **1974,** *14,* 76–78.
3. Frisch, H. L.; Wasserman, E. *J. Am. Chem. Soc.* **1961,** *83,* 3789–3795.
4. Prager, S.; Frisch, H. L. *J. Chem. Phys.* **1967,** *46,* 1475–1483.
5. Klempner, D.; Frisch, H. L.; Frisch, K. C. *J. Polym. Sci.* **1970,** A-2, *8,* 921–935.
6. Frisch, K. C.; Klempner, D.; Migdal, S.; Frisch, H. L. *J. Polym. Sci.* **1974,** A-1, *12,* 885–896.
7. Kim, S. C.; Klempner, D.; Frisch, K. C.; Frisch, H. L.; Ghiradella, H. *Polym. Eng. Sci.* **1975,** *15,* 339–342.
8. Sperling, L. H.; Thomas, D. A.; Covitch, M. J.; Curtius, A. J. *Polym. Eng. Sci.* **1972,** *12,* 101–108.
9. Sperling, L. H.; Arnts, R. R. *J. Appl. Polym. Sci.* **1971,** *15,* 2317–2319.
10. Pernice, R.; Frisch, K. C.; Navare, R. *J. Cell. Plast.* **1982,** *18,* 121–128.
11. Frisch, K. C. *Rubber Chem. Technol.* **1980,** *53,* 126–140.
12. Lee, L. J. *Rubber Chem. Technol.* **1980,** *53,* 542–599.
13. Harper, R. C.; Reber, D. H. *37th SPE ANTEC* **1979,** *35,* 673–680.
14. Von Hassell, A. *Plast. Tech.* **1979,** *25,* 57–61.
15. Tucker, III, C. L.; Suh, N. P. *Polym. Eng. Sci.* **1980,** *20,* 875–886.
16. Nguyen, L. T.; Suh, N. P. presented at the Annu. AIChE Meet., American Chemical Society 75th Jubilee, Washington, D.C., October, 1983.
17. Suh, N. P.; Tucker III, C. L. "Proc. Int. Conf. on Polymer Processing", MIT Press, Cambridge, 1978.
18. Richter, E. B.; Macosko, C. W. *Polym. Eng. Sci.* **1978,** *18,* 1012–1018.
19. Malguarnera, S. C.; Suh, N. P. *Polym. Eng. Sci.* **1977,** *17,* 111–115; *ibid,* 116–121.
20. Tucker, III, C. L., Ph.D. Thesis, MIT, Cambridge, MA, 1978.
21. Suh, N. P., *ASME Trans.* **1982,** Paper 82-Prod-18.
22. Suh, N. P., U.S. Patent 4,279,843, 1981.
23. Suh, N. P.; Tsuda, N.; Moon, M. G.; Saka, N. *ASME Trans.* **1982,** Paper 82-Prod-19.
24. Lee, L. J.; Ottino, J. M.; Ranz, W. E.; Macosko, C. W. *Polym. Eng Sci.* **1980,** *20,* 868–874.
25. Rice, D. M.; Dominguez, R. J. G.; Lloyd, R. F. In "ASTM Special Technical Publication 788"; Ashe, W. A.; Dunleavy, R. A., Eds.; 1981, 46–55.
26. Davies, W. E. A. *J. Phys. D* **1971,** *4,* 1325–1339.
27. Allen, G.; Bowden, M. J.; Todd, S. M.; Blundell, D. J.; Jeffs, G. M.; Davies, W. E. A. *Polymer* **1974,** *15,* 28–32.
28. Kerner, E. H. *Proc. Phys. Soc. London* **1956,** *B69,* 808–813.
29. Halpin, J. C. *J. Compos. Mater.* **1969,** *3,* 732–734.
30. Nielsen, L. E. *J. Appl. Phys.* **1970,** *41,* 11, 4626–4627.
31. Nielsen, L. E. *Rheol. Acta* **1974,** *13,* 86–92.
32. Lipatov, Y. *J. Polym. Sci.: Polym. Symp.* **1977,** *61,* 369–388.
33. Nguyen, L. T.; Suh, N. P. Proc. Symp. Multiphase Polymers and Composites, 12th Ann. NATAS, Williamsburg, Va., September 25–29, 1983; *Thermochim. Acta* **1984,** *76,* 265–271.

RECEIVED for review November 18, 1983. ACCEPTED April 2, 1984.

20

Impact Properties of Rubber-Modified Epoxy Resin–Graphite-Fiber Composites

WILLIAM J. GILWEE and ZOHAR NIR[1]

Chemical Research Projects Office, NASA Ames Research Center, Moffett Field, CA 94035

To improve the impact resistance of graphite-fiber composites, a commercial and an experimental epoxy resin were modified with liquid reactive rubber and a brominated epoxy resin. The commercial epoxy was a tetrafunctional resin, and the experimental epoxy was a trifunctional resin. The reactive rubber was a carboxyl-terminated butadiene–acrylonitrile copolymer. The rubber content was varied from 0 to 25% (wt). The brominated epoxy resin was used at Br levels of 4, 19, and 38% of the resin. Composites were prepared with woven graphite cloth reinforcement. The composites were evaluated by using flexural strength in the dry state and at an elevated temperature after saturation with water. The impact properties were determined by measuring shear strength after falling-ball impact and instrumented impact. The rubber-modified, trifunctional resin exhibited better properties, when tested in hot–wet conditions in a heated oven at 366 K (after boiling the material for 2 h in demineralized water), than the tetrafunctional resin. Improved impact resistance was observed with the addition of the reactive rubber to the epoxy resin. Further improvement was observed with the addition of the brominated epoxy resin.

THE POTENTIAL USE OF GRAPHITE-FIBER–RESIN MATRIX composites to achieve weight savings in aircraft and space applications is well documented (*1*). The design requirements for this type of composite include high strength, stiffness, impact resistance, and resistance to burning. The use of reactive liquid rubber to improve the toughness (impact resistance) of epoxy resin composites has been reported by a number of investigators (*2–5*). The two-phase system of a brittle

[1] Permanent address: Makhteshim Chemical Works, P.O. Box 60, Beer Sheeva, Israel.

0065-2393/84/0208-0321/$06.00/0

epoxy resin phase and rubber phase is believed to increase the impact strength by means of a crack-terminating mechanism.

In this investigation, which is part of a screening program, we have studied the properties of a rubber-modified experimental epoxy resin as well as a standard epoxy as the matrix in a composite. In addition, a brominated epoxy resin was used in varying quantities to improve the fire resistance of the composite. The experimental resin was tris(hydroxyphenyl)methane triglycidyl ether, known as tris epoxy novolac (TEN). The standard epoxy resin used was tetraglycidyl 4,4′-diaminodiphenylmethane (TGDDM). The brominated epoxy was a diglycidyl bisphenol A (DGBA) polymer with 50% by weight of Br.

The above resins were modified with carboxyl-terminated butadiene–acrylonitrile (CTBN) rubber. The rubber was added as a prereacted concentrate containing 50% CTBN rubber and 50% epoxy resin. Two different concentrates were used. For the nonbrominated formulation, the concentrate was prepared by reacting 50% CTBN rubber (w/w) with TGDDM. For the brominated formulations, a prereacted concentrate of the brominated DGBA-type epoxy with 50% Br content was used. Both concentrates were synthesized by following recommended procedures (2). The chemical structures for the various resins are shown on page 323.

Composites were made with these resins by using woven graphite-fiber reinforcement. The composites were tested for glass transition temperature (T_g), moisture absorption, flexural strength and modulus, and impact–shear; a limited number were submitted to instrumented impact testing.

Experimental

Materials Used. The graphite cloth used for reinforcement was an eight harness satin weave (Hexcel F3T–584–42, 300, 3K) with an epoxy resin sizing. The resins used included the following: TEN, an experimental resin (6) with an epoxide equivalent weight (EEW) of 162 (The Dow Chemical Company, XD 7342.00L); F2001P, a brominated DGBA epoxy resin with an EEW of 545 and 50 wt% Br (Makhteshim Chemical Works, Israel); TGDDM (Ciba-Geigy, MY–720); and a CTBN rubber (The BFGoodrich Company, HYCAR 1300 × 13). The epoxy resin was cured with 4,4′-sulfonylbisbenzamine (4,4′-diaminodiphenyl sulfone) (DDS). The curing agent was used at a ratio of 88% of the stoichiometric amount of amino hydrogen per equivalent of the epoxy group. Table I shows the components of the various resins used.

Modification of Epoxy Resin with CTBN Rubber. The epoxy resin and the CTBN rubber were placed in a resin kettle and heated to 353 K; mixing was effective at that temperature. Triphenylphosphine (0.15%) was added under N_2 atmosphere. The reaction continued for 2 h at temperatures of 403–423 K. The reaction advancement was monitored by 0.1 *N* KOH–EtOH titration to determine the equivalent per hundred grams (EPHR) of the carboxyl group. The reaction continued to 1% of the initial EPHR.

Composite Preparation. The epoxy formulations were prepared by dissolving the epoxy components in methyl ethyl ketone (MEK) in a tumbler for 4 h. DDS was added and the solution was tumbled for an additional hour.

1. Tris(hydroxyphenyl)methane Triglycidyl Ether

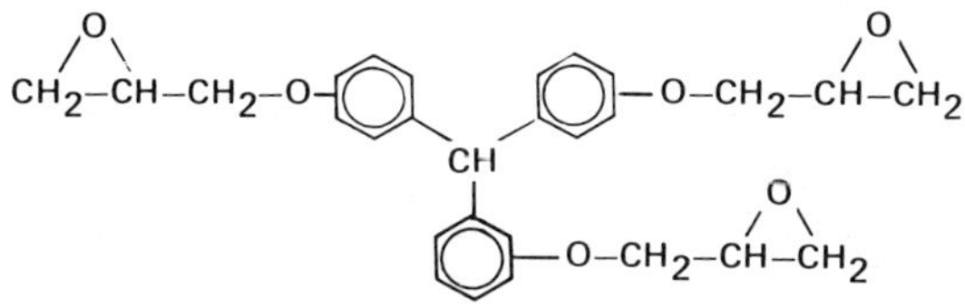

2. Diaminodiphenyl Sulfone (DDS)

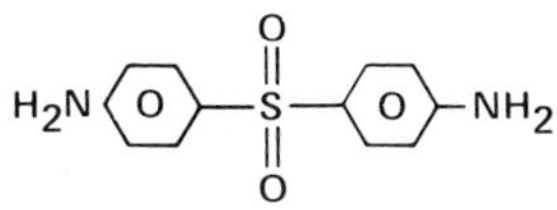

3. Tetraglycidyl 4,4′-Diaminodiphenylmethane (TGDDM)

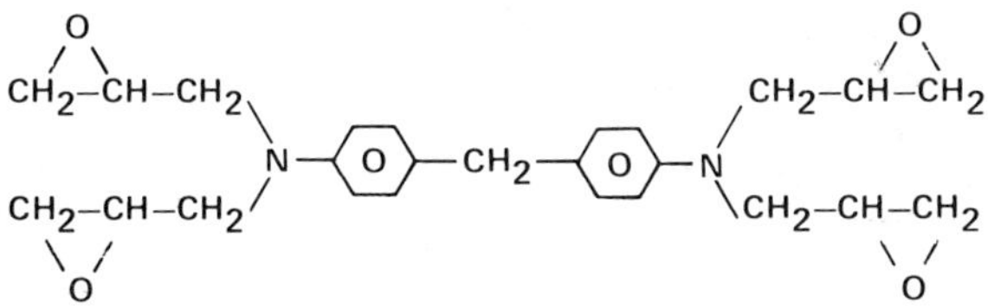

4. Carboxyl-Terminated, Liquid Copolymer of Butadiene and Acrylonitrile (CTBN)

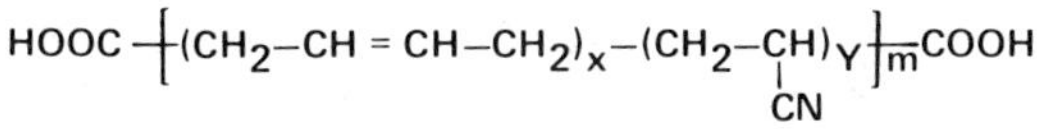

5. Brominated Polymeric Additive (BPA)

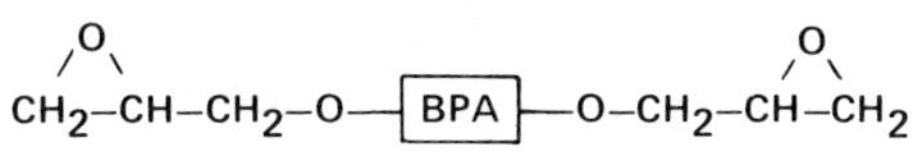

6. BPA-CTBN Prereacted Concentrate

The graphite cloth was preimpregnated (prepregged) with the resin solvent varnish and staged for 5 min at 358 K followed by 8 min at 398 K to remove the solvent. The prepregged cloths were trimmed to 15 × 30 cm and nine plies were stacked together. Curing was accomplished by increasing the temperature from ambient (room) temperature to 450 K at 4 deg/min, holding at 450 K for 2 h while applying 0.35-MPa pressure (50 psi), and cooling under pressure to ambient temperature for 600 min; the material was postcured at 465 K for 4 h.

Mechanical Tests. The mechanical tests were performed on an Instron testing machine with a constant rate of crosshead movement. The flexural strength and modulus were determined by using the ASTM D790–70 method under two conditions: The first was at room temperature and the second (hot–

Table I. Formulation (by Weight) of Composites

Composite	*Resin*	*F2001P*	*CTBN*	*DDS*	*MEK*
TGDDM					
Control	100	—	—	43	95
+ 2% CTBN	97	—	6	43	97
+ 5% CTBN	93	—	15	43	101
+ 10% CTBN	84	—	32	43	106
+ 15% CTBN	76	—	96	43	143
TEN					
Control	100	—	—	34	89
+ 2% CTBN	97	—	6	34	91
+ 5% CTBN	93	—	14	34	94
+ 10% CTBN	85	—	30	34	99
+ 15% CTBN	76	—	48	34	105
+ 25% CTBN	55	—	90	34	119
+ 19% Br	100	89	—	43	154
+ 38% BR	100	653	—	100	959
+ 4% Br + 10% CTBN	100	—	32	35	111
+ 19% Br + 10% CTBN	127	101	72	56	237
+ 38% Br + 10% CTBN	11	222	67	29	219

NOTE: Data are given in grams.

wet) occurred at 366 K after boiling the material for 2 h in demineralized water. The short-beam shear was determined by using the ASTM D2344 method. The T_g was determined on a dynamic mechanical analyzer (DMA, Du Pont Model 980) and a thermal analyzer (Du Pont Model 1090). The impact–shear test is similar to the method reported by Susman (*7*), Harper–Trevet (*8*), and Nir et al. (*9*). In this procedure, the test specimens are impacted at center point with a hemispherical impactor (Gardner impact apparatus). After the specimens have been subjected to impact, the shear strength of the specimens is determined.

Instrumented Impact Tests. The instrumented impact testing was performed by using two methods. The first method (A) used the General Research (formerly Effects Technology) Dynatup impact tester (Model 8200). In this procedure a 16-mm hemispherical head impacts a test specimen at a constant velocity. This instrumented impact technique provides a complete record of the impact event. Automated data analysis provides a record of the applied load and energy absorbed during impact. The second method (B) used a Rheometrics impact tester. This method also uses a microprocessor to collect, calculate, and display impact data. For both tests, the test specimen is supported by a metal frame. In method A the specimen is mounted horizontally, whereas in method B the specimen is mounted vertically. The use of both methods for testing the impact resistance of composites is documented in the literature (*8, 10, 11*).

Measurements Taken. The limiting oxygen index (LOI) was measured according to the ASTM–2863 method. The resin content was determined by digesting the composite laminate specimen in boiling HNO_3 for 2 h. Comparative water absorption was measured by subjecting one set of composite specimens to room-temperature demineralized water for 90 days [room-temperature (RT) specimens] and another set of specimens to boiling water for 2 h [boiling-water (BW) specimens] and then weighing the specimens for weight gain.

Results and Discussion

The TGDDM and TEN epoxy resins were modified with up to 25% rubber. The brominated composites were formulated with 10% rubber and Br contents of 4, 19, and 38% (all percent by weight).

The LOI and the T_g are given in Table II. The LOI values for the modified TGDDM and TEN epoxy resins decrease (higher flammability) with the increase in rubber content, a result of the more aliphatic character of the rubber. Composites prepared with Br had higher LOI values than the nonbrominated composites. The LOI of rubber-modified TEN epoxy resin composites increased in direct relation to the Br content. For example, with 4% Br, the LOI was 48.9%; with 19% Br, the LOI was 69.3%; and, with 38% Br, the LOI was 86.6%. Composites with Br but no rubber had even higher LOI values.

The modification of TGDDM and TEN epoxy resins with rubber and brominated resin does not have a deleterious effect on the T_g of the composites. This result has been seen by other investigators (*8*) when working in the range of 5 to 10% rubber modification of epoxy resin. However, other investigators (*12*) have observed an additive effect of modification of epoxy resin.

The short-beam shear strength and the flexural strength and modulus of the various composites are shown in Table III. The flexural strength was determined at room temperature and at 366 K after 2 h in boiling water. The shear strength is shown as the original shear strength and also the shear-strength retention after an impact of 0.56 J.

Table II. Glass Transition Temperatures and Limited Oxygen Indexes of Composites

Composite	*LOI*	T_g *(K)*
TGDDM		
Control	35.9	482
+ 2% CTBN	32.6	470
+ 5% CTBN	34.5	488
+ 10% CTBN	34.6	481
+ 25% CTBN	33.1	542
TEN		
Control	40.3	544
+ 2% CTBN	37.4	553
+ 5% CTBN	38.5	559
+ 10% CTBN	39.6	540
+ 15% CTBN	37.5	555
+ 25% CTBN	34.2	596
+ 19% Br	75.3	528
+ 38% Br	91.3	487
+ 4% Br + 10% CTBN	48.9	477
+ 10% Br + 10% CTBN	69.3	534
+ 38% Br + 10% CTBN	86.6	544

Table III. Shear Strength, Shear Impact, and Flexural Strength and Modulus of Composite Properties

Composite	*Shear Strength*[a] *(MPa)*	*Shear Impact*[b]	*Flexural Strength*[c] *(MPa)*		*Flexural Modulus*[d] *(GPa)*	
			RT Conditions	*BW Conditions*	*RT Conditions*	*BW Conditions*
TGDDM						
Control	46.12 (6704)	73	632.3 (91.9)	544.9 (79.2)	57.38 (8.34)	54.83 (7.97)
+ 2% CTBN	42.52 (6180)	81	785.7 (114.2)	466.5 (67.8)	54.63 (7.94)	46.51 (6.76)
+ 5% CTBN	53.33 (7752)	70	677.8 (98.5)	262.1 (38.1)	51.32 (7.46)	45.82 (6.66)
+ 10% CTBN	64.19 (9330)	103	800.9 (116.41)	317.2 (46.1)	52.7 (7.66)	41.00 (5.96)
+ 25% CTBN	62.29 (9054)	82	714.14 (103.8)	263.5 (38.3)	48.16 (7.0)	38.39 (5.58)
TEN						
Control	39.39 (5726)	48	676.3 (98.3)	627.4 (91.2)	59.58 (8.66)	61.78 (8.98)
+ 2% CTBN	28.72 (4175)	82	538.70 (78.3)	479.5 (69.7)	63.30 (9.20)	65.70 (9.55)
+ 5% CTBN	35.38 (5142)	67	649.47 (94.4)	436.2 (63.4)	62.13 (9.03)	56.83 (8.26)

+ 10% CTBN	50.76 (7378)	90	765.1 (111.2)	646.0 (93.9)	54.97 (7.99)	59.17 (8.60)
+ 15% CTBN	43.67 (6348)	85	730.7 (106.2)	558.0 (81.1)	57.38 (8.34)	57.86 (8.41)
+ 25% CTBN	27.88 (4053)	82	527.7 (76.7)	377.7 (54.9)	54.0 (7.85)	51.32 (7.46)
+ 19% Br	50.02 (7271)	53	763.0 (110.9)	623.3 (90.6)	64.53 (9.38)	64.40 (9.36)
+ 38% Br	54.27 (7888)	74	828.3 (120.4)	635.7 (92.4)	66.87 (9.72)	69.49 (10.1)
+ 4% Br + 10% CTBN	35.12 (5104)	54	551.1 (80.1)	424.5 (61.7)	60.89 (8.85)	59.72 (8.68)
+ 19% Br + 10% CTBN	39.98 (5811)	75	688.0 (100.0)	470.6 (68.4)	76.37 (11.1)	72.24 (10.5)
+ 38% Br + 10% CTBN	34.55 (5022)	74	566.2 (82.3)	344.7 (50.1)	58.82 (8.55)	58.41 (8.49)

NOTE: Values normalized to 62% by volume of fibers.
[a] Psi values in parentheses.
[b] Percent shear retained with 0.56-J impact.
[c] Ksi values in parentheses.
[d] 10^6 psi values in parentheses.

The original shear strength and the shear after impact for the rubber-modified TGDDM and TEN epoxy resins peak at 10% modification with the rubber. This 10% level of rubber in the brominated system produces a decline in original shear strength; however, the modification does produce an improvement over the TEN epoxy resin control in shear-strength retention after impact.

The flexural strength and modulus (Table III) of the composites made with rubber-modified TGDDM and TEN epoxy resins tested at room temperature, with two exceptions, increased in flexural strength with the addition of rubber. The brominated systems had better flexural strength without the rubber addition. The flexural modulus was closely maintained or improved with the addition of rubber. Even with the 25% rubber content, the TGDDM and TEN epoxy resins retained 85 and 91%, respectively, of the original flexural modulus.

The flexural samples modified with rubber and measured in hot–wet conditions showed more loss in original flexural strength than the TGDDM and TEN epoxy resin controls. However, the brominated and nonbrominated TEN epoxy resins were consistently higher than the modified TGDDM resin. For example, the TGDDM resin with 10% rubber had a flexural strength retention of 39% when tested in hot–wet conditions. The nonbrominated TEN epoxy resin with 10% rubber retained 84%, and the brominated TEN epoxy resin with 10% rubber (19% Br) retained 68% of the original flexural strength when tested in hot–wet conditions.

The flexural moduli of the rubber-modified, brominated and nonbrominated TEN epoxy resins measured in hot–wet conditions were uniformly higher than the modified TGDDM composites. In fact, they were equal to or better than the TGDDM control resin. For example, the hot–wet measured modulus of the TGDDM control resin was 54.8 GPa (96% retention) and the hot–wet measured modulus of the TEN epoxy resin with 10% rubber was 50.2 GPa (108% retention).

Moisture Absorption. Table IV shows moisture-absorption values for the composites from the two procedures described earlier. The RT specimens were 12.7 × 19.1 mm, and the edges were coated with wax (carnauba) to eliminate moisture absorption through the machined edge. Because of the different sample surface areas and sizes, we observed no direct correlation between the two values. In fact, the relative weight pick-up in one test does not correspond with the other test. For example, the TEN epoxy resin with 10% rubber composite had one of the lower moisture-absorption values in the BW specimens but exhibited one of the highest moisture-absorption values in the RT specimens.

Instrumented Impact. The data for instrumented impact are given in Table V. This test was performed only with composites made

Table IV. Moisture-Absorption Values of Boiling Water and Room Temperature Specimens

Composite	*BW*	*RT*
TGDDM		
Control	0.4%	1.68%
+ 2% CTBN	0.4	—
+ 5% CTBN	—	0.46
+ 10% CTBN	0.5	0.91
+ 25% CTBN	0.5	1.21
TEN		
Control	2.0	1.34
+ 2% CTBN	1.3	—
+ 5% CTBN	1.3	1.16
+ 10% CTBN	0.3	0.79
+ 15% CTBN	0.4	0.64
+ 25% CTBN	0.9	0.60
+ 19% Br	0.80	0.50
+ 38% Br	0.32	0.58
+ 4% Br + 10% CTBN	—	1.02
+ 19% Br + 10% CTBN	0.15	0.51
+ 38% Br + 10% CTBN	0.98	1.74

Table V. Instrumented Impact Measurements

	Maximum Force (P_F) *(N)*		*Total Energy Absorbed (J/cm)*	
Composite	*Method A*	*Method B*	*Method A*	*Method B*
TEN				
Control	1922	—	52.6	—
+ 5% CTBN	2326	2926	59.1	55.7
+ 10% CTBN	2023	2491	48.1	56.4
+ 15% CTBN	2194	2996	52.9	57.1
+ 25% CTBN	2153	3721	60.1	67.0
+ 19% Br	2989	5446	68.9	89.3
+ 38% Br	2475	3811	65.9	73.2
+ 4% Br + 10% CTBN	2027	3557	51.0	58.8
+ 19% Br + 10% CTBN	2904	5332	72.0	82.2
+ 38% Br + 10% CTBN	3297	5211	83.6	106.8

NOTE: Values normalized to 62% by volume of fibers. *See* text for discussion of methods A and B.

with the TEN epoxy resin. No instrumented impact was run on the TGDDM composites because of the low values exhibited by the hot–wet flexural-test specimens. The instrumented-impact test specimens in all cases were 10.2 × 10.2 cm, and thickness varied from 23 to 28 mm. The impact tests were run at room temperature.

Two values were reported for each test method. The first value is the maximum force (expressed in newtons) necessary for penetration of the test specimen. This penetration is characterized as the delamination and fiber breaking of the composite specimen. The second value is the total energy absorbed during the impact event [expressed in joules per centimeter (J/cm) thickness]. The impact velocity of method A was 1737 mm/s and that of method B was 1219 mm/s. The maximum-force and energy-absorbed test results were higher for method B than for method A. However, the relative rankings of the samples by the two methods were, in general, in close agreement.

To determine the effect of impact velocity on the test specimens, one sample (TEN epoxy resin with 10% rubber) was tested with method B at different impact velocities. The impact velocities were 1219, 2438, and 3048 mm/s. The maximum force for the samples was 3721, 3425, and 3425 N respectively. The total energy absorbed for each of the samples was 67 J/cm. Therefore, within the range of velocities tested, we observed no significant variation in test values. This result agrees with prior work by Sykes and Stoakley (*10*) in which they impacted samples in the range of 400 to 1300 mm/s and found that the penetration energy was insensitive to impact velocity.

The TEN epoxy resin composites were made with rubber modification of up to 25%. The highest values for maximum load and energy absorbed were obtained with method B for the 25% rubber TEN epoxy composites. The load value was 3721 N and total energy was 67 J/cm. The brominated samples gave excellent values without the addition of rubber. The 19% Br (zero rubber) gave close to the best values in both test methods and each of the criteria (maximum load and energy absorbed).

Figures 1 and 2 present the energy absorbed in graphic form. In both methods little change is observed until the rubber level reaches 25% in the nonbrominated TEN epoxy resin. The best values are seen with the brominated system. The three levels of Br (4, 19, and 38%) and 10% rubber show the effect of Br on the impact resistance of the composites. The best values were obtained with the 38% Br and 10% rubber-modified composite.

Conclusions

Six definite conclusions were formulated from this experiment: (1) Modification of epoxy resin with a reactive rubber compound improves the impact resistance of graphite fabric–resin composites. To

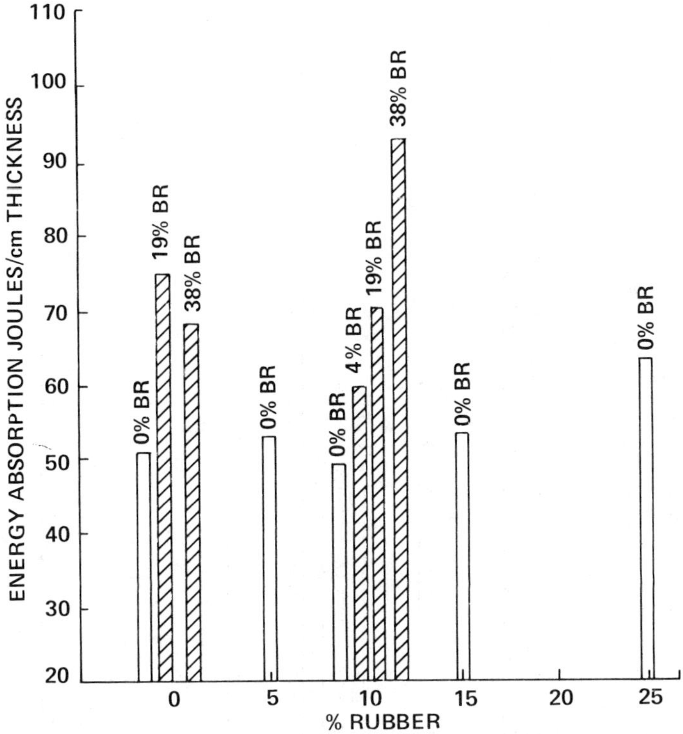

Figure 1. Instrumented-impact method A with TEN–woven graphite cloth composites, impact velocity 1737 mm/s. Clear bars indicate composites with no Br added.

obtain a significant increase in impact resistance, a level of 25% rubber is necessary. (2) Modification of the TEN epoxy resin with brominated epoxy resin gave higher impact resistance than the control or the rubber-modified epoxy resin composites. (3) The amount of brominated resin present in the rubber-modified TEN epoxy resin has a direct relationship to the impact resistance of the composite. (4) When tested in hot–wet conditions, the experimental TEN epoxy resin had better flexural strength and modulus than the TGDDM resin. (5) The impact velocity does not significantly affect the energy absorbed by the test sample. (6) The T_g of the composites did not decline significantly with the rubber modification.

Acknowledgments

The authors gratefully acknowledge the contributions of various personnel at General Research Corporation and Rheometrics, Inc. for the impact tests. We also thank the following personnel at NASA Ames Research Center: E. J. Fontes, S. J. Crane, and Trinh Tran for their dedicated technical assistance.

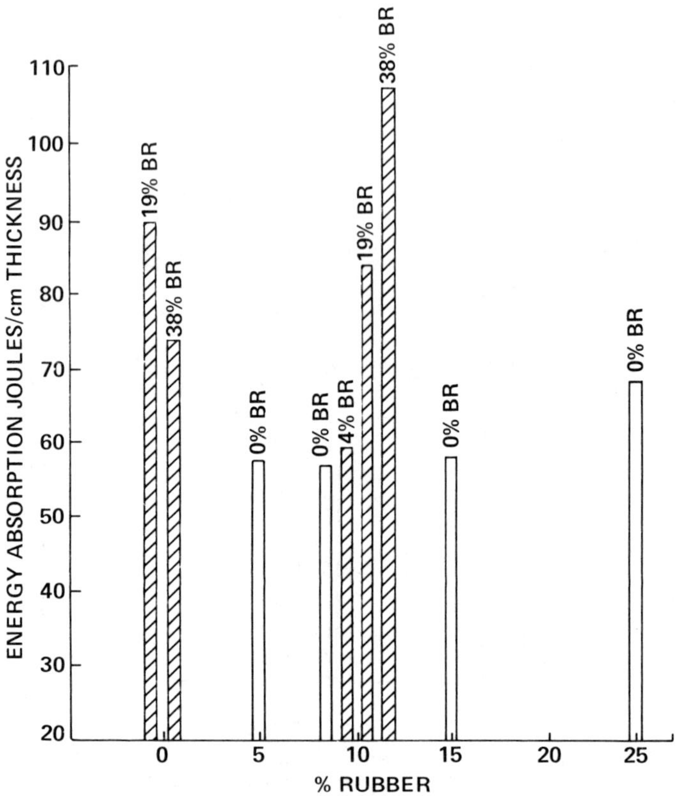

Figure 2. Instrumented-impact method B with TEN–woven graphite cloth composites, impact velocity 1219 mm/s. Clear bars indicate composites with no Br added.

The use of trade names of manufacturers in this report does not constitute an official endorsement of such products or manufacturers, either expressed or implied, by the National Aeronautics and Space Administration.

Literature Cited

1. Kourtides, D. A. *Proc. 21st AIAA Structures, Structural Dynamics and Mater. Conf.*, May 12–14, 1980.
2. Drake, H.; Siebert, R. A. *SAMPE Q.*, 1975, *6*, No. 4.
3. Rowe, E. H.; Siebert, R. A. in "Toughness and Brittleness of Plastics"; Deanin, R. D.; Crugnola, A. M., Eds.; Advances in Chemistry Series No. 154; American Chemical Society, Washington, D.C., 1976; pp. 326–346.
4. Lee, B. L.; Lizak, C. M.; Riew, C. K.; Moulton, R. J. *Proc. 12th Natl SAMPE Tech. Conf.*; Seattle, Oct. 7–9, 1980.
5. Nir, Z.; Gilwee, W. J.; Kourtides, D. A.; Parker, J. A. *Proc. 41st Annu. Tech. Meet. (ANTEC) SPE;* Chicago, May 2–5, 1983.
6. Hawthorne, K. L.; Henson, F. C.; Pinzelli, R. *Org. Coat. Appl. Polym. Sci. Proc.;* Las Vegas, March 25–April 2, 1982, Am. Chem. Soc. *46*, 493–497.

7. Susman, E. S., *Proc. 12th Natl SAMPE Tech. Conf.;* Seattle, Oct. 7–9, 1980.
8. Harper-Trevet, F. *Proc. 27th Natl SAMPE Symp.;* San Diego, May 4–6, 1982.
9. Nir, Z.; Gilwee, W. J.; Kourtides, D. A.; Parker, J. A. *SPI Proc. 38th Annu. Conf., Reinf. Plast. Compos. Inst.*, Feb. 7–11, 1983.
10. Sykes, G. F.; Stoakley, D. M. *Proc. 12th Natl SAMPE Tech. Conf.;* Seattle, Oct. 7–9, 1980.
11. Miller, A. G.; Hertzberg, P. E.; Rantala, V. W. *Proc. 12th Natl SAMPE Tech. Conf.;* Seattle, Oct. 7–9, 1980.
12. Lee, G.; Hartman, B. *J. Appl. Polym. Sci.* 1983, *28*, 823–830.

RECEIVED for review November 18, 1983. ACCEPTED May 16, 1984.

21

Influence of Matrix Toughening on the Impact Properties of Injection-Molded Sheet Molding Compound

J. T. ENGLEHARDT[1], and C. S. NICHOLS[2]

[1] Research and Development Division, Owens–Corning Fiberglas Corporation Technical Center, Granville, OH 43023.

[2] The BFGoodrich Research and Development Center, Brecksville, OH 44141.

A study was conducted to determine optimum processing techniques for the addition of a reactive liquid polymer in the preparation of reinforced-polyester, injection-molding compound to achieve improved impact properties. Significant improvements in the physical properties of the modified elastomer over the control compounds were noted. The microscopic morphology of the injection-molded composites provided information on the lack of influence of processing conditions on the degree of improvement in physical properties among the elastomer-modified samples.

THE TECHNOLOGY OF BRITTLE-RESIN TOUGHENING through the dispersion of discrete elastomeric particles in the resin matrix is well established. The toughening mechanism in thermosetting epoxy resins has been defined by the use of carboxyl-terminated butadiene–acrylonitrile copolymers (CTBN) (*1, 2*). Optimum property improvements have been found to be a function of elastomer functionality and acrylonitrile (AN) content, in situ elastomer particle formation during cure, and ultimate particle size.

A logical extension of the work on toughening epoxy resins was to investigate the modification of unsaturated polyesters. The degree of toughening in unsaturated polyester resins is dependent on the nature of the elastomeric modifier and the chemical structure of the polyester resin (*3–4*). An effective modifier for this class of resin is a vinyl-terminated butadiene–acrylonitrile copolymer with pendent vinyl groups (VTBNX). This material increases the crack resistance

0065-2393/84/0208-0335/$06.00/0

of unsaturated polyester resins by up to 150% (*4*). Although this increase is significant, it is much lower than the degree of improvement observed in epoxy resins. Proposed explanations for these smaller increases in polyester crack resistance include the inherent incompatibility of VTBNX in these resins. This incompatibility places a lower limit of approximately 10 μm on elastomer domain size in the cured matrix. Consequently, domains in polyesters are at least an order of magnitude larger in diameter than the optimum particle sizes for epoxy resin toughening. However, no definitive answer for the smaller degree of improvement is available.

The influence of resin properties such as toughness on the properties of polyester-based, fiber-reinforced composites has been an active area of research. The initial stages of composite failure (i.e., matrix microcracking) are related to the properties of the base resin in sheet molding compound (SMC) and bulk molding compound (BMC) (*5, 6*). This microcracking near the resin–fiber interface occurs at low strain and inflicts permanent damage. Deterioration of part modulus, strength, and fatigue properties are all results of this matrix failure. Consequently, toughening of the polyester matrix resin in a fiber-reinforced composite with VTBNX has been extensively evaluated.

Reactive liquid polymer (RLP) modification of SMC has demonstrated a distinct relationship between increase in the crack resistance of a base polyester resin and increase in SMC properties such as critical knee strain, ultimate strain, critical knee stress, and secondary modulus retention (*7*). The increases in the critical stress and strain properties of SMC by elastomeric toughening of the matrix resin have been correlated with the initiation of matrix microcracking; the elastomer reduces the rate or extent of this initial failure process (*8*). These factors certainly combine to explain the increased fatigue resistance of VTBNX-modified SMC (*9*).

The advent of injection molding for the production of polyester composite parts has placed an even greater emphasis on the properties of the matrix resin. Physical properties of injection-molded parts compared to compression-molded parts are significantly lower, as shown in Table I.

Table I. Properties of Polyester Composites

Property	*Compression Molding*	*Injection Molding*
Unnotched Izod impact (J/cm)	3.8	1.4
Tensile strength (MPa)	76	36
Flexural strength (MPa)	148	92
Flexural modulus (GPa)	11.4	10.6

The lower physical properties of injection-molded composites are due to the shorter, average fiber length in these parts as compared to a corresponding compression-molded part. The shorter fiber length is a result of the compound passing through the injection-molding machine (frequently with an extruder) and flowing in the mold. Consequently, the properties of the base resin have a much greater influence on the properties of the finished composite. The injection molding of SMC can offer advantages in raw material costs, part-to-part consistency, secondary finishing operations, and greater automation (*10*).

This work was designed to investigate the influences of VTBNX modification and process conditions on typical injection-molding formulations.

Experimental

By using two formulations that typify current industry technology, studies were conducted to compare the effect of contrasting mixing methods, namely low-shear and high-shear techniques. This study was accomplished through the use of a Cowles mixer, to provide a high-shear condition, and a Littleford mixer, which functions on a low-shear mixing principle.

The two formulations selected for this work were essentially BMC formulations prepared to run in an SMC process. Both formulations are based on an E–4297–5 (Owens–Corning Fiberglas) resin system. Formulation A (Table II) is typical of certain automotive molders and is significant because it contains a viscosity reducer and low filler level. Formulation B (Table III) is also used by some automotive molders. It differs from Formulation A in that it contains a polyethylene (PE) surface modifier (Microthene FN 510). In addition, it contains a higher filler level and has no viscosity reducer. Hence, the two formulations represent a contrast in viscosities.

An elastomeric modifier (The BFGoodrich Chemical Company, HYCAR VTBNX 1300x24) was added to certain batches of each formulation as indicated in Tables II and III. The investigation was designed to assess the influence of the order of introduction of elastomeric modifier into the system. In one version,

Table II. Formulation A Compositions

		HYCAR	
Compound	*Control*	*Prior to Filler*	*After Filler*
E–4297–5	98.04	98.04	98.04
P–543	2.9	2.9	2.9
Vinyl toluene	1.7	1.7	1.7
Peroxide catalyst	1.5	1.5	1.5
Viscosity reducer	1.96	1.96	1.96
Calcium stearate	1.4	1.4	1.4
VTBNX 1300x24	—	10.0	—
Camelwite	165	165	165
VTBNX 1300x24	—	—	10.0
Thickener	0.83	0.83	0.83

Table III. Formulation B Compositions

Compound	Control	HYCAR Prior to Filler	HYCAR After Filler
E-4297-5	100	100	100
Peroxide catalyst	1.2	1.2	1.2
Calcium stearate	4.2	4.2	4.2
Polyethylene surface modifier	5.0	5.0	5.0
VTBNX 1300x24	—	10.0	—
Camelwite	180	180	180
VTBNX 1300x24	—	—	10.0
Thickener	0.85	0.85	0.85

the modifier was added prior to the addition of the filler. In the other, modifier was added immediately following filler addition (Figure 1).

For the actual production of the paste, we prepared two batches (40 lb each) of each formulation. The first four compounds in Table II were blended for 1 min and the first two compounds in Table III were blended for 1 min. Then, the remaining components for the formulation, up to but not including the filler, were added and mixed 1 min. The filler was added with mixing. After the addition was complete, mixing was continued for 2 min (control and RLP prior to filler) or 1 min (RLP after filler). HYCAR was then added, and the formulation was mixed an additional minute so that total mixing time for all formulations was the same. At this point, past samples were removed from the A formulations for the preparation of unreinforced castings. The thickeners were finally added and blended in for 1 min. Mixing in the high-shear mixer produced a paste temperature of approximately 95 °F, and the low-shear mixer produced essentially no heat buildup.

The prepared paste was subsequently used in the preparation of SMC with

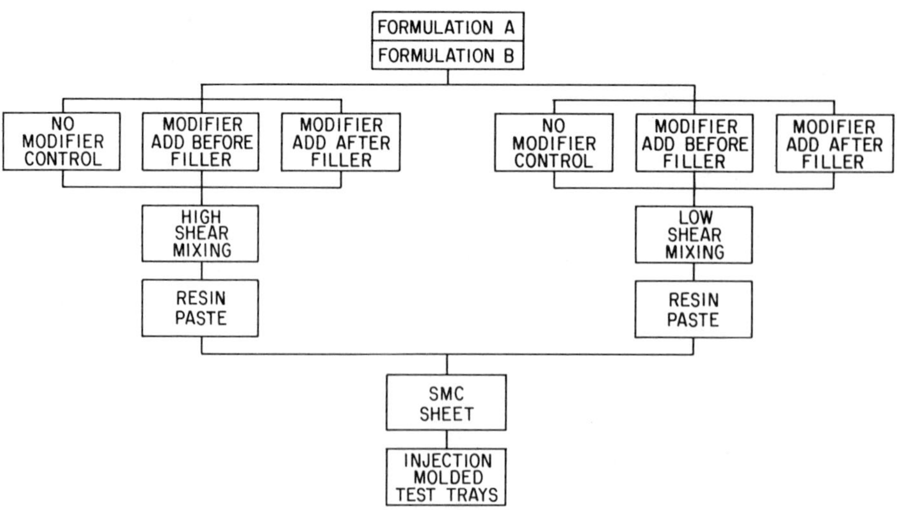

Figure 1. SMC paste mixing scheme.

conventional SMC processing techniques. Each formulation was compounded into a 16-oz/ft^2 sheet product containing 20 wt%, 0.5-in., OCF 951 glass reinforcement.

Following a 72-h maturation period, the SMC compound was injection molded into a 12 × 12 in. shallow tray. We began the molding process by introducing the SMC, which had been slit and rolled into compact charges, into the stuffer of the injection molding machine.

The trays were molded by using a Hull 200-ton injection-molding machine equipped with a 50-mm reciprocating screw. The mold was single cavity with sprue gating at the center of the part. Conventional molding parameters consisting of low screw speed, low back pressure, and medium injection rate were employed. Mold temperatures of 275 and 300 °F were maintained for formulations A and B, respectively, because of the catalyst used in each system.

The molded trays or plaques were subsequently submitted to testing laboratories for physical property determinations. Flexural, tensile, and Izod impact testing were carried out according to ASTM specifications D790, D638, and D256, respectively. Rheometrics impact testing (Rheometrics impact tester) was conducted at a test speed of 5 mph by using a ½-in. diameter ram and a 3-in. diameter clamp. Fracture-surface energy determinations were conducted on the unreinforced paste casting by using a double cantilever beam specimen that was pulled at the rate of 1.25 mm/min.

Results

Results of the performance evaluations are discussed in the following sections and compiled in Tables IV, V, VI, and VII.

Table IV. Physical Properties of High-Shear-Mixed Formulation A Series

		RLP	
Property	*Control*	*Prior to Filler*	*After Filler*
Flexural			
Strength (MPa)	103.5	123.4	124.9
Strain (%)	1.50	1.86	1.80
Modulus (GPa)	8.28	8.03	8.41
Tensile			
Strength (MPa)	43.7	42.2	38.5
Strain (%)	1.02	1.21	0.95
Modulus (GPa)	7.97	7.12	7.51
Izod Impact			
Notched (ft lbs/in.)	2.6	3.0	3.2
Unnotched (ft lbs/in.)	3.1	4.3	4.0
Rheometrics Impact			
Force (lbs)	87	99	104
Energy (ft lbs)	8	12	12
Fracture Surface			
Energy (J/m^2)	35.9	52.4	45.5

Table V. Physical Properties of Low-Shear-Mixed Formulation A Series

Property	Control	RLP Prior to Filler	RLP After Filler
Flexural			
Strength (MPa)	110.4	124.1	117.5
Strain (%)	1.65	1.92	1.78
Modulus (GPa)	8.15	7.98	7.94
Tensile			
Strength (MPa)	48.1	38.1	45.1
Strain (%)	1.18	1.11	1.24
Modulus (GPa)	7.58	6.45	6.98
Izod Impact			
Notched (ft lbs/in.)	2.4	3.3	3.0
Unnotched (ft lbs/in.)	3.3	3.8	4.1
Rheometrics Impact			
Force (lbs)	72	92	92
Energy (ft lbs)	7	12	10
Fracture Surface			
Energy (J/m^2)	39.3	50.5	55.2

Table VI. Physical Properties of High-Shear-Mixed Formulation B Series

Property	Control	RLP Prior to Filler	RLP After Filler
Flexural			
Strength (MPa)	110.2	130.4	110.7
Strain (%)	1.69	1.90	1.76
Modulus (GPa)	7.83	8.42	7.84
Tensile			
Strength (MPa)	36.1	36.3	41.5
Strain (%)	0.97	0.89	1.25
Modulus (GPa)	7.91	7.75	6.67
Izod Impact			
Notched (ft lbs/in.)	2.5	3.2	3.3
Unnotched (ft lbs/in.)	3.2	4.1	4.5
Rheometrics Impact			
Force (lbs)	99	100	98
Energy (ft lbs)	8	10	10

Table VII. Physical Properties of Low-Shear-Mixed Formulation B Series

Property	*Control*	*RLP* Prior to Filler	*RLP* After Filler
Flexural			
Strength (MPa)	94.9	127.0	131.0
Strain (%)	1.57	1.89	1.89
Modulus (GPa)	7.39	8.24	8.66
Tensile			
Strength (MPa)	38.9	38.7	35.5
Strain (%)	1.01	0.94	0.91
Modulus (GPa)	7.49	7.67	7.31
Izod Impact			
Notched (ft lbs/in.)	2.7	3.2	3.2
Unnotched (ft lbs/in.)	3.4	4.3	3.9
Rheometrics Impact			
Force (lbs)	78	98	103
Energy (ft lbs)	10	11	11

Formulation A. Notched and unnotched Izod impact results (Figure 2) indicate significant impact improvement resulting from the addition of VTBNX modifier. Impact values of the two modified systems are statistically the same; this result indicates that the point of RLP addition is insignificant. Also, we saw no evidence that the degree of shear in mixing had a significant influence on Izod impact.

Rheometrics impact results (Figure 3) are expressed in terms of

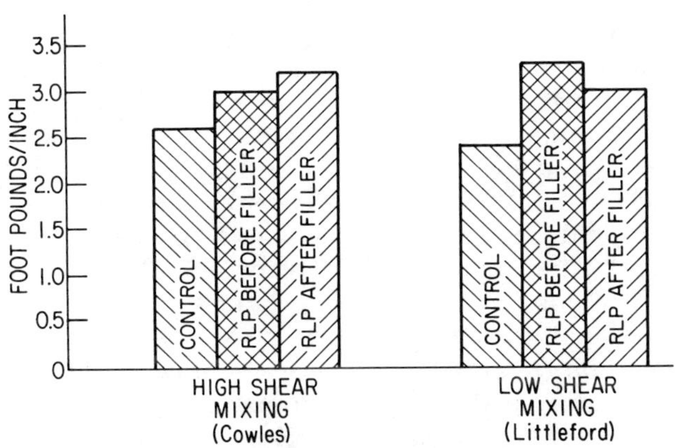

Figure 2. Notched Izod impact data for formulation A.

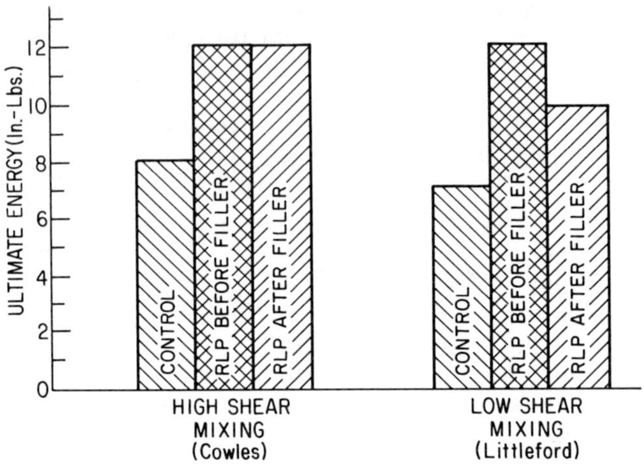

Figure 3. Rheometrics impact data at 5280 fpm for formulation A.

ultimate energy. This term relates to the area under the stress–strain curve at the point of failure. Results with formulation A confirm the Izod findings and indicate a 50% increase in energy absorption when HYCAR RLP is added. Once again, no influence on impact performance is indicated by the amount of shear in mixing, and the point of modifier addition produces no effect on high-shear mixing.

Flexural strength, as well as strain, was improved by elastomer addition, and no deterioraton of flexural modulus was noted.

Tensile properties show a slight reduction in strength and modulus with essentially no change in elongation when elastomer is present in the system.

Fracture-surface energy data show significant improvement in crack resistance with HYCAR modification. However, the point in the process at which modifier is added appears to be insignificant, regardless of the degree of shear applied in mixing.

Photomicrographs of the paste castings indicate a wide range of rubber particle sizes in the system (Figures 4–6). No significant difference in elastomer particle sizes is evident when the HYCAR is added prior to the filler in either mixer. A clear differentiation is observed when the HYCAR is added after the filler, and the low-shear mixer produces the largest domains. These particle sizes do not translate to the molded parts (Figures 7–9), and the domains become indistinguishable from the resin matrix. Sizable rubber particles can be found in regions of high glass content.

Formulation B. Impact response, both Izod and Rheometrics, associated with formulation B was similar to that of formulation A; a distinct improvement was shown where the system contained RLP (Figures 10 and 11). However, in this formulation, no preference in

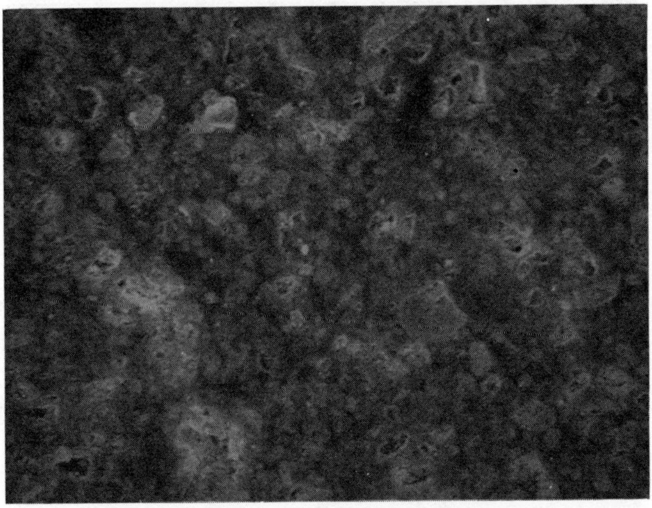

Figure 4. Unreinforced formulation A casting containing no modifier (magnification, 510×).

impact performance was shown for either order of modifier addition or character of mixing.

In formulation B, flexural strength of the high-shear version showed an increase when the RLP was added prior to the filler. The addition of RLP after the filler produced no enhancement in flexural properties. Improvements were evident in both low-shear systems

Figure 5. Unreinforced formulation A casting with modifier added prior to filler (magnification, 510×).

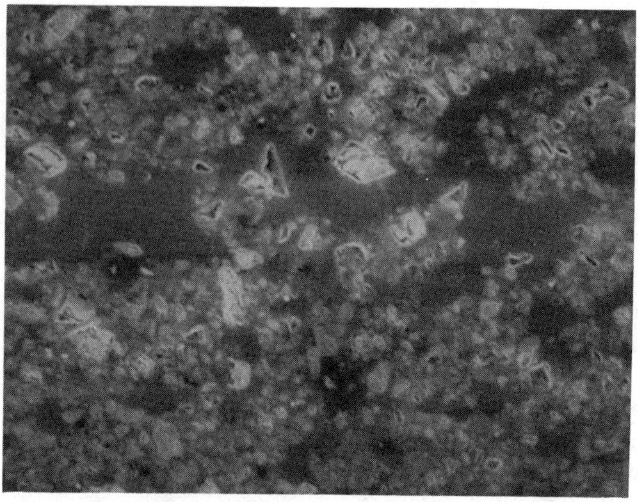

Figure 6. Unreinforced formulation A casting with modifier added after filler (magnification 510×).

containing HYCAR. This result would indicate a preference for adding RLP prior to the filler when high-shear mixing is practiced.

Tensile properties associated with formulation B generally are unaffected by elastomeric modification, except for a slight decrease in modulus when the RLP was added after the filler in the high-shear mixing experiment.

Because of the high viscosity associated with formulation B, we

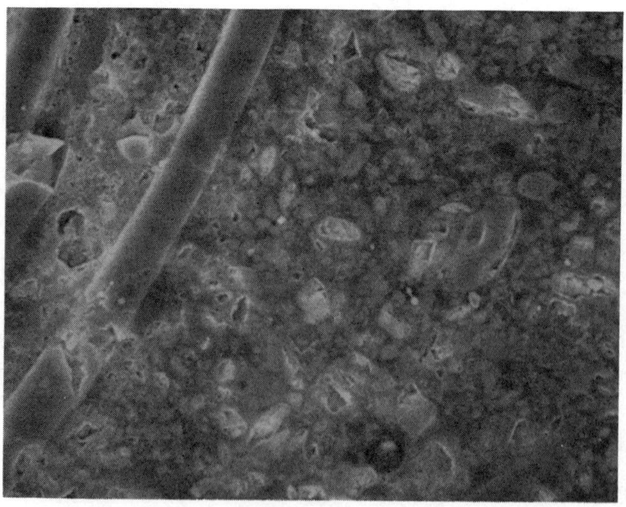

Figure 7. Glass-reinforced formulation A composite containing no modifier (magnification, 510×).

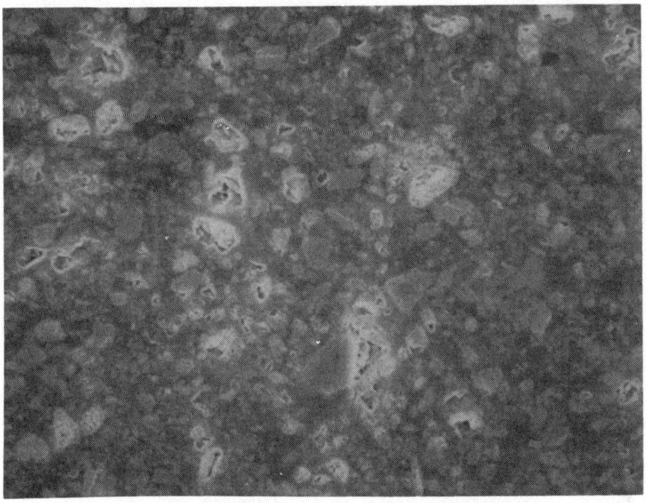

Figure 8. Glass-reinforced formulation A composite with modifier added prior to filler (magnification 510×).

could not prepare paste castings. Hence, measurement of fracture-surface energy was not conducted.

A microscopic analysis (Figure 12) indicates an affinity between the PE surface modifier and the elastomer, as well as the previously noted affinity for the elastomer and the glass fibers. The PE also serves as a carrier to move the "elastomer-coated" PE into the resin matrix away from the glass.

Figure 9. Glass reinforced formulation A composite with modifier added prior to filler (magnification 510×).

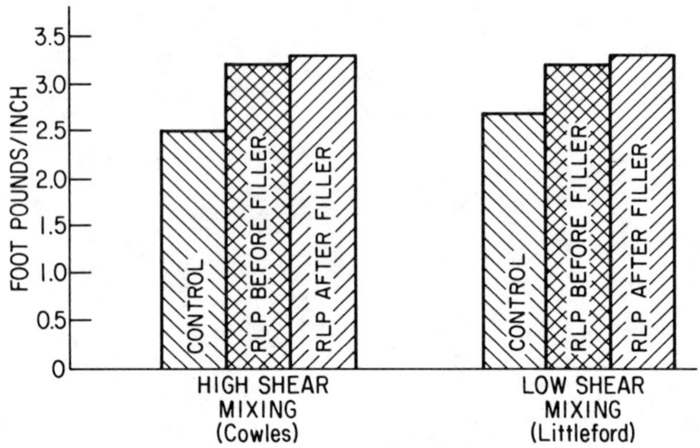

Figure 10. Notched Izod impact data for formulation B.

Discussion

The physical property improvements of the injection-molded SMC indicate that the HYCAR VTBNX 1300x24 can be used to toughen injection-molded composites based on the OCF E–4297–5 resin system. The Izod impact values show increases similar to those obtained in production situations with impact increases of 20–30%. The Rheometrics impact data for formulation A show significant increases in both ultimate force and energy absorbed at ultimate force. This result would indicate a part that has a higher load to failure and the ability to absorb a greater amount of energy prior to failure. This modification is particularly important in production where shipping, handling, and assembly operations contribute to part damage. The

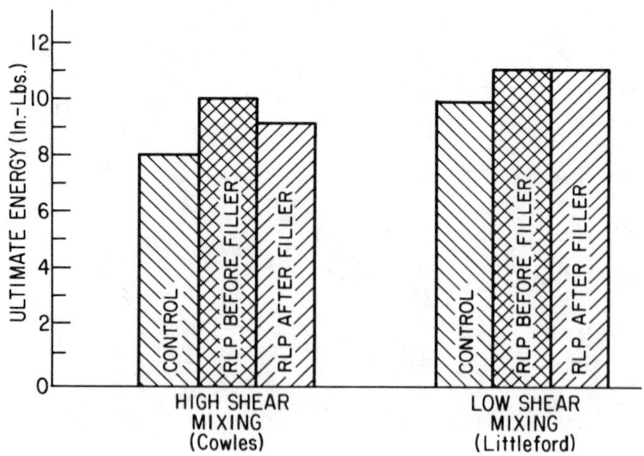

Figure 11. Rheometrics impact data at 5280 fpm for formulation B.

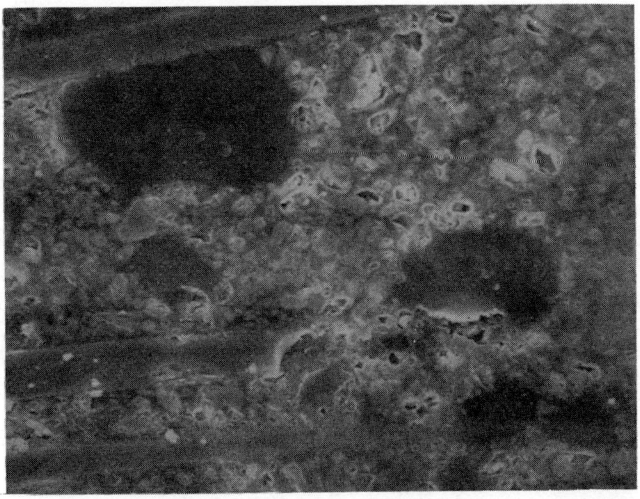

Figure 12. Glass-reinforced formulation B composites with modifier added prior to filler (magnification 510×).

formulation B series does not show as significant an increase in Rheometrics impact as the formulation A. The major difference in the two formulations is the presence of the PE in formulation B. The association of the VTBNX with the PE, which will be discussed in more detail later, may account for these smaller increases.

The influence of the HYCAR on the flexural properties of both material groups was encouraging. The increases in flexural strength with no loss of modulus, always a concern during elastomeric modification, was a highlight of this work. However, the tensile properties did generally show small decreases in strength and modulus.

Although physical property improvements were the objective of this project, an interesting facet of the results came from the morphology study. The distribution of the elastomer particles in the injection-molded plaques was a complete surprise.

As was desired, a wide range of elastomer particle sizes was achieved in the various formulation A pastes as seen in Figures 4–6. These distributions were directly related to the shear of mixing of the paste. However, these particle distributions did not carry over to the injection-molded plaques (Figures 7–9). Sizable HYCAR domains generally were not found in the resin matrix of the composite in contrast to the corresponding casting. Potentially, the additional shear and/or turbulence of the injection-molding process produces particles of a smaller size. Distortion or elongation of the elastomeric domains may occur to the extent that they are no longer distinguishable in the resin matrix. The exception is in regions of high glass content. In these areas, it is not difficult to identify elastomer domains. Often these domains are in contact with a glass fiber. The high-modulus glass may be protecting the elastomer particles from the shear–turbulence of the injection-molding process; this protection would maintain the integrity of the HYCAR domains. The consistent association of these elastomer domains with the surface of the glass fibers implies an association between the butadiene–acrylonitrile (BAN) backbone of VTBNX and the sizing on the fibers. This concept certainly warrants additional investigation. Changes in the type of glass or the AN content of the elastomer could influence the degree of this association.

These observations could explain the lack of differentiation in physical properties of the rubber-modified systems prepared under different shear conditions. The additional shear–turbulence of the compound going through the extruder or flowing in the mold largely diminishes the initial difference in paste particle size. Thus, these homogenized systems show similar responses in toughening. However, the absence of significant rubber particles in the resin matrix and their association with glass fibers cause one to consider a new potential mechanism for toughening injection-molded composites. Traditionally, we have believed that VTBNX toughens the resin-rich regions in a polyester composite and, thus, makes the formation and propagation of the initial crack(s) more difficult. With the absence of rubber particles of the size generally associated with polyester toughening in the resin-rich regions, the toughening mode may be associated with fiber debonding. The elastomer domains on the glass fibers could retard crack propagation at the resin–fiber interface; this retardation would make fiber debonding and pull-out more energy intensive. Both toughening modes could be functioning in this system, but testing to determine the actual toughening mechanism is difficult.

Formulation B is morphologically different as a result of the presence of the Microthene. HYCAR-modified formulations show an association of the elastomer with the PE (Figure 12). When the VTBNX is added prior to the filler, only rarely is the elastomer not associated with the PE. This association allows the HYCAR to be carried into the resin-rich regions with the Microthene. So, unlike the formulation A series where elastomer was not present in the resin-rich matrix, it is present in these regions in the B series. This association is reasonable in relationship to solubility parameters. The polyester resin and low profile additive are both fairly polar polymers with solubility parameters of approximately 10. The nonpolar PE has a solubility parameter of approximately 7, so the components of this formulation have no particular affinity. The BAN-based VTBNX has a solubility parameter of approximately 8.8, a value between the resin and the PE. The elastomer should, therefore, distribute itself at the interface of the two components that differ so greatly in polarity.

Realistically, much of the VTBNX is present as HYCAR-coated

PE. This coating of soft elastomer on harder plastic has implications for the physical properties of the composite. The harder, RLP-coated Microthene particle may not be as efficient in preventing cracks or retarding crack propagation in the polyester matrix. Thus, the Rheometrics impact was not as improved in the B formulation series as it was in the A series.

When the HYCAR is added after the filler, we find RLP associated with Microthene and free RLP domains that are often associated with glass fibers. This type of system, when mixed on a high-shear mixer, does not show the flexural improvements of the other formulations. Although the explanation is not clear, RLP addition prior to the filler in the preparation of SMC–BMC appears to be the best procedure.

Conclusions

The addition of HYCAR VTBNX 1300x24 to injection-molding compounds based on OCF E–4297 produces significant improvements in the physical properties of the composite. The improvement in Izod impact was 20-30%, flexural properties improved, and tensile properties slightly decreased.

Part performance is not generally affected by the degree of shear in the paste preparation nor is mixing order of great significance. However, where mixing is performed under high-shear conditions, preference might be given to adding HYCAR prior to filler addition. Although dramatic differences in paste rubber morphology were observed, these differences did not translate into particle differences in the parts. The increased shear–turbulence of the injection-molding process appears to homogenize the different morphologies of the elastomer-modified systems so that no differentiation in properties can be noted.

The elastomer is attracted to various components in the compound such as the glass or PE. This attraction probably influences the properties of the system in a way we are not able to predict at this time. The influence of this attraction, especially to glass fibers, should certainly be investigated further. The effect of the AN content of the RLP and the use of different types of glass reinforcement would be good starting points.

Finally, no direct correlation could be found between the fracture-surface energy of the unreinforced SMC paste and any properties of the injection-molded parts.

Acknowledgment

This work was originally presented at the Society of Plastics Engineers NATEC '83 in Detroit, Michigan.

Literature Cited

1. McGarry, F. J.; Willner, A. M. "Toughening of an Epoxy Resin by an Elastomer Second Phase", Research Report R68-8, School of Engineering, MIT, Cambridge, MA, 1968.
2. Riew, C. K.; Rowe, E. H.; Siebert, A. R. In "Toughness and Brittleness of Plastics"; Deanin, R. D., Crugnola, A. M., Eds.; ADVANCES IN CHEMISTRY SERIES NO. 154; American Chemical Society: Washington, D.C., 1976; pp. 326-343.
3. Rowe, E. H. "Developments in Improved Crack Resistance of Thermoset Resins and Composites", presented at the Proc. 37th Annu. Tech. Conf. Soc. Plast. Eng., 1979; pp. 598-600.
4. Nichols, C. S.; Horning, G. J. "Rubber Toughening of Polyesters: The Effect of Isophthalic Acid Content", presented at the Proc. 38th Annu. Tech. Conf. Reinf. Plast./Comp. Inst., Soc. Plast. Ind., 1983, Session 19-B.
5. Owen, G. E. "Assessing the Effects of Resin Type and Glass Content on FRP Impact Resistance", presented at the Proc. 35th Annu. Tech. Conf. Reinf. Plast./Comp. Inst., Soc. Plast. Ind., 1980, Session 10-G.
6. Myers, F. A. "Impact Response of SMC/BMC Composites", presented at the Proc. 37th Annu. Tech. Conf. Reinf. Plast./Comp. Inst., Soc. Plast. Ind., 1982, Session 1-C.
7. Lee, B. L.; Howard, F. H. "Effect of Matrix Toughening on the Strength and Strain Properties of Sheet Molding Compound", presented at the Proc. 36th Annu. Tech. Conf. Reinf. Plast./Comp. Inst., Soc. Plast. Ind., 1981, Session 23-E.
8. Lee, B. L.; Howard, F. H.; Rowe, E. H. "Effect of Matrix Toughening on the Crack Resistance of SMC Under Static Loading", presented at the Proc. 38th Annu. Tech. Conf. Reinf. Plast./Comp. Inst., Soc. Plast. Ind., 1983, Session 9-A.
9. Rowe, E. H.; McGarry, F. J. "Improving Damage Resistance of SMC and BMC", presented at the Proc. 35th Annu. Tech. Conf. Reinf. Plast./Comp. Inst., Soc. Plast. Ind., 1980, Session 19-E.
10. Mali, D. T. J. *Plast. Technol.*, **1982.**

RECEIVED for review November 18, 1983. ACCEPTED April 19, 1984.

22

Flame-Retardant Composition of Epoxy Resins with Phosphorus Compounds

JOHN A. MIKROYANNIDIS[1] and DEMETRIUS A. KOURTIDES
NASA Ames Research Center, Moffett Field, CA 94035

*Flame-retardant compositions were prepared by using 1-[bis(2-chloroethoxyphosphonoyl)methyl]-2,4-diaminobenzene and 1-[bis(2-chloroethoxyphosphonoyl)methyl]-2,6-diaminobenzene (together as DCEPD) as curing agents for three typical epoxy resins. In addition, compositions of the three test epoxy resins cured with agents such as 1,3-benzenediamine (*m*-phenylenediamine, MPD) and 4,4′-sulfonylbisbenzamine (4,4′-diaminodiphenyl sulfone, DDS) were studied to compare their reactions with those of DCEPD-cured test resins. The reactivity of the three curing agents toward the epoxy resins, measured by differential scanning calorimetry (DSC), was MPD > DCEPD > DDS. The polymerization heat increases with the increase of the epoxy functionality of the resin. The polymers obtained were studied by DSC and thermogravimetric analysis. The polymers of DCEPD showed a relatively lower polymer decomposition temperature and a higher char yield than the polymers of the regular curing agents. The flame retardancy of the polymers was evaluated by determining their limiting oxygen index value. The DCEPD polymer with polyfunctional epoxy resin showed a significantly higher flame retardancy as compared with the polymers of regular curing agents.*

FLAMMABILITY REDUCTION OF EPOXY RESIN COMPOSITIONS is conveniently accomplished by using a flame-retardant cross-linking agent to impart fire-retardant properties. Brominated poly(*p*-vinylphenol) (*1*), hydroxy alkylphosphines (*2*), phosphate and pyrophosphate esters (*3*), organic phosphonates (*4*), and complex compounds of boron trifluoride with phosphines (*5*) have been proposed as curing agents for epoxy resins.

[1] Permanent address: Chemical Technology Laboratory, University of Patras, Patras, Greece.

The purpose of this study was to improve the flammability characteristics of epoxy resin compositions by using a phosphorus-containing curing agent. Specifically, the polymerization of some commercially available epoxy resins with 1-[bis(2-chloroethoxyphosphonoyl)methyl]-2,4-diaminobenzene and 1-[bis(2-chloroethoxyphosphonoyl)methyl]-2,6-diaminobenzene (together as DCEPD) was investigated.

Materials and Method

The DCEPD was prepared by nitration and subsequent hydrogenation of bis(2-chloroethoxyphosphonoyl)methylbenzene. From IR and ^{1}H NMR data, the product consists of about 90% 2,4-diamino and 10% 2,6-diamino isomer (Structure **I**).

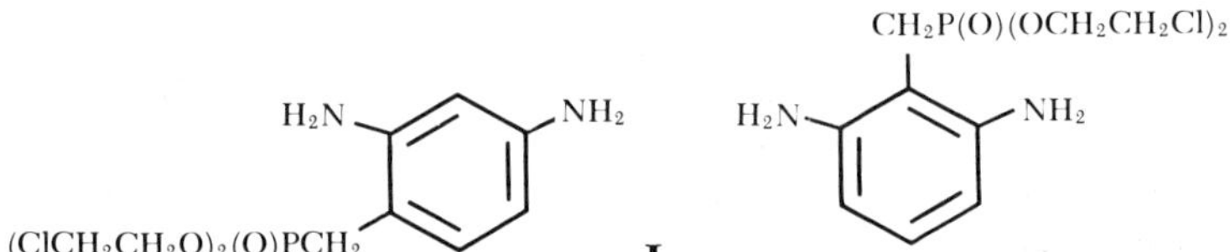

I

The polymerizations of epoxy resins with DCEPD and with curing agents such as 1,3-benzenediamine (*m*-phenylenediamine, MPD) and 4,4′-sulfonylbisbenzamine (4,4′-diaminodiphenyl sulfone, DDS) were compared.

Three test epoxy resins, EPON 828 (Shell), XD 7342 (Dow), and MY 720 (Ciba-Geigy) were used. The chemical structures, as well as the epoxy equivalent weights (EEW) of these resins, are shown in Table I.

The epoxy resins and curing agents were dissolved in acetone to obtain a homogeneous mixture of the reactants. The solvent was stripped off in a vacuum and the resinous residue obtained was directly used for differential scanning calorimetry (DSC) measure-

Table I. Chemical Structures of Epoxy Resins

Epoxy Resin	*EEW*	*Structure*
EPON 828	185–192	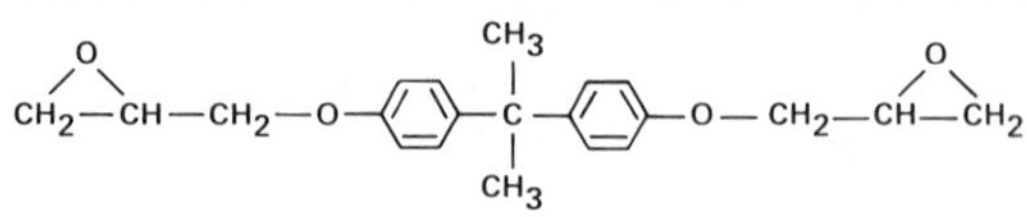
XD 7342	162	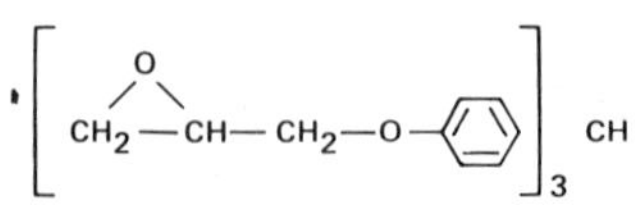
MY 720	105.5	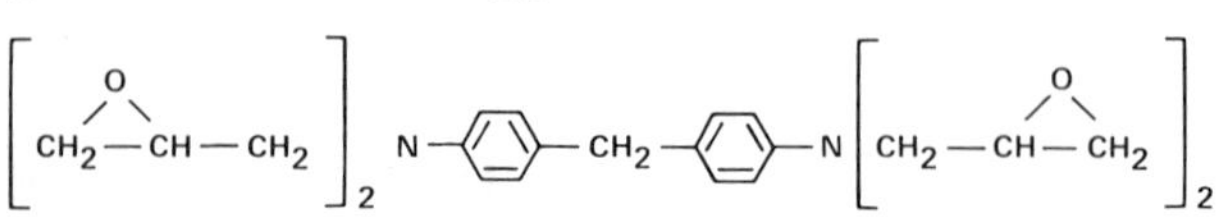

ments. Mixtures cured to a temperature at which the first curing heat was released were studied to determine their thermal characteristics. Specifically, the mixtures were cured for 4 h at 120 °C (MPD), 160 °C (DCEPD), and 190 °C (DDS). DSC and thermogravimetric analysis (TGA) measurements were performed on a thermal analyzer (Dupont 1090) at a heating rate of 20 °C/min and at a nitrogen flow of 80 mL/min. The limiting oxygen index (LOI) of the compositions was determined with a flammability unit (Stanton Redcroft) (ASTM D2863–74).

Results and Discussion

Some thermal characteristics of the epoxy resin–curing agent compositions uncured and cured are presented in Table II. Specifically, the onset temperature (T_1) calculated by the intersection of the tangent to the front side of the exotherm curve with the baseline, the exotherm peak temperature (T_2), and the evolved polymerization or pyrolysis heat (ΔH_{pol} or ΔH_{pyr}) are shown in Table II. To obtain a highly cross-linked polymer with good thermal stability, 1 EEW of the epoxy resins was polymerized with 0.25 mol of curing agent.

Typical DSC thermograms of the uncured epoxy resin–curing agent compositions are shown in Figure 1. The systems start to release heat at 133–216 °C. No exotherms can be detected in the DSC thermograms of the epoxy resins without a curing agent or in the completely cured epoxy resins in the same temperature range. Therefore, the peaks are attributed to the exothermic curing reactions. The influence of the chemical reactivity of the curing agents on cure behavior is apparent in Figure 1. For all the epoxy resins used, the progression of exotherm starting temperature is MPD < DCEPD < DDS. A curing agent that shows lower exotherm starting temperature under the same set of curing conditions is more reactive

Table II. Thermal Characteristics of Epoxy Resin–Curing Agent Compositions

		Polymerization					*Pyrolysis*		
				ΔH_{pol}					
Epoxy Resin	*Curing Agent*	T_1 (°C)	T_2 (°C)	(J/g)	(J/g · epoxide)	(J/mol, epoxide)	T_1 (°C)	T_2 (°C)	ΔH_{pyr} (J/g)
EPON 828	DCEPD	156.9	201.0	145	—	—	270.1	288.9	94
EPON 828	MPD	140.4	177.3	308	353	131	364.0	386.6	70
EPON 828	DDS	191.8	243.7	224	—	—	352.5	389.0	102
XD 7342	DCEPD	151.4	190.7	172	—	—	278.2	307.1	210
XD 7342	MPD	133.7	165.3	320	373	181	360.0	378.1	121
XD 7342	DDS	186.9	233.8	276	—	—	359.9	395.9	97
MY 720	DCEPD	174.7	213.7	208	—	—	221.7	269.1	214
MY 720	MPD	153.2	177.8	356	440	186	268.6	291.0	85
MY 720	DDS	216.3	249.7	331	—	—	304.8	352.7	100

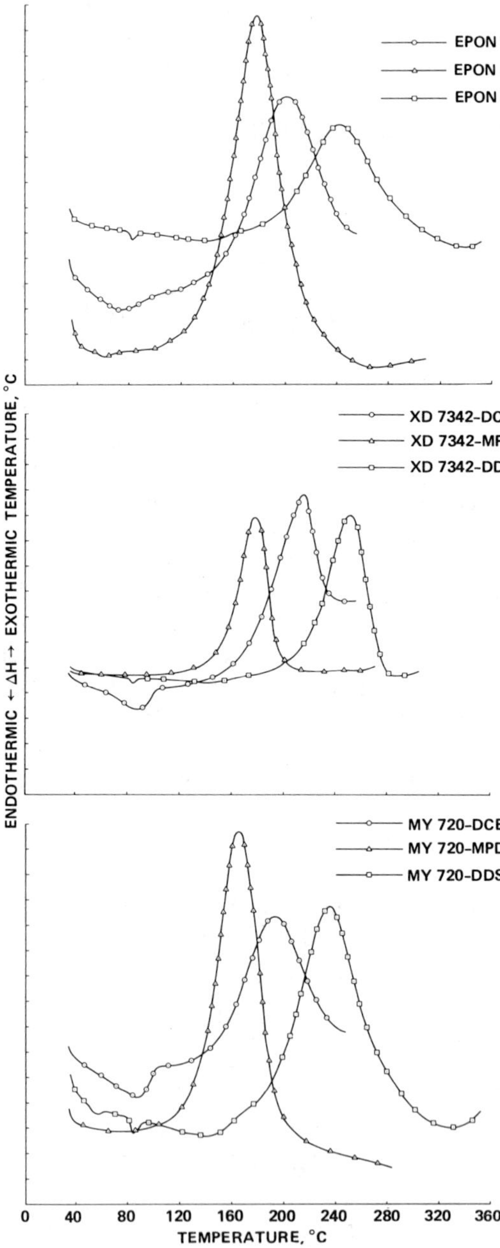

Figure 1. DSC thermograms of the uncured epoxy resin–curing agent compositions.

Table III. Thermal Stability of Epoxy Resin–Curing Agent Compositions in Nitrogen Atmosphere

Epoxy Resin	*Curing Agent*	*PDT (°C)*	PDT_{max} *(°C)*	*TCP (°C)*	*Char Yield (%, 650 °C)*
EPON 828	DCEPD	287.0	351.6	500.0	42.0
EPON 828	MPD	399.2	436.3	476.9	18.5
EPON 828	DDS	416.7	459.1	503.3	15.0
XD 7342	DCEPD	265.0	317.7	550.0	56.0
XD 7342	MPD	388.0	431.5	483.3	31.5
XD 7342	DDS	401.7	443.3	491.0	35.0
MY 720	DCEPD	235.0	297.7	550.0	53.0
MY 720	MPD	361.4	424.7	492.9	23.5
MY 720	DDS	381.7	439.2	502.7	29.5

toward the epoxy resin (6). Therefore, we conclude that the progression of chemical reactivity of the three different curing agents toward the epoxy resins is MPD > DCEPD > DDS. Furthermore, the ΔH_{pol} of the same curing agent with the test epoxy resins increases as the epoxy functionality of the resins increases.

TGA data in nitrogen atmosphere are shown in Table III. Specifically, the polymer decomposition temperature (*PDT*), the maximum polymer decomposition temperature (PDT_{max}), the temperature of complete pyrolysis (*TCP*), and the char yield in percent at 650 °C are given. The DCEPD compositions show higher char yields (42–56%) than the other curing agents. Char formation is important to the prediction of the flammability of polymers (7).

The flammability (tendency to burn) of the compositions of DCEPD, as well as of the regular curing agents, was evaluated by determining their LOI value. The results are shown in Table IV. The estimated phosphorus, chlorine, and nitrogen content of the compositions is also given in the table. The compositions of DCEPD show

Table IV. Limited Oxygen Index Values of the Epoxy Resin–Curing Agent Compositions

Epoxy Resin	*Curing Agent*	*P (%)*	*Cl (%)*	*N (%)*	*LOI*
EPON 828	DCEPD	2.87	5.64	2.59	28.1
EPON 828	MPD	0	0	3.25	27.2
EPON 828	DDS	0	0	2.79	22.9
XD 7342	DCEPD	3.18	6.26	2.87	29.6
XD 7342	MPD	0	0	3.70	28.0
XD 7342	DDS	0	0	3.12	25.3
MY 720	DCEPD	4.14	8.14	7.47	59.3
MY 720	MPD	0	0	10.57	30.8
MY 720	DDS	0	0	8.36	29.0

somewhat significantly higher fire resistance than the compositions of regular curing agents because of the phosphorus–chlorine–nitrogen synergism. This observation was especially evident with the MY 720 resin. However, the LOI is only an indication of fire performance and not an absolute measure of it. The content of the compositions in these three elements increases as the epoxy functionality of the resin increases; therefore, the increase in LOI follows the same trend. The MY 720–DCEPD composition shows an increase in LOI because of its relatively higher phosphorus, chlorine, and nitrogen contents. Compositions with approximately 2% phosphorus and 6% chlorine are known to produce self-extinguishing epoxy resins (*8*); therefore, the DCEPD compositions studied are expected to be self-extinguishing as well.

Literature Cited

1. Takahasi, A.; Shimizu, M. *Fire Retard. Proc. Int. Symp. Flammability Fire Retard.* **1979,** *212.*
2. Lyapichev, V. E.; Tuzhikov, O. I. *Functs. Org. Soedin. Polim.* **1972,** *57.*
3. Hosoda, Minoru; Kashiwagi, Eiichi *Japan Kokai* 73 56, 229 1973; *Chem. Abstr.* **1973,** *79,* 116421y.
4. David, F. W. (Ross, Ger. Offen. 2,212,390, 1972); *Chem. Abstr.* **1973,** *78,* 30806v.
5. Sadykova, E. M.; Chelnokova, Z. B.; Kolli, I. D.; Rodionov, R. A.; Spitsyn, V. I. *Izv. Akad. Nauk SSSR, Ser. Khim.* **1972,** *4,* 786.
6. May, C. *SPE Trans.* **1963,** *3,* 251.
7. Van Krevelen, D. W. *Polymer* **1975,** *6.*
8. Lewin, M.; Atlas, S, M.; Pearce, E. M. "Flame-Retardant Polymeric Materials"; Plenum: New York and London, 1975; p. 362.

RECEIVED for review November 18, 1983. ACCEPTED March 12, 1984.

INDEXES

AUTHOR INDEX

SUBJECT INDEX

A

G

K

L

M

R

T

Editing and indexing by Susan Robinson
Production by Anne Riesberg and Deborah Corson
Jacket design by Pamela Lewis
Managing Editor: Janet S. Dodd

Typeset by The Sheridan Press, Hanover, Pa. and
Hot Type Ltd., Washington, D.C.
Printed and bound by Maple Press Co., York, Pa.